Laser Shock Processing and Related Phenomena

Laser Shock Processing and Related Phenomena

Editors

José Luis Ocaña
Janez Grum

MDPI • Basel • Beijing • Wuhan • Barcelona • Belgrade • Manchester • Tokyo • Cluj • Tianjin

Editors
José Luis Ocaña
Polytechnical University of
Madrid
Spain

Janez Grum
University of Ljubljana
Slovenia

Editorial Office
MDPI
St. Alban-Anlage 66
4052 Basel, Switzerland

This is a reprint of articles from the Special Issue published online in the open access journal *Metals* (ISSN 2075-4701) (available at: https://www.mdpi.com/journal/metals/special_issues/LSP_technology).

For citation purposes, cite each article independently as indicated on the article page online and as indicated below:

LastName, A.A.; LastName, B.B.; LastName, C.C. Article Title. *Journal Name* **Year**, *Article Number*, Page Range.

ISBN 978-3-03936-798-6 (Hbk)
ISBN 978-3-03936-799-3 (PDF)

Contents

About the Editors

José L. Ocaña is Chair Professor in Mechanical Engineering at the Polytechnical University of Madrid (Spain), where he earned his Ph.D. in 1982. He has developed scientific collaborations and stages in most relevant worldwide experimental laser and nuclear fusion facilities, including Kernforschungszentrum Karlsruhe (Karlsruhe, Germany), P.N. Lebedev Physical Institute (Moscow, Russia), Institut für Hochleistungsstrahltechnik (ISLT) TU Wien (Vienna, Austria) and others. Prof. Ocaña was the Founder and Director of the UPM Laser Center at the Polytechnical University of Madrid (Spain) (1999–2016). He has been an active participant in national (Spain) and European RTDI initiatives, leading projects in the field of scientific and industrial applications of high power lasers. He has authored and co-authored numerous papers in national and international journals and congresses. He is a member of many different specialized committees, associations, and networks. He was the former chair of the EUREKA EULASNET II network on Laser Technology and Applications (2006–2010), and is member of the Executive Board for the European Laser Institute (Vice-President since 2013). He hosted the 4th International Conference on Laser Peening and Related Phenomena in Madrid on May 6th–10th 2013.

Janez Grum is retired Professor of Materials Science at the Faculty of Mechanical Engineering, University of Ljubljana (Slovenia). He is the founder and Editor-in-Chief of a new journal, *International Journal of Microstructure and Materials Properties (IJMMP)*, and since 1994 served as the Editor for *Journal: News of Society for Nondestructive Testing* by the Slovenian Society. He was the organizer of the international conferences and the Editor of the 16 conference proceedings and Guest Editor of 30 special issues in various journals, author of 28 book chapters published at ASM, Taylor&Francis, CRC Press, Marcel Dekker, Springer, Kluwer and Academic Press and 16 books with several reprints. He has also published in more than 300 refereed journals and more than 450 conference papers on heat treatment, laser materials processing, and materials testing including nondestructive testing. He is scientific board member of various journals and member of internastional associations. Prof. Grum is a Fellow of the American Society for Materials (ASM), for his "sustained contributions in metallurgical research and technologies, including nondestructive testing, failure analysis, and laser processing of steel and other engineering alloys". He is also a Fellow of the British Institute for Non-Destructive Testing.

Editorial

Laser Shock Processing and Related Phenomena

José Luis Ocaña [1,*] and Janez Grum [2]

[1]   Polytechnical University of Madrid, UPM Laser Centre, 28031 Madrid, Spain
[2]   Faculty of Mechanical Engineering, University of Ljubljana, 1000 Ljubljana, Slovenia; janez.grum@fs.uni-lj.si
*   Correspondence: jlocana@etsii.upm.es

Received: 28 May 2020; Accepted: 3 June 2020; Published: 16 June 2020

1. Introduction and Scope

Laser Shock Processing (LSP) is continuously developing as an effective technology for improving the surface and mechanical properties of metallic alloys and is emerging in direct competition with other established technologies, such as shot peening, both in preventive manufacturing treatments and maintenance/repair operations.

The level of maturity of Laser Shock Processing has been increasing during the last few years, and several thematic international conferences have been organized (the 7th ICLPRP held in Singapore, June 17–22, 2018, being the last reference), where different developments on a number of key aspects have been discussed, i.e.:

- Fundamental laser interaction phenomena;
- Material behavior at high deformation rates/under intense shock waves;
- Laser sources and experimental processes implementation;
- Induced microstructural/surface/stress effects;
- Mechanical and surface properties experimental characterization and testing;
- Numerical process simulation;
- Development and validation of applications;
- Comparison of LSP to competing technologies;
- Novel related processes.

All these aspects have been recursively treated by well-renowned specialists, providing a firm basis for the further development of the technology in its path to industrial penetration.

However, the application of LSP (and related technologies) to different types of materials, envisaging different types of applications (ranging from the always demanding aeronautical/aerospatial field to the energy generation, automotive, and biomedical fields), still requires extensive effort in the elucidation and mastering of different critical aspects, thus deserving a great research effort as a necessary step prior to its industrial readiness level.

The present Special Issue of *Metals* in the field of "Laser Shock Processing and Related Phenomena" aims, from its initial launching date, to collect (especially for the use of LSP application developers in the different target sectors) a number of high-quality and relevant papers representing present state-of-the-art technology also useful to newcomers in realizing its wide and relevant prospects as a key manufacturing technology.

Consequently, and in an additional and complementary way to papers presented at the thematic ICLPRP conferences, a call was made to those authors willing to prepare a high-quality and relevant paper for submission to the journal, with the confidence that their work would become part of a fundamental reference collection providing the present state-of-the-art LSP technology.

The result is now available and the Special Issue has been completed, with two review and nine full research papers, really setting reference knowledge for LSP technology and covering the practical totality of open issues leading the present-day research at worldwide universities, research centers, and industrial companies.

2. Contributions

As a first section, two review articles are included, representing, on one side, the previous history of developments of LSP technology [1] and, on the other, an analysis based on such developments of the prospects for the industrial implementation of the LSP technique in critical reliability applications [2]. It is needless to say that these two review papers were written by two of the most renowned experts in the LSP field—i.e., Dr. Clauer was one of the original inventors of the LSP technique at Batelle Columbus Labs. (USA) in the 1970s, and Dr. Sano was the scientist responsible for one of the most impressive research programs on the application of the LSP technique to the nuclear industry in Japan in the last 25 years.

In the second section (comprising nine full research papers), we aimed to compile as representative as possible coverage of the different key aspects leading the present-day research in LSP technology and related disciplines. The result has been a collection of articles ranging from the study of fundamental physics aspects (mostly laser–plasma interaction diagnosis and plasma pressure development, respectively represented by the articles of Colón et al. [3] and Sadeh et al. [4]); passing through the application of numerical modelling to the predictive assessment of the results of the application of LSP to the most relevant present-day materials (represented by the articles of Langer et al. [5] and Angulo et al. [6]); continuing on to the theoretical and experimental analysis of the parametric space of LSP in view of realistic applications (represented by the articles of Kallien et al. [7], Troiani and Zavatta [8], and Petan et al. [9]); and, finally, arriving at two of the most advanced developments at present day in the industrial application of LSP (i.e., the articles of Le Bras et al. [10] and T. Sano et al. [11]). In short, a collection of first-rank articles covering fundamental processes, numerical modelling, microstructural and material-related issues, materials and standard specimens testing, parametric applications design, advanced LSP applications, and implementation issues has been obtained.

3. Conclusions and Outlook

According to the initial spirit of the Special Issue, it is desired and hoped that this collection results in an useful reference tool, complementing and updating previous similar issues of the journal and forming a solid and reliable basis for further thematic research in the field

Conflicts of Interest: The authors declare no conflict of interest.

References

1. Sano, Y. Quarter century development of laser peening without coating. *Metals* **2020**, *10*, 152. [CrossRef]
2. Clauer, A.H. Laser shock peening, the path to production. *Metals* **2019**, *9*, 626. [CrossRef]
3. Colón, C.; de Andrés-García, M.I.; Moreno-Díaz, C.; Alonso-Medina, A.; Porro, J.A.; Angulo, I.; Ocaña, J.L. Experimental Determination of Electronic Density and Temperature in Water-Confined Plasmas Generated by Laser Shock Processing. *Metals* **2019**, *9*, 808. [CrossRef]
4. Sadeh, S.; Gleason, G.H.; Hatamleh, M.I.; Sunny, S.F.; Yu, H.; Malik, A.S.; Qian, D. Simulation and Experimental Comparison of Laser Impact Welding with a Plasma Pressure Model. *Metals* **2019**, *9*, 1196. [CrossRef]
5. Langer, K.; Spradlin, T.J.; Fitzpatrick, M.E. Finite Element Analysis of Laser Peening of Thin Aluminum Structures. *Metals* **2020**, *10*, 93. [CrossRef]
6. Angulo, I.; Cordovilla, F.; García-Beltrán, Á.; Porro, J.A.; Díaz, M.; Ocaña, J.L. Integrated Numerical-Experimental Assessment of the Effect of the AZ31B Anisotropic Behaviour in Extended-Surface Treatments by Laser Shock Processing. *Metals* **2020**, *10*, 195. [CrossRef]
7. Kallien, Z.; Keller, S.; Ventzke, V.; Kashaev, N.; Klusemann, B. Effect of Laser Peening Process Parameters and Sequences on Residual Stress Profiles. *Metals* **2019**, *9*, 655. [CrossRef]

8.   Troiani, E.; Zavatta, N. The Effect of laser peening without coating on the fatigue of a 6082-T6 aluminum alloy with a curved notch. *Metals* **2019**, *9*, 728. [CrossRef]

9.   Petan, L.; Grum, J.; Porro, J.A.; Ocaña, J.L.; Šturm, R. Fatigue Properties of Maraging Steel after Laser Peening. *Metals* **2019**, *9*, 1271. [CrossRef]

10.   Le Bras, C.; Rondepierre, A.; Seddik, R.; Scius-Bertrand, M.; Rouchausse, Y.; Videau, L.; Fayolle, B.; Gervais, M.; Morin, L.; Valadon, S.; et al. Laser Shock Peening: Toward the Use of Pliable Solid Polymers for Confinement. *Metals* **2019**, *9*, 793. [CrossRef]

11.   Sano, T.; Eimura, T.; Hirose, A.; Kawahito, Y.; Katayama, S.; Arakawa, K.; Masaki, K.; Shiro, A.; Shobu, T.; Sano, Y. Improving Fatigue Performance of Laser-Welded 2024-T3 Aluminum Alloy Using Dry Laser Peening. *Metals* **2019**, *9*, 1192. [CrossRef]

Review

Quarter Century Development of Laser Peening without Coating

Yuji Sano [1,2]

1   Institute for Molecular Science, National Institutes of Natural Sciences, Okazaki 444-8585, Japan; yuji-sano@ims.ac.jp or yuji-sano@sanken.osaka-u.ac.jp
2   Institute of Scientific and Industrial Research, Osaka University, Ibaraki 567-0047, Japan

Received: 4 January 2020; Accepted: 17 January 2020; Published: 19 January 2020

Abstract: This article summarizes the development of laser peening without coating (LPwC) during the recent quarter century. In the mid-1990s, the study of LPwC was initiated in Japan. The objective at that time was to mitigate stress corrosion cracking (SCC) of structural components in operating nuclear power reactors (NPRs) by inducing compressive residual stresses (RSs) on the surface of susceptible components. Since the components in NPRs are radioactive and cooled underwater, full-remote operation must be attained by using lasers of water-penetrable wavelength without any surface preparation. Compressive RS was obtained on the top-surface by reducing pulse energy less than 300 mJ and pulse duration less than 10 ns, and increasing pulse density (number of pulses irradiated on unit area). Since 1999, LPwC has been applied in NPRs as preventive maintenance against SCC using frequency-doubled Q-switched Nd:YAG lasers (λ = 532 nm). To extend the applicability, fiber-delivery of intense laser pulses was developed in parallel and has been used in NPRs since 2002. Early first decade of the 2000s, the effect extending fatigue life was demonstrated even if LPwC increased surface roughness of the components. Several years ago, it was confirmed that 10 to 20 mJ pulse energy is enough to enhance fatigue properties of weld joints of a structural steel. Considering such advances, the development of 20 mJ-class palmtop-sized handheld lasers was initiated in 2014 in a five-year national program, ImPACT under the cabinet office of the Japanese government. Such efforts would pave further applications of LPwC, for example maintenance of infrastructure in the field, beyond the horizons of the present laser systems.

Keywords: fatigue; handheld laser; nuclear power reactor; residual stress; stress corrosion cracking

---

1. Introduction

Progress in laser science and technology has realized advanced processes and applications in industries. Development of laser peening without coating (LPwC) is a landmark to deploy high-power lasers for maintenance work of infrastructure in the field. LPwC has advantage because of inertia-less process over mechanical treatment in operating nuclear facilities [1,2]. LPwC introduces compressive residual stresses (RSs) on metallic materials by simply irradiating successive laser pulses to the bare surface of components covered with water [3]. A remote processing system of LPwC was developed and has been applied to components of existing nuclear power reactors (NPRs) to mitigate stress corrosion cracking (SCC) since 1999 [1].

In the earliest system of LPwC for NPRs, laser pulses travel 50 m from laser units to the reactor components through waterproof guide pipes with reflecting mirrors at corners [1]. A technology for delivering 20 MW (100 mJ, 5 ns) laser pulses using optical fiber was also developed to increase the flexibility and extend the applicability of LPwC [4,5]. A miniaturized optical head with a diameter of 10 mm was developed with a fast-responding focusing function [6,7] that controls the focal point just on the surface within an accuracy required for fiber-delivered LPwC, namely less than ±0.5 mm. By integrating these technologies, fiber-delivery has been utilized in NPRs since 2002 [2].

Regarding fatigue issues, LPwC has positive effects to improve mechanical properties of various materials including ceramics [8,9]. LPwC significantly enhanced the fatigue strength and prolonged the fatigue life of steels [10–12], aluminum alloys [13] titanium alloys [14,15], etc. Recently, Sakino et al. confirmed the effect enhancing fatigue properties of HT780 (780 MPa grade high-strength steel) by low-energy LPwC with pulse energies down to 20 and 10 mJ [16]. Considering these advances, the Japanese government launched a five-year national program, ImPACT (Impulsing PAradigm Change through Disruptive Technologies) in 2014 to develop compact high-power pulsed lasers including 20 mJ-class palmtop-sized handheld lasers [17], which brings about further applications beyond the horizons of the present LPwC by realizing a portable system with the handheld lasers, for example applications to infrastructure in the field such as bridges, windmills, etc.

In this article, the development of LPwC in the recent quarter century is reviewed including the perspective brought by palmtop-sized handheld lasers.

2. Fundamental Process of LPwC

The fundamental process of LPwC is illustrated in Figure 1a [18]. When the high-power laser pulse with a duration of several nanoseconds is focused on the material, the top surface immediately transforms into plasma through ablative interaction with the laser pulse. If the surface of the material is covered with water, the pressure of the plasma significantly increases because the inertia of the water prevents expansion of the plasma. Under certain conditions, the peak pressure becomes 10 to 100 times higher than that in air, reaching several GPa which exceeds the yield strength of most metals. A shock wave is generated by this sudden pressure rise, propagates toward inside the material and attenuates to induce plastic deformation of the material. After passage of the shock wave, compressive RS generates in the surface layer due to elastic constraint from the surrounding part.

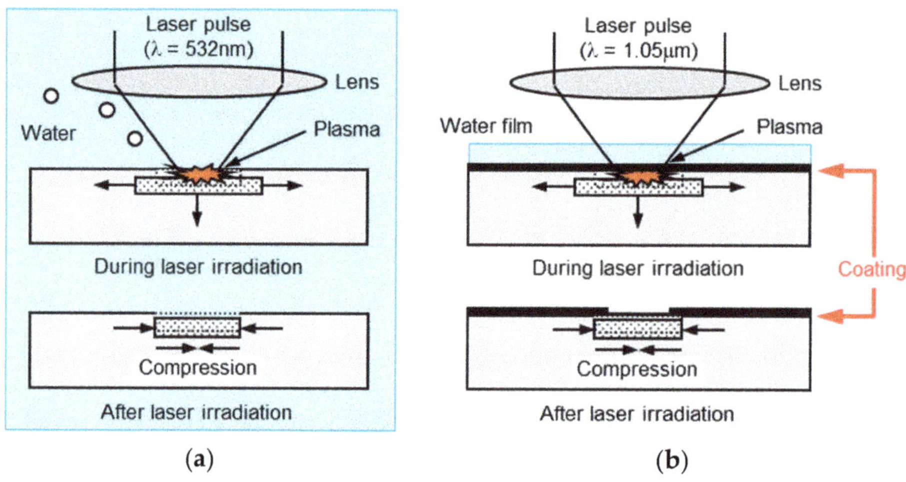

Figure 1. Fundamental process: (**a**) Laser peening without coating (LPwC); (**b**) Laser peening with coating (sacrificial overlay).

LPwC usually employs Q-switched Nd:YAG lasers. In our development, the wavelength was halved to water-penetrable visible light (λ = 532 nm) to apply to water-immersed objects. Surface RSs become compressive by increasing the number of pulses irradiated in unit area (pulses/m^2) [18], in spite of intense heat input due to the direct interaction of laser pulses with the bare surface of the objects. To make the heat input negligible the interaction time was reduced, i.e., the laser pulse duration was decreased to several nanoseconds from tens of nanoseconds in the laser peening with coating [19–22]. The pulse energy was also reduced to around 200 mJ from several tens of Joules.

In the mid-1990s, we attained surface compression by LPwC for the first time in the world [18]. This achievement is a landmark for the maintenance of NPRs because LPwC doesn't require drainage of cooling water used for radiation shielding but only irradiates laser pulses to bare components underwater without any preparation on the surface of the components.

In case of laser peening with coating, sacrificial overlay (coating) is pasted on material [19–22], which controls laser energy absorption and prevent the surface from melting. This scheme of laser peening uses high energy Nd:glass lasers with near infrared wavelength (λ = 1.05 µm) and black polymer tape or metal foil as the coating which is pasted prior to laser irradiation and removed after the treatment. The details of the process described elsewhere [22].

3. Effects of LPwC

3.1. Effects on Residual Stress

The effect of LPwC on RS was studied through experiments. As shown in Figure 2, a sample was immersed in water and driven two-dimensionally with an X-Y stage during consecutive irradiation of laser pulses. Samples were cut out from a type-304 austenitic stainless steel plate after 20% cold-working which simulated the irradiation hardening due to fast neutrons during long-term operation of NPRs (2×10^{25} neutrons/m^2, neutron energy >1 MeV). The size of the samples was 40 mm × 60 mm with 10 mm thickness and an area of 20 mm × 20 mm was processed. Laser irradiation conditions were 200 mJ pulse energy, 8 ns pulse duration, 0.8 mm focal spot diameter and 36 pulses/mm^2 pulse density. This corresponds to 50 TW/m^2 laser peak power density on the sample. Prior to LPwC, the sample surface was ground in the rolling direction of the original plate to introduce a tensile RS on the surface. X-ray diffraction (XRD; sin$^2\Psi$ method) was used to measure the surface RS, and the in-depth profile was estimated by alternately repeating the XRD and electrolytic polishing.

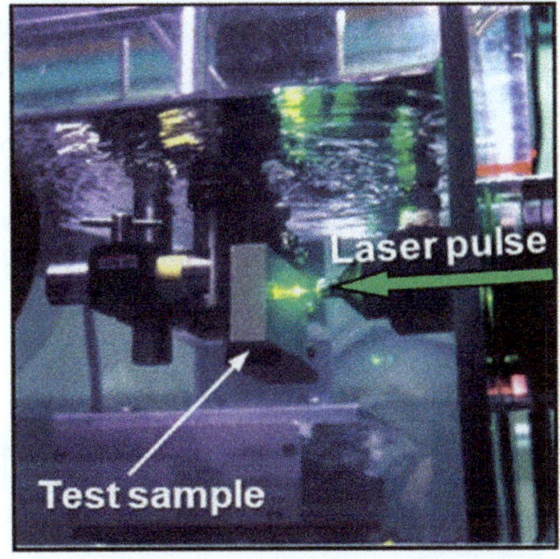

Figure 2. Experiment of underwater LPwC.

Figure 3 exhibits the RS in-depth profiles with and without LPwC together with those predicted by time-dependent elasto-plastic simulation based on finite element method (FEM) [23–25], which reproduced experimental results quite well in terms of magnitude and profile. It is obvious that LPwC can induce compressive RSs in the surface layer of material, typically up to around 1 mm depth.

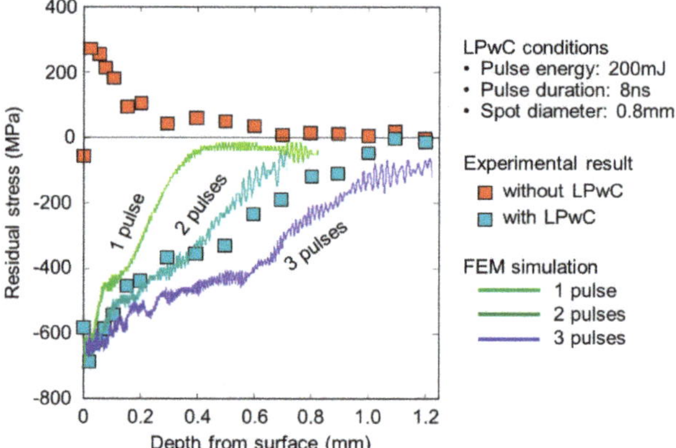

Figure 3. Residual stress in-depth profiles of 20% cold-worked type-304 austenitic stainless steel. Time-dependent elasto-plastic simulation based on a finite element method (FEM) well reproduces the experimental result.

The simulation of LPwC was made in two steps. The first one is to calculate temporal evolution of plasma pressure based on Fabbro's model [26] in which the plasma was assumed to be an ideal gas. To calibrate the plasma pressure, we measured the expansion velocity of the plasma generated on the sample surface underwater [3], then the velocity was converted to the pressure with Fabbro's model. The second step is to calculate the RS in-depth profile by using a home-made FEM program SAFFRON developed in a framework of a non-linear displacement-based incremental scheme [27]. The calculation system was discretized with 20-node isoparametric solid elements [23–25]. The plasma pressure calculated in the first step of the simulation was used as the time-dependent external load working on the sample. Stress-strain relation was modeled by the data obtained from static tensile test of the sample material. The Poisson's ratio was assumed to be 0.28. The von Mises yield criterion and a combined hardening rule were applied in the second step of the simulation.

3.2. Effects on Fatigue Properties

Fatigue test samples were prepared from a low carbon type austenitic stainless steel (type-316L) as shown in Figure 4 [28]. Two types of heat treatments were applied to the samples before LPwC, namely full heat (FH; 1373 K, 3600 s in vacuum) treatment and stress relieving (SR; 1173 K, 3600 s). Figure 5 shows the microstructure of the materials after the heat treatments. The grain sizes of the materials after FH and SR treatments were 88 μm and 24 μm, respectively. LPwC was made with 200 mJ pulse energy, 0.8 mm spot diameter and 36 pulses/mm^2 pulse density. Then, rotating bending fatigue testing (R = −1) were made with a frequency of 2820 rpm. During fatigue loading, the samples were cooled by flowing distilled water. The micro-vickers hardness (Hv) and RS were measured for the samples with and without LPwC [28].

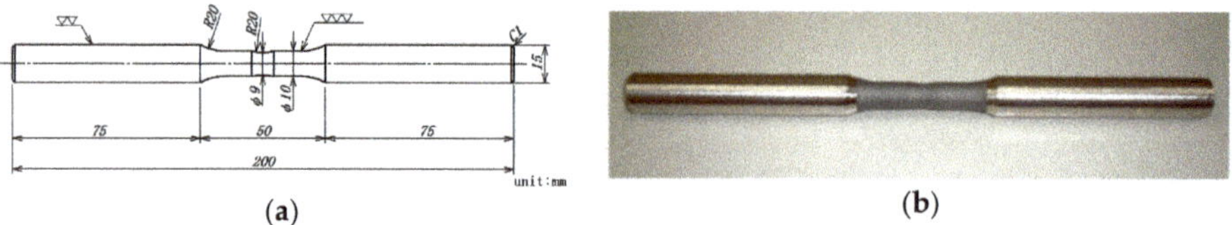

Figure 4. Type-316L austenitic stainless steel sample: (a) Dimensions; (b) External appearance. The color of the center part changed from metallic to grayish due to direct laser irradiation.

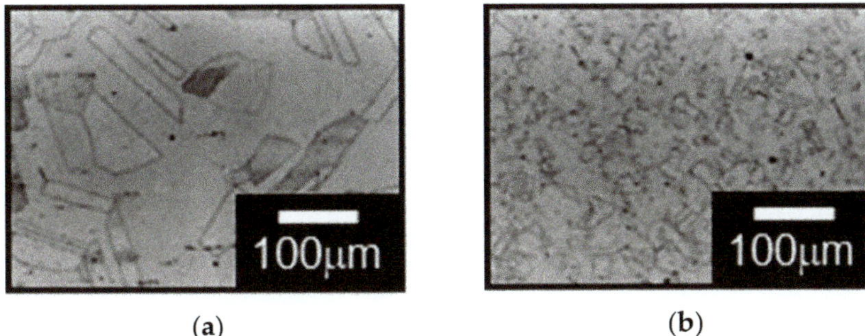

(a)   (b)

Figure 5. Microstructure of type-316L austenitic stainless steel: (**a**) Full heat-treated (FH); (**b**) Stress-relieved (SR).

The results showed that LPwC hardened the surface of both FH and SR materials down to about 0.6 mm from the surface. The hardness of both materials was increased by about 140 Hv with LPwC and reached about 300 Hv at just below the surfaces. The RS in-depth profiles exhibited anisotropy between longitudinal (z) and circumferential (θ) directions; σ_z on the surface was about −400 MPa, on the other hand σ_θ was about −200 MPa. The maximum compressive RSs were about −600 MPa (σ_z) and −400 MPa (σ_θ) at 60–100 μm depth.

Figure 6 shows the fatigue test results. Fatigue strengths of FH and SR materials with LPwC were 300 MPa and 340 MPa at 10^8 cycles, respectively, i.e., LPwC enhanced the fatigue strengths by 1.7 and 1.4 times as great as those of the reference materials. Fatigue properties enhancement was also confirmed in uniaxial fatigue of steel [16,29,30], aluminum alloy [31] and titanium alloy [15].

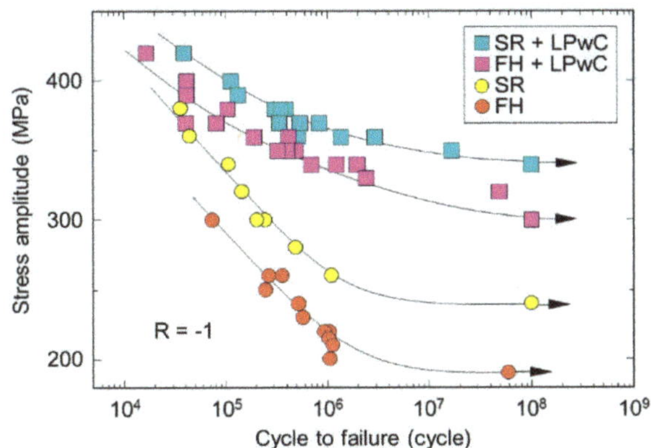

Figure 6. Rotating bending fatigue test results of type-316L austenitic stainless steel. Fatigue strengths of FH and SR materials were increased by LPwC by 70% and 40%, respectively.

3.3. Effects on SCC Susceptibility and Application to NPRs

Creviced bent beam (CBB) type testing was performed to evaluate the effect of LPwC on SCC susceptibility [24]. Samples of 10 mm × 50 mm and 2 mm thick were cut out from a plate of type-304 austenitic stainless steel with thermal sensitization (893 K, 8.64×10^4 s) followed by 20% cold working. As shown in Figure 7, samples were bent to make 1% tensile strain on the surface by using a curved fixture. After LPwC on the sample surface, crevices were made with graphite wool, and then the samples were immersed in 561 K water with 8 ppm dissolved oxygen and 10^{-4} S/m electrical conductivity for 1.8×10^6 s duration by using autoclaves.

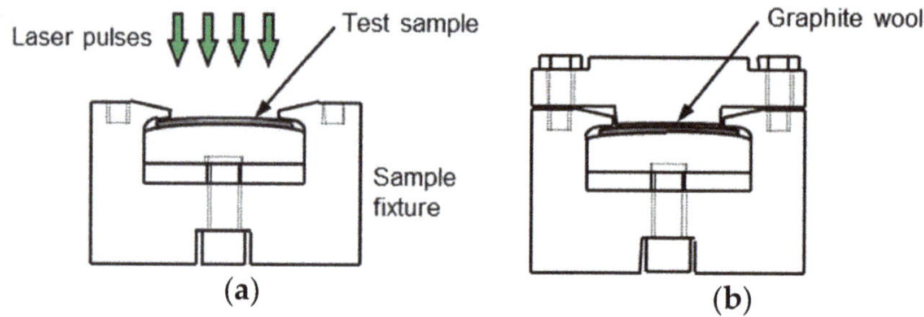

Figure 7. Procedure of accelerating stress corrosion cracking (SCC) test: (**a**) Sample setting and LPwC; (**b**) Preparation of crevices on sample surface for immersion in autoclave.

After the immersion, the surface and cross-section of all samples were precisely observed with microscopes. Inter-granular type SCC appeared in all reference samples, however no cracks were found out in samples with LPwC. Typical cross-sectional micrographs are shown in Figure 8. LPwC induced compressive RSs on the surfaces of austenitic stainless steels, nickel-based alloys and their weld metals, and prevented SCC in all tested materials [32].

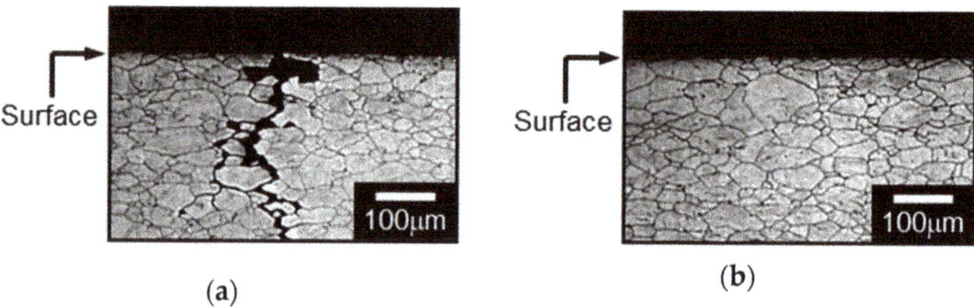

Figure 8. SCC test results of type-304 austenitic stainless steel: (**a**) Cross-section of reference material (unpeened); (**b**) Material with LPwC.

Figure 9 illustrates LPwC in a boiling water reactor (BWR) [1]. Laser pulses are delivered from the laser system on the top floor of the reactor building to weld lines of the reactor core shroud with waterproof guide pipes and mirrors at corners of the piping. An elaborate beam tracking/alignment system with a fast-responding anti-vibration function was developed and implemented to control laser irradiation point within accuracy of 0.1 mm at about 50 m away from the laser system.

Fiber delivery technology was also developed to extend the applicability of LPwC [4,5]. The intense laser pulses sometimes cause damage on the inlet surface of optical fiber and, if not, the incoming laser pulses tend to converge and lead to damage inside the optical fiber due to reflection at the curved boundary between core and cladding and/or the non-linear effect of refractive index. To avoid this situation, an inlet optics with a homogenizer consist of micro lens arrays was developed, which flatten the spatial distribution of laser intensity and eliminated conceivable hot spots. Thus, the technology was established for delivering frequency-doubled Nd:YAG laser pulses with 100 mJ energy and 5 ns duration with a single optical fiber, which improves the applicability to 3D structures, together with a tiny optical head as presented in Figure 10.

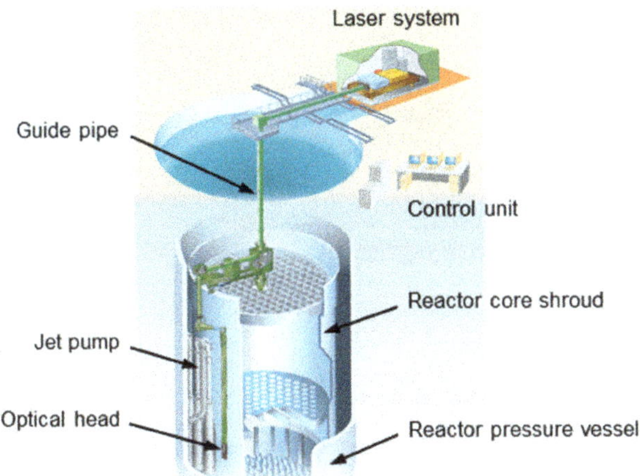

Figure 9. Schematic of LPwC for weld lines of a reactor core shroud in a boiling water reactor (BWR).

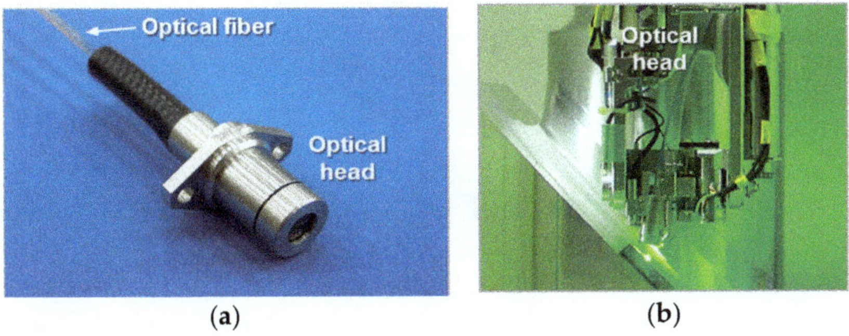

(a)   (b)

Figure 10. Fiber-delivered LPwC: (**a**) Optical head; (**b**) Mockup experiment for the bottom of a BWR.

After the completion of the system and personnel training, LPwC has been applied to reactor core shrouds, bottom-mounted nozzles, etc. of NPRs since 1999 [1,2].

4. Palmtop-Sized Handheld Laser Development

The effect of low-energy LPwC on fatigue properties was investigated for HT780 welded joints around 2013. In the course of the investigation, the pulse energy was reduced from 200 mJ to 100 mJ and then 50 mJ, the fatigue lives were significantly prolonged nevertheless [33]. Further experiments showed LPwC with the pulse energy even down to 20 mJ or 10 mJ has sufficient effects to enhance fatigue properties as shown in Figure 11 [16].

Considering such progress on the low-energy LPwC, the development of 20 mJ-class palmtop-sized handheld lasers was initiated in 2014 in a five-year Japanese national program, ImPACT [17]. A near-infrared (λ = 1.06 µm), sub-nanosecond (<1 ns) and passively Q-switched Nd:YAG laser with a weight of less than 1 kg was developed in IMS (Institute for Molecular Science) led by Prof. Taira [34,35], as shown in Figure 12.

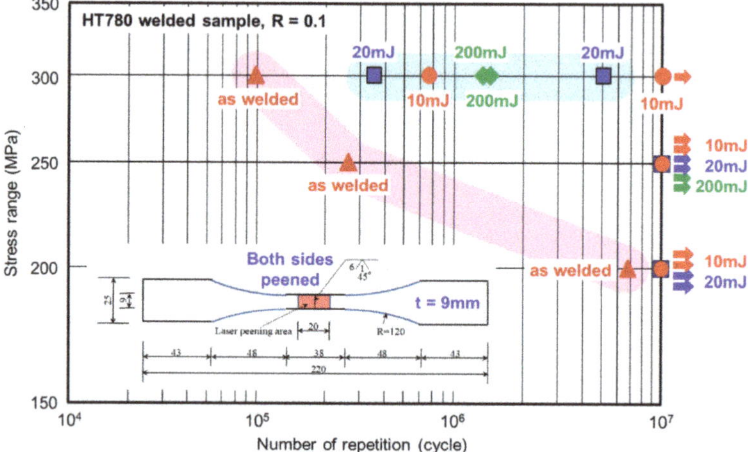

Figure 11. Fatigue test results of 780 MPa grade high-strength steel (HT780) welded joints. LPwC with 10 mJ and 20 mJ pulse energies significantly extends the fatigue life.

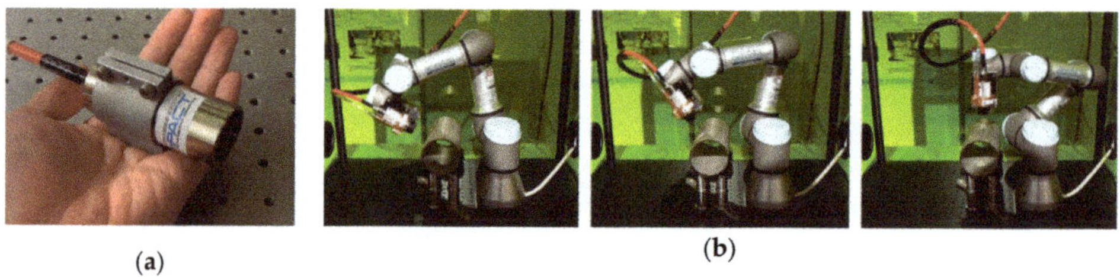

(a)   (b)

Figure 12. Palmtop-sized handheld laser: (**a**) External appearance; (**b**) Handheld laser manipulated by a robotic arm along a pipe object. Neither the movement nor vibration affects the function of the handheld laser.

A concept of LPwC system with the handheld laser is illustrated in Figure 13. A miniaturized optical head containing the laser is manipulated by a multi-axes robotic arm. Such a simple LPwC system could certainly extend the applicability and drastically reduce the time required in all phases of applications, i.e., designing, manufacturing, system integration, testing, training, transportation, installation, operation, quality assurance and dismantling.

Compared to earlier LPwC systems with current massive lasers, the system proposed above would be much smaller and simpler taking full advantage of ultra-compact handheld lasers. The pronounced characteristics expected are as follows:

- Higher reliability and operability can be expected due to simplicity of the system, which requires fewer personnel for the operation and maintenance.
- The system is much tolerant toward ambient conditions, i.e., temperature change, vibration, etc., resulting from the smaller system volume and number of parts.
- Required laser power can be decreased due to smaller transmitting loss of laser energy resulting from the shorter optical path and simpler optics.
- Application to infrastructure such as NPRs, bridges, windmills, etc. could be easier due to the smaller and simpler system.

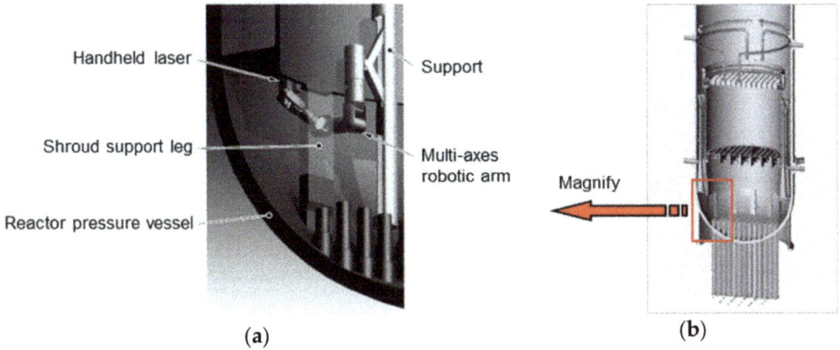

Figure 13. Schematic of LPwC using a handheld laser for SCC mitigation in a BWR: (**a**) Concept to apply LPwC to hidden weld lines; (**b**) Cutaway view of a reactor pressure vessel in outage. The concept reduces the scale of LPwC system and laser transmission distance from tens of meters (~50 m) to tens of millimeters (~0.05 m).

5. Concluding Remarks

The processes, effects, and applications of laser peening without coating (LPwC) were reviewed. A series of experimental studies clearly demonstrated that LPwC improves fatigue properties and reduces the susceptibility to stress corrosion cracking (SCC) through the impartment of compressive residual stresses (RSs) in the near surface layer of objects. LPwC has been applied to nuclear power reactors (NPRs) as a preventive maintenance against SCC of structural components since 1999 [1].

Low-energy LPwC was applied to welded joints of HT780 (780 MPa grade high-strength steel) structural steel with pulse energies down to 10 mJ. Fatigue testing revealed that the fatigue lives were sufficiently prolonged by LPwC even if 10 mJ pulse energy was used [16].

The Japanese government launched a five-year national program, ImPACT in 2014 [17], which was designed to trigger off disruptive innovation for changes in society. In the program, compact high-power pulsed lasers including 20 mJ-class palmtop-sized handheld lasers has been developed. Due to the simplicity and robustness of handheld lasers, the application including LPwC necessarily expands in various fields, for example field maintenance of infrastructure such as bridges, windmills, power plants, etc.

Funding: This work was partially supported by ImPACT Program of Council for Science, Technology and Innovation (Cabinet Office, Government of Japan).

Conflicts of Interest: The author declares no conflict of interest.

References

1. Sano, Y.; Kimura, M.; Sato, K.; Obata, M.; Sudo, A.; Hamamoto, Y.; Shima, S.; Ichikawa, Y.; Yamazaki, H.; Naruse, M.; et al. Development and Application of Laser Peening System to Prevent Stress Corrosion Cracking of Reactor Core Shroud. In Proceedings of the 8th International Conference on Nuclear Engineering (ICONE-8), Baltimore, MD, USA, 2–6 April 2000.
2. Yoda, M.; Chida, I.; Okada, S.; Ochiai, M.; Sano, Y.; Mukai, N.; Komotori, G.; Saeki, R.; Takagi, T.; Sugihara, M.; et al. Development and Application of Laser Peening System for PWR Power Plants. In Proceedings of the 14th International Conference on Nuclear Engineering (ICONE-14), Miami, FL, USA, 17–20 July 2006.
3. Sano, Y.; Mukai, N.; Okazaki, K.; Obata, M. Residual stress improvement in metal surface by underwater laser irradiation. *Nucl. Instrum. Methods Phys. Res. B* **1997**, *121*, 432–436. [CrossRef]
4. Schmidt-Uhlig, T.; Karlitschek, P.; Marowsky, G.; Sano, Y. New simplified coupling scheme for the delivery of 20 MW Nd: YAG laser pulses by large core optical fibers. *Appl. Phys. B* **2001**, *72*, 183–186. [CrossRef]
5. Schmidt-Uhlig, T.; Karlitschek, P.; Yoda, M.; Sano, Y.; Marowsky, G. Laser shock processing with 20 MW laser pulses delivered by optical fibers. *Eur. Phys. J. AP* **2000**, *9*, 235–238. [CrossRef]

6. Sano, Y.; Tamura, M.; Chida, I.; Suezono, N. Underwater Maintenance and Repair Technologies for Reactor Components by Laser Material Processing. In Proceedings of the 7th International Welding Symposium (7WS), Kobe, Japan, 20–22 November 2001.

7. Sano, Y.; Kimura, M.; Yoda, M.; Mukai, N.; Sato, K.; Uehara, T.; Ito, T.; Shimamura, M.; Sudo, A.; Suezono, N. Development of Fiber-Delivered Laser Peening System to Prevent Stress Corrosion Cracking of Reactor Components. In Proceedings of the 9th International Conference on Nuclear Engineering (ICONE-9), Nice, France, 8–12 April 2001.

8. Akita, K.; Sano, Y.; Takahashi, K.; Tanaka, H.; Ohya, S. Strengthening of Si_3N_4 ceramics by laser peening. *Mater. Sci. Forum* **2006**, *524–525*, 141–146. [CrossRef]

9. Saigusa, K.; Takahashi, K.; Sibuya, N. Evaluation of surface properties of silicon nitride ceramics treated with laser peening. *Int. J. Peen. Sci. Technol.* **2019**, *1*, 221–232.

10. Sano, Y.; Obata, M.; Kubo, T.; Mukai, N.; Yoda, M.; Masaki, K.; Ochi, Y. Retardation of crack initiation and growth in austenitic stainless steels by laser peening without protective coating. *Mater. Sci. Eng. A* **2006**, *417*, 334–340. [CrossRef]

11. Sakino, Y.; Sano, Y.; Kim, Y.-C. Application of laser peening without coating on steel welded joints. *Int. J. Struct. Integ.* **2011**, *2*, 332–344. [CrossRef]

12. Masaki, K.; Ochi, Y.; Matsumura, T.; Ikarashi, T.; Sano, Y. Effects of laser peening treatment on high cycle fatigue and crack propagation behaviors in austenitic stainless steel. *J. Power Energy Syst.* **2010**, *4*, 94–104. [CrossRef]

13. Sano, Y.; Masaki, K.; Gushi, T.; Sano, T. Improvement in fatigue performance of friction stir welded A6061-T6 aluminum alloy by laser peening without coating. *Mater. Des.* **2012**, *36*, 809–814. [CrossRef]

14. Maawad, E.; Sano, Y.; Wagner, L.; Brokmeier, H.-G.; Genzel, C. Investigation of laser shock peening effects on residual stress state and fatigue performance of titanium alloys. *Mater. Sci. Eng. A* **2012**, *536*, 82–91. [CrossRef]

15. Altenberger, I.; Sano, Y.; Nikitin, I.; Scholtes, B. Fatigue Behavior and Residual Stress State of Laser Shock Peened Materials at Ambient and Elevated Temperatures. In Proceedings of the 9th International Fatigue Congress (FATIGUE 2006), Atlanta, GA, USA, 14–19 May 2006.

16. Sakino, Y.; Sano, Y. Investigations for lowering pulse energy of laser-peening for improving fatigue strength. *Q. J. Jpn. Weld. Soc.* **2018**, *36*, 153–159. [CrossRef]

17. Ubiquitous Power Laser for Achieving a Safe, Secure and Longevity Society under ImPACT Program. Available online: https://www.jst.go.jp/impact/sano/index.html (accessed on 31 December 2019).

18. Mukai, N.; Aoki, N.; Obata, M.; Ito, A.; Sano, Y.; Konagai, C. Laser Processing for Underwater Maintenance in Nuclear Plants. In Proceedings of the 3rd JSME/ASME International Conference on Nuclear Engineering (ICONE-3), Kyoto, Japan, 23–27 April 1995. S404-3.

19. Fabbro, R.; Peyre, P.; Berthe, L.; Scherpereel, X. Physics and applications of laser-shock processing. *J. Laser Appl.* **1998**, *10*, 265–279. [CrossRef]

20. Peyre, P.; Chaieb, I.; Braham, C. FEM calculation of residual stresses induced by laser shock processing in stainless steels. *Model. Simul. Mater. Sci. Eng.* **2007**, *15*, 205–221. [CrossRef]

21. Fairand, B.P.; Clauer, A.H.; Jung, R.G.; Wilcox, B.A. Quantitative assessment of laser-induced stress waves generated at confined surfaces. *Appl. Phys. Lett.* **1974**, *25*, 431–433. [CrossRef]

22. Sokol, D.W.; Clauer, A.H.; Ravindranath, R. Applications of Laser Peening to Titanium Alloys. In Proceedings of the ASME/JSME 2004 Pressure Vessels and Piping Division Conference, San Diego, CA, USA, 25–29 July 2004.

23. Sano, Y.; Kimura, M.; Mukai, N.; Yoda, M.; Obata, M.; Ogisu, T. Process and Application of Shock Compression by Nano-Second Pulses of Frequency-Doubled Nd: YAG Laser. In Proceedings of the International Forum on Advanced High-Power Lasers and Applications (AHPLA'99), Osaka, Japan, 1–5 November 1999.

24. Sano, Y.; Mukai, N.; Yoda, M.; Ogawa, K.; Suezono, N. Underwater laser shock processing to introduce residual compressive stress on metals. *Mater. Sci. Res. Int.* **2001**, *2*, 453–458.

25. Sano, Y.; Yoda, M.; Mukai, N.; Obata, M.; Kanno, M.; Shima, S. Residual stress improvement mechanism on metal material by underwater laser irradiation. *J. Atom. Energy Soc. Jpn.* **2000**, *42*, 567–573. [CrossRef]

26. Fabbro, R.; Fournier, J.; Ballard, P.; Devaux, D.; Virmont, J. Physical Study of Laser-produced Plasma in Confined Geometry. *J. Appl. Phys.* **1990**, *68*, 775–784. [CrossRef]

27. Sano, Y. A Finite Element Method for Contact Problems between Three-Dimensional Curved Bodies. *J. Nucl. Sci. Technol.* **1996**, *33*, 119–127. [CrossRef]

28. Ochi, Y.; Masaki, K.; Matsumura, T.; Wakabayashi, Y.; Sano, Y.; Kubo, T. Effects of Laser Peening on High Cycle Fatigue Properties in Austenitic Stainless Steel. In Proceedings of the 12th International Conference on Experimental Mechanics (ICEM12), Bari, Italy, 29 August–2 September 2004.

29. Sakino, Y.; Sano, Y.; Sumiya, R.; Kim, Y.-C. Major factor causing improvement in fatigue strength of butt welded steel joints after laser peening without coating. *Sci. Technol. Weld. Join.* **2012**, *17*, 402–407. [CrossRef]

30. Sano, Y.; Sakino, Y.; Mukai, N.; Obata, M.; Chida, I.; Uehara, T.; Yoda, M.; Kim, Y.-C. Laser peening without coating to mitigate stress corrosion cracking and fatigue failure of welded components. *Mater. Sci. Forum* **2008**, *519*, 580–582. [CrossRef]

31. Adachi, T.; Takehisa, H.; Nakajima, M.; Sano, Y. Effect of Laser Peening on Fatigue Properties for Aircraft Structure Parts. In Proceedings of the 10th International Conference on Shot Peening (ICSP10), Tokyo, Japan, 15–18 September 2008.

32. Sano, Y.; Obata, M.; Yamamoto, T. Residual stress improvement of weldment by laser peening. *Weld. Int.* **2006**, *20*, 598–601. [CrossRef]

33. Sakino, Y.; Yoshikawa, K.; Sano, Y.; Sumiya, R.; Kim, Y.-C. A basic study for application of laser peening to large-scale steel structure. *Q. J. Jpn. Weld. Soc.* **2013**, *31*, 231–237. [CrossRef]

34. Zheng, L.; Kausas, A.; Taira, T. Drastic thermal effects reduction through distributed face cooling in a high power giant-pulse tiny laser. *Opt. Mater. Exp.* **2017**, *7*, 3214–3221. [CrossRef]

35. Ubiquitous Power Laser for Achieving a Safe, Secure and Longevity Society under ImPACT Program. Available online: https://www.youtube.com/watch?v=nMsOkkEPK5I (accessed on 31 December 2019).

Review

Laser Shock Peening, the Path to Production

Allan H. Clauer

LSP Technologies,6161 Shamrock Court, Dublin, OH 43016-1284, USA; ahhc1936@gmail.com;
Tel.: +1-614-718-3000

Received: 9 May 2019; Accepted: 28 May 2019; Published: 29 May 2019

Abstract: This article describes the path to commercialization for laser shock peening beginning with the discovery of the basic phenomenology of the process through to its implementation as a commercial process. It describes the circumstances leading to its invention, the years spent on exploring and defining characteristics of the process, and the journey to commercialization. Like many budding technologies displaying unique characteristics, but no immediately evident application, i.e., "a solution looking for a problem", there were several instances where its development may have been delayed or ended except for an unanticipated event that enabled it to move forward. An important contributor to the success of laser peening, is that nearly 15 years after its invention, universities world-wide began extensive research into the process, dramatically broadening the knowledge base and increasing confidence in, and understanding of its potential. Finally, a critical problem in need of a solution, laser peening, appeared, culminating in its first industrial application on aircraft turbine engine fan blades.

Keywords: laser peening; fatigue; residual stress; laser shock waves; laser peening history

---

1. Introduction

New technologies are invented, developed and applied following many different paths. It is often difficult to accurately describe these paths in hindsight, particularly the events critical to sustaining interest and support for the technology in the early and middle stages where its proponents are few and the ultimate use not certain. Fortunately, laser peening offers the opportunity to describe such a path clearly and definitively. This is possible because its invention and early development occurred within a single organization, and relatively few people and organizations were instrumental in taking it to commercial use. The insights into the phenomena vital to the success of laser shock peening can be traced to a few basic research investigations performed in the 1960s, followed by its invention in the early 1970s. It took another 40 years to become an accepted industrial process to treat metal surfaces for increasing fatigue strength and fatigue life. Along the way several critical, key events are identified. Without these events progress would have been significantly delayed or stopped completely. If any one or more of these events had not occurred, the use of laser shocks to modify material properties would still have been recognized at some point in the future, but the path would have been much different. While under development, the technology was referred to as laser shock processing. It was lacking a defined target application until further understanding and development of the technology would bring one or more into focus. The first application became laser shock peening, or laser peening, to increase the fatigue strength and fatigue life of metal alloys. This was followed by laser peen forming. In the last two decades, investigations into laser shock processing have reached beyond laser peening, to include the use of laser-induced stress waves to evaluate adhesive bond strength in bonded structures and coatings, metal die forming, surface imprinting and other possible uses.

2. The Phenomenological Origins of Laser Processing

After the invention of the laser, the first of the key events leading to laser shock processing was provided by Askaryan and Moroz at the P.N. Lebedev Physics Institute in 1962 [1]. In an experiment to measure the pressure exerted on a metal surface by a high intensity photon beam, they discovered that the pressure was at least several orders of magnitude greater than the calculated photon pressure. They rightfully concluded that they actually measured the vaporization recoil pressure produced by vaporization of material from the target surface by the laser beam. They further speculated that it was large enough to possibly be used to steer space vehicles.

Two years later, Neuman investigated the magnitude of the momentum transfer at constant and varying beam intensities for a number of different metals, at the NASA Ames Research Center [2]. He noted that a short, 50 ns "giant" laser pulse produced a greater recoil pressure than a "normal" 1 ms laser pulse with five times the energy of the giant pulse. An observation that would later be recognized as peak pressure increasing with power density. Soon after, these findings were expanded by a number of investigators, both experimental and theoretical, pursuing studies of the creation of stress waves using lasers [3–7]. All these experiments were performed with the target residing in a vacuum chamber to avoid dielectric breakdown in the beam in air at the high power densities necessary to achieve increasing pressure. While generating high pressure laser shock waves in a vacuum was acceptable for research purposes, it would not be acceptable for industrial applications.

The path to removing this obstacle was demonstrated by the second key event, a discovery made by N.C. Anderholm at Sandia Laboratories in 1968 [8,9]. He vapor-deposited an aluminum film onto a 6 mm-thick quartz disk, irradiated this aluminum film through the 6 mm-thick quartz disk and measured the pressure profile using a piezoelectric quartz gauge pressed against the aluminum film. Irradiating the aluminum film with a 1.9 GW/cm^2, 12 ns laser pulse, he measured 3.4 GPa peak pressure. Although this experiment, too, was performed in a vacuum, it clearly demonstrated that with a transparent overlay, significant shock pressures could be achieved at beam power densities not causing dielectric breakdown in air. This breakthrough observation would open the door a few years later to exploring the potential for using laser-induced shock waves as a materials processing tool.

These previous investigations were focused on studying the surface effects produced by the pulsed laser irradiation. Soon, investigators began looking at the effects of the laser-induced shock waves within the metals. In 1970, Mirkin at the M.V. Lomonosov Moscow State University realized that the higher energy, short laser pulses were capable of driving a relatively high pressure shock wave into the metal surface [10]. This suggested that the known effects of explosive or plate driven shock waves on metals' microstructure and hardness should also occur with laser-induced shock waves. He was the first to report the effects of laser-induced shocks on metal microstructure, observing twinning in steel ferrite grains located only below the laser-irradiated crater, down to a depth greater than 0.5 mm. The next year, Metz and Schmidt at the U.S. Naval Research Laboratory, investigated the effects of mild laser shocks, 0.18 GW/cm^2, 35 ns pulse width, on annealed, 50 μm-thick nickel and vanadium foils [11]. After again annealing the irradiated foils after laser shocking, they observed vacancy voids in the nickel foils and vacancy loops in the vanadium foils. Although this irradiation condition was relatively mild, these loops were evidence of a high density of lattice vacancies created by the shock wave.

During this same period, 1968–1972, other investigators were investigating the important issue of the effect of varying the transparent overlay on the pressure enhancement observed by Anderholm. O'Keefe and Skeen at TRW Systems Group explored the use of thin volatile coatings of RTV (Room Temperature Vulcanizing) silicone adhesive and Duco cement as transparent overlays on 76 μm-thick 1100-0 aluminum targets [7]. For a 50 ns pulse of 1.8 GW/cm^2, the peak pressure of the stress wave with a coating of 25 μm of the silicone adhesive was eight times higher than without the silicone coating. A 63 μm-thick coating of Duco cement increased the pressure about 15 times compared to the bare surface. With these overlays, both the plasma confinement and the vaporization of the overlay contributed to the pressure pulse. The contribution of vaporization of the overlay was deduced from the observation that increasing the curing time of the RTV, i.e., decreasing its volatility, also decreased the pressure.

3. The Transition to Laser Shock Processing

3.1. Setting the Stage

To this point, all the research was understandably dedicated to exploring the science of laser induced shock waves. There was as yet no coherent effort to define how or for what purpose they might be used. However, the third key event would both enable and foster this effort. It was the decision in 1968 by Battelle Memorial Institute in Columbus, Ohio, to purchase and install a large Compagnie Gènèrale Electrique (CGE) VD-640 Q-switched, Nd-glass laser system imported from France for the purpose of initiating work in laser fusion. Philip Mallozzi and Barry Fairand of the Laser Physics Group were members of the team setting up and operating the laser, which became operational in 1970. The system consisted of six linearly aligned amplifying stages, each supported by a large wall cabinet containing the capacitors to operate the flash lamps energizing the Nd-glass rods as shown in Figure 1.

(a)   (b)

Figure 1. The Battelle Compagnie Gènèrale Electrique (CGE) VD-640 Q-switched laser became operational in 1970 and was used for laser shock investigations to the mid-1990s: (**a**) capacitor banks; (**b**) laser rod amplifiers.

After the system became operational, Fairand and Mallozzi sought to expand its use within the laboratory. To pursue one possibility, Fairand approached Benjamin Wilcox in Battelle's Metals Science Group in early 1972, proposing that using laser-induced shock waves to modify metal properties might provide useful benefits. This was suggested by the known effects of flyer plate impacts on metals. Wilcox agreed and suggested laser shocking 7075 aluminum alloy tensile specimens to determine whether there was sufficient change in strength to warrant a further look. This first experiment consisted of clamping a 1 mm-thick glass slide against the gauge section of small, 1.35 mm-thick, dog bone specimens using sodium silicate as a coupling layer between the glass slide and aluminum surface. The 10 mm × 5 mm gage length of the specimens was shocked on each side consecutively with one shot at a power density of 1.2–2.2 GW/cm^2, 32 ns, Gaussian pulse. The specimens were backed by a 3.2 mm-thick brass plate. After laser shocking, the yield strength increased 18% for the solution treated condition, 28% for the over-aged T73 temper and a slight decrease for the peak-aged T6 temper. In this latter condition, precipitation hardening dominated the strain hardening effect of the shock wave. Transmission electron microscopy confirmed the increase in yield strength was due to the substantial increase in dislocation density in the microstructure, i.e., cold work hardening. These results were presented in the very first publication reporting an improvement in mechanical properties and the associated microstructural changes after laser shocking [12]. Based on these results, the National Science Foundation (NSF) supported a proposal to investigate the primary parameters influencing the magnitude of the in-material and property changes associated with laser shock processing of metals. The possibility that this might develop into a process that could be used for treating metals was recognized, but how and for what would play out in the years ahead.

In January, 1973, the NSF program was initiated. At that same time, Allan Clauer returned to Battelle after a year's absence at Denmark's Risø National Laboratory and Wilcox left Battelle soon after. Clauer and Fairand immediately began the journey to explore laser shock processing with this program and others to follow. In 1974 Fairand and Mallozzi were awarded the first patent for laser shock processing, "Altering Material Properties Using Confined Plasma" [13].

The NSF program had two major objectives: (1) investigate the distribution, depth, and intensity of laser shock-induced plastic strain, and (2) initiate modeling of the peak pressure and shape of the pressure pulse. The distributions of plastic strain formed by the passage of the shock wave were investigated using the etch pitting technique in specimens fabricated from Fe-3Si steel. This method had been used extensively in fracture studies at Battelle by George Hahn and coworkers to study the plastic zone size and shape at the tip of a crack [14]. A large number of disks of different diameters and thicknesses were irradiated with a range of power densities and laser spot diameters. During shocking, the back surface of the disks was a free surface except where supported on the outer rim or pressed against a quartz pressure gauge. After laser shocking, the disks were sectioned along a diameter and the sectioned surface was polished and chemically etched. Since each etch pit on the surface corresponded to a dislocation intersecting the surface, the local density of the etch pits represented the local density of dislocations and thereby the magnitude of the local plastic strain. The relative dislocation density could be easily discerned up to about 3–4% plastic strain, where the etch pits overlapped extensively. Fortunately, the plastic strains were generally below this level.

A variety of deformation patterns were observed depending on the overlay conditions, disk thickness and spot size relative to the disk diameter [15]. Generally, if the beam diameter was significantly less than the disk diameter, or the disk was 5 mm thick, the strain gradient was highest at the surface and decreased with depth as expected. By comparison, if the spot diameter was the same as or larger than the disk diameter and the thickness was about 3 mm or less, the patterns were more complex as shown in Figure 2. This was attributed to strong release waves reflected from the circumferential surface of the disk with the passage of the shock wave. These waves focused along the disk centerline and interacted with the planar shock and reflected waves traveling between the front and back surfaces. Periodically these waves constructively interfere, causing the local stress to rise above the yield strength either in tension or compression, creating various symmetrical, radial patterns like those seen in Figure 2.

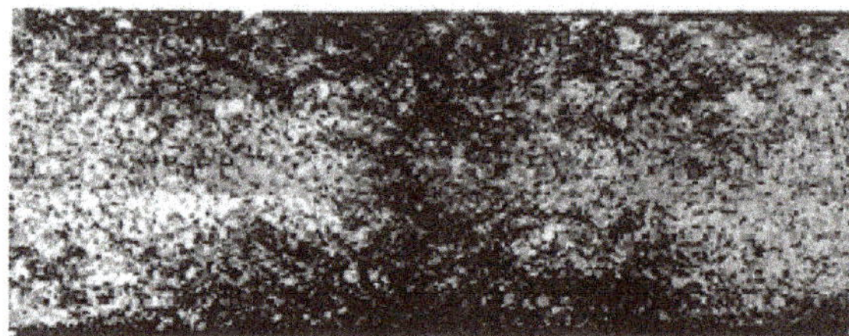

Figure 2. Etched cross-section of a laser shocked 19 mm-diameter, 3 mm-thick Fe-3Si disk showing the plastic strain distribution. 3 mm-thick quartz + 10 μm-thick lead overlays, 27 mm diameter spot, 5.64×10^8 GW/cm^2, 30 ns pulse width. Reproduced with permission from [15], The Minerals, Metals & Materials Society and ASM International, 1977.

Shock wave pressure measurements were also made to relate the intensity of the observed deformation to the incident shock pressures. The pressure was measured on the back surface of Fe-3Si disks of different thicknesses using different overlays, i.e., bare surface, quartz and quartz plus lead. In addition, modeling of the pressure pulse on the target surface and shock wave propagation into the target was undertaken to support understanding of the experimental results [16]. The pressure profiles in Figure 3 demonstrated that the Hugoniot Elastic Limit (HEL), above which plastic yielding occurs in the shock front, was easily visible in shock wave. In 0.2 mm-thick disks, plastic deformation occurred through the entire cross section producing an increase in hardness of nearly 25% after laser peening [15].

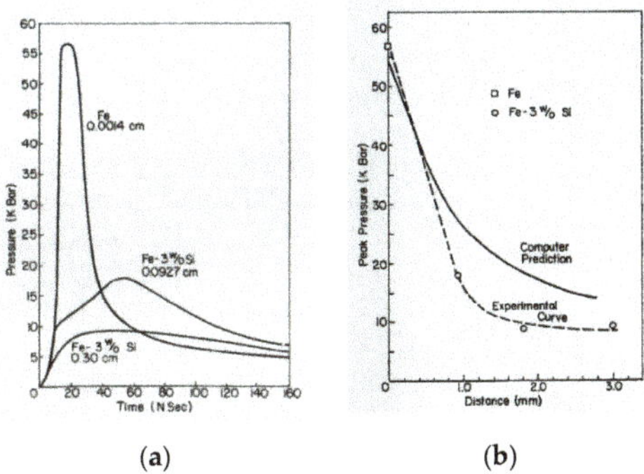

(a)   (b)

Figure 3. Pressure profiles and experimental and modeled peak pressure attenuation in Fe-3Si disks with a quartz overlay, 30 J/cm^2, 30 ns pulse width: (**a**) measured pressure profiles through different thicknesses; (**b**) peak pressure attenuation through iron [16].

At this early stage it was desirable to have the capability to predict the surface pressure for various overlay and target combinations of interest, and the in-material behavior of the shock wave. A one-dimensional radiation hydrodynamics code was written based on the PUFF computer program [17] to model the laser-material interaction for predicting the surface pressure, and a hydrodynamic code to predict the shock wave attenuation in the disks. This model was first applied to laser shocking the Fe-3Si disks. Figure 3b shows that the predicted surface pressure was close to the experimental pressure. The attenuation of the peak pressure appears to be largely hydrodynamic through the first 0.5–0.6 mm in depth. Beyond this, the attenuation is faster than the hydrodynamic code predicts due to microstructure-related damping effects such as plastic deformation. Lastly, beyond 2 mm the wave is elastic and only weakly attenuated. It should be noted that all pressure measurements using a thin metal foil, vapor deposited film or black paint on a quartz gauge is the pressure developed in the quartz [16].

The research up to early 1975 used only quartz as a transparent overlay. However, it was understood that while quartz was convenient in the laboratory, it was not a viable transparent overlay for a commercial process. Using a quartz overlay required firmly pressing it against a flat, smooth target surface. It could not adapt to curved surfaces without expensive custom design and fabrication of the overlay. In 1973, Fox had used water and paint overlays when investigating spallation of metal samples by laser induced shocks, and observed pressure increases with water and paint overlays [18]. Considering this, it was obvious that water as a transparent overlay had many desirable characteristics. It was transparent to the laser beam and due to the short pressure pulse durations of tens of nanoseconds, a thin, 1 mm layer effectively confined the plasma to the target surface to produce useful shock pressures. It had highly desirable properties for practical use, it was easily applied and removed, and easily accommodated curved surfaces. It was also inexpensive. In our investigation using water as an overlay, the first pressure measurements were made for three setups using a 2 mm-thick layer of still water: on 25 μm-thick aluminum foil, with and without black paint, and on 3 μm-thick aluminum film vapor deposited onto a quartz gauge. The tests demonstrated that water did provide the same pressure enhancement, nominally 2 GPa at 1.2 GW/cm^2, on both aluminum and black paint surfaces. In addition, pressure attenuation profiles similar to those in Figure 3a were also observed in 5086 aluminum using a water overlay [19]. Although these results confirmed the value of a water overlay, subsequent experiments continued to use quartz overlays when necessary to compare results to previous work.

It was also important during this early stage to understand the temporal relationship between the laser pulse and the pressure pulse. A direct comparison of a set of laser and pressure pulse profiles from the same shot is shown in Figure 4. It clearly shows that the rise time of the pressure pulse coincided with that of the laser pulse, and the pressure pulse was nominally twice the width of the laser pulse [20]. Since most of the beam energy initially goes into heating the plasma, driving the pressure, the leading portions of the laser and pressure pulses are similar. After the peak of the laser pulse, the pressure decays, but more slowly than the laser pulse, at a rate determined by the work against the confining materials by the continued expansion of the plasma and loss of thermal energy to the colder surroundings.

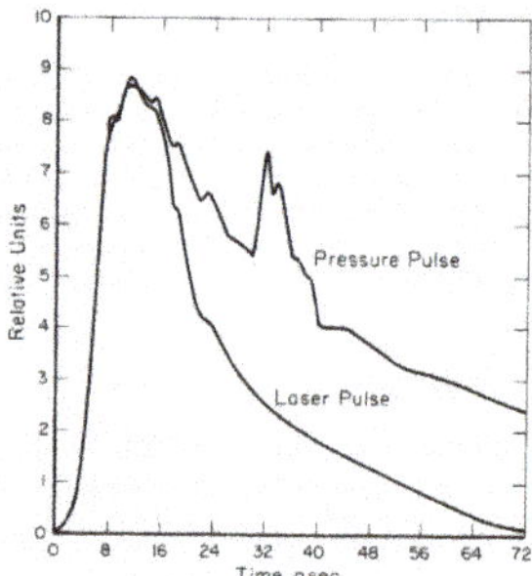

Figure 4. Comparison of laser and pressure pulse profile for the same shot measured for aluminum vapor deposited on a quartz gauge with water overlay, 1.2×10^9 W/cm^2 [20].

The code used for the first predictions of the shock pressures, shown in Figure 3, was of limited use. To support better understanding of the of the laser shock process going forward, the first robust model of laser induced confined plasmas was developed. A one-dimensional model named LILA, based on the method of finite differences, was written in the mid-1970s to model the laser induced pressure on a confined surface. LILA was then used for all subsequent pressure predictions.

Following development of this model, a number of pressure measurements and predictions were performed to investigate various combinations of transparent and opaque overlays, including iron with lead and quartz overlays, aluminum with water overlay, zinc with water overlay, black paint on aluminum and other combinations [21,22]. An example of water overlay on aluminum foil is shown in Figure 5 [21]. There is good agreement between the peak pressures, although the calculated rise time at the front of the shock wave is slower. The model for zinc foil with a water overlay showed similar agreement, but with the experimental trailing pressure much lower than calculated.

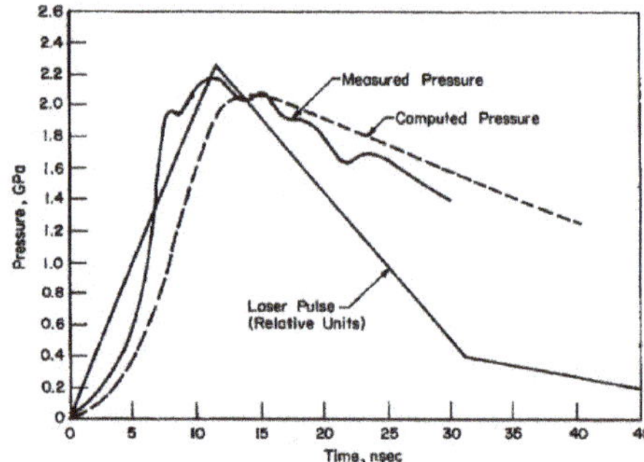

Figure 5. First modeling of pressure pulse for water overlay over a 3 µm foil of aluminum against a quartz gauge [21].

The first investigation of the dependence of peak pressure on power density, both experimental and predicted, is shown Figure 6. The pressures were measured using quartz pressure gauges with either a 3 μm-thick metallic film vapor deposited directly onto the front electrode surface of the quartz gauges, or with 8–10 μm of ultraflat black Krylon paint sprayed onto the surface of the gauges. For transparent overlays, the films were covered with either 3 mm-thick disks of fused quartz, or 3 mm thickness of distilled water. The laser spot size was several times the gauge inner electrode diameter to ensure one-dimensional strain conditions in the gauge [21].

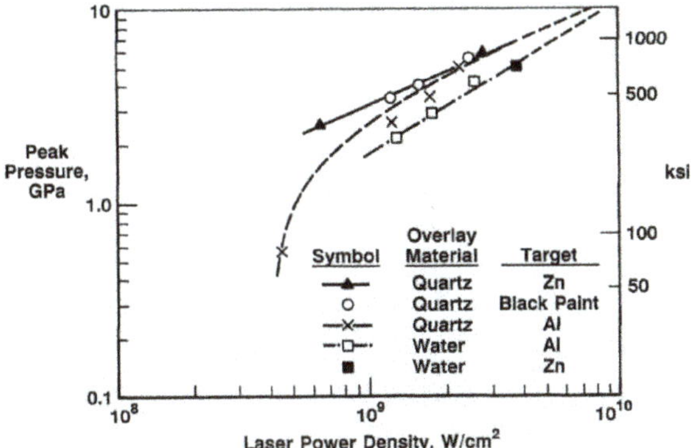

Figure 6. Comparison of predicted and measured pressures for aluminum, zinc and black paint confined by quartz and water overlays. The data points are experimental measurements. The curves are predicted by the LILA code [21].

The figure clearly shows the higher peak pressures reached using quartz overlays compared to water overlays due to the much higher acoustic impedance of quartz relative to water. The pressures created by the zinc and black paint are higher than for aluminum when using quartz overlays at the lower power densities. This was attributed to the higher thermal conductivity of aluminum conducting thermal energy from the plasma into the target. The lower thermal conductivities of zinc and black paint minimize this effect. This effect disappears at higher power densities. The agreement between the experimental and predicted pressures is very good. This series of experiments demonstrated that black paint would make an ideal opaque overlay. It could be easily applied to and removed from any surface to both protect the surface and provide a consistent surface for processing.

During this same time period, 1971–1974, others were also pursuing investigations of laser shock-induced material effects. O'Keefe et. al. investigated the laser shock-induced deformation modes in thin 6061-T6 aluminum and stainless steel targets using a Nd-glass laser and fused quartz or Plexiglass for confining the plasma [23]. They attributed the time sequence of events during bulging and puncturing the thin targets to the interplay of the dilatational and shear waves generated by the pressure pulse. Fox examined the effects of water and paint overlays on cracking and spalling of plexiglass, 6061-T6 aluminum and lead [18]. In addition, he also investigated the overlays' effects on the peak pressure at the back surface of 1 mm-thick 6061-T6 aluminum coupons. The peak pressure increased as the surface condition was varied between bare, paint only, water only, and water plus paint. At the same time, Yang reported on an extensive study to determine the sensitivity of the peak pressure generated by a confined plasma to target composition, target thickness, and energy density [24]. He found that the peak pressure was relatively insensitive to the target material, and discussed the results in terms of various aspects of plasma generation and thermal effects.

This program helped to understand in general terms the dependence of peak pressure on power density, the pressure pulse relationship to the laser pulse, the use and selection of viable overlays and the in-material plastic strain patterns. The plastic strain distributions observed in the etch pitted Fe-3Si demonstrated that depending on the target geometry, the interactions of the shock wave from internal surfaces could create different strain distributions.

3.2. Exploring the Effects of Laser Shocks on Material Properties

By the mid-1970s, although there remained much to learn about the characteristics of laser shocks, how to produce them, and how to adapt the means to produce them to achieve a desired result, the salient features of laser shock waves and how to apply them were beginning to take shape and define a process for application to metals. However, to maintain essential funding for developing a laser shock process it was necessary to begin identifying potential commercial uses for the process. The question was, what material properties driving commercial applications, if any, would be most affected in a positive, beneficial way by laser shocking as a process? Could it be developed into a commercially viable process? After all, flyer plate, explosive, and other similar methods had been around for years and had very limited commercial success, and then only in niche applications, such as welding. Laser shocking did have advantages over these earlier technologies. A big advantage was that it was non-contact and treatment could be limited to only the location on a part where it was needed. It appeared that with the use of black paint and water or water only, seldom would other special surface preparations be necessary. Additionally, the shock delivery system could be physically separated from the part manipulation system. The part could be manipulated to the beam by a robot or other tooling already widely used in manufacturing. The Battelle team was confident that a laser facility with sufficient power and processing speed could be reduced to a size compatible with safe processing in a manufacturing environment. It remained to convince others this was a promising, new metal treatment that had strong potential to be developed into a manufacturing process. To do this, it would have to be demonstrated that the effects of laser shock processing on commercial metal alloys would potentially increase strength and/or service life beyond the reach of existing technologies.

In the mid-1970s, one possible area of interest was the strengthening of weld joints in welded aluminum structures. Dogbone-shaped tensile specimens, 3 mm thick, of 5086-H32 and 6061-T61 aluminum alloys containing a transverse weld were laser shocked over the weld and heat affected zones simultaneously from both sides [25]. In the welded condition, both alloys have the same strength i.e., the weld was neither work hardened or precipitation strengthened. After laser shocking, the yield strength of the welded joint in 5086, a strain hardened alloy, was increased to nearly that of the parent alloy by laser shock induced work hardening. By comparison, the yield strength of the welded joint in 6061, a precipitation hardenable alloy, was increased to only midway between the welded and parent levels, at about the same strength as the shocked 5086 alloy. Figure 7 shows the sequential changes in microstructure: before welding, at the edge of the heat affected zone (HAZ) and after laser shocking. The initial microstructure of the 5086-H32 alloy has a fine-grained recrystallized microstructure. The edge of the HAZ has a coarse-grained annealed microstructure with few dislocations. The laser shocked weld zone has the dislocation clusters and tangles of a cold worked microstructure. By comparison, the initial microstructure of the 6061-T6 alloy contains fine lathe-like magnesium silicide precipitates and larger manganese-rich precipitates for strength, but few dislocations. The edge of the HAZ shows the magnesium silicide precipitates have dissolved. The laser shocked microstructure shows a somewhat higher and more tangled dislocation density than the 5086 alloy. The microstructures after shocking showed dislocation densities typical of cold working. In the 6061 alloy, the precipitates responsible for the strength in the T-6 condition had dissolved in the weld and HAZ zones and the laser-induced work hardening was unable to fully compensate for the absence of the precipitate strengthening. For both alloys the relative increases in ultimate tensile strength and hardness were smaller than the increases in the yield strength. It was also found that shocking both sides simultaneously increased the strength more than sequentially shocking both sides. This was expected from observations that simultaneous shocking significantly increased the hardness at the mid-plane of thin cross sections due to increased cold working from the superposition of the opposing shock waves. In addition, a set of shock wave attenuation curves for different thicknesses of 5086 aluminum were very similar to those shown in Figure 3 [26].

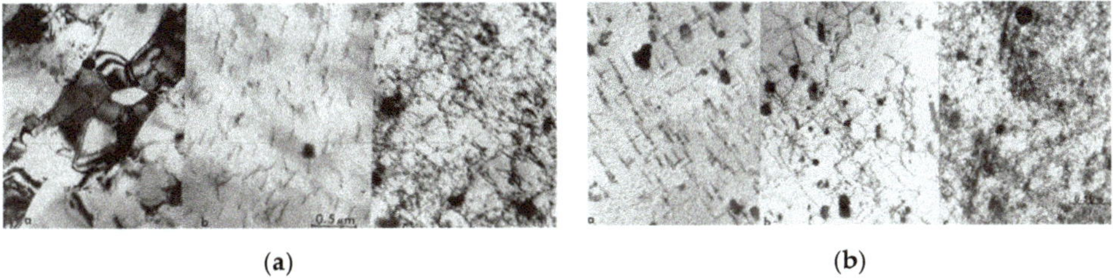

(a)   (b)

Figure 7. TEM micrographs of the microstructures of the welded and shocked aluminum alloys: (**a**) 5086-H32 alloy, left to right: as-received, weld heat affected zone (HAZ), laser shocked; (**b**) 6061-T6 alloy, left to right: as-received, weld HAZ, laser shocked. Reproduced with permission from [25], The Minerals, Metals & Materials Society and ASM International, 1977.

About this same time, the National Aeronautics and Space Administration (NASA) agreed to support an investigation on alloys and properties of interest to them. These included the effect of laser shocking on hardness and tensile strength, and stress corrosion and stress corrosion cracking resistance of 2024 and 7075 aluminum alloys [27]. The 2024 alloy was treated in the lower strength T351 temper and the higher strength, slightly overaged T851 temper. The 7075 alloy was treated in the peak aged T651 and overaged T73 tempers. There were several parts to this investigation. One was intended to compare the hardness response of 2024 to laser shocks and flyer plate shocks to determine whether there were any significant differences that may be related to the different shape of the shock waves. Concurrently laser shocking for tensile strengthening would be examined including transmission electron microscopy of the shocked microstructures. The program would also survey stress corrosion cracking behavior by polarization curves and corrosion crack initiation tests.

The hardness response in each alloy was examined over a range of peak pressure with longer pulse lengths than generally used today. With laser shocks applied with increasing shock peak pressure, the surface hardness of the 2024-T351 condition began increasing at about 1 GPa consistent with an HEL less than 1 GPa (Figure 8a). The T851 condition did not show any hardening with increasing pressure up to 5 GPa, the highest laser shock pressure (Figure 8b). For comparison, Herring and Olsen treated this same alloy in similar aged conditions with flyer plate shocks of 150 ns shock duration at increasing pressure [28]. The initial hardness of the comparable alloys is in good agreement. Despite differences in the shape of the shock wave between the two methods, the data appear to blend together well. The combined data show that the T351 condition reaches a saturation level of hardening at about 5 GPa, and the T851 condition does not show hardness increasing until about 5–6 GPa as defined by the flyer plate data. Figure 8b shows that the laser shocking and flyer plate shocking data are in good agreement at 5 GPa. Although the initial hardness of the two tempers differs by about 15 DPH (Diamond Pyramid Hardness), the saturation hardness level is the same, about 180 DPH. This suggests that the hardness of the T851 temper did not increase until the cold work hardening component exceeded the age hardening component. Then, however, with further increasing peak pressure the hardness increased at a rate similar to the T351 temper to saturation. This may also be related to the lower strain hardening rate for T851 observed in tensile tests. For comparison, a heavily hammered surface gave a hardness of 165–178 DPH [26].

To investigate effects on tensile strength, the test specimens were 1 mm thick and laser shocked either on one side only or on both sides simultaneously to increase the plastic strain at the mid-thickness where the two shock waves superpose. After laser shocking, the yield strength of 2024-T351 did increase, but the ultimate strength remained the same. The total elongation decreased, but the reduction in area increased by a factor of two or more. From limited testing, the yield and ultimate tensile strength of 2024-T851 were relatively unchanged, the total elongation slightly reduced and the reduction in area slightly increased. These changes in yield strength with laser shocking are consistent with the observed changes in surface hardness. For 7075-T651 the changes were similar to 2024-T851. The yield and ultimate strengths increased for 7075-T73, but the total elongation and reduction in area were relatively unchanged.

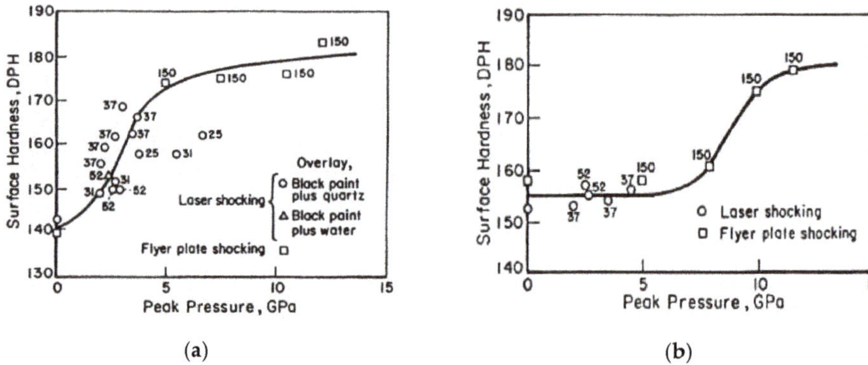

Figure 8. Shock-induced surface hardening dependence on peak pressure. The data point numbers are the pulse width for laser shocks and shock wave width for flyer plate shocks: (**a**) 2024-T351; (**b**) 2024-T851. Reproduced with permission from [26], ASM International, 2019. [27].

Transmission electron microscopy of the slightly over aged 2024-T851 and peak aged 7075-T651 coupons showed lower and more uniform dislocation densities, whereas the natural aged 2024-T351 showed dense dislocation tangles and overaged 7075-T73 showed dense dislocation bands. This is consistent with no discernable hardening in the peak aged conditions and the obvious hardening response in the non-peak aged conditions [26].

Polarization curves were measured in aerated 3.5% NaCl solution for both alloys, on sheet cut both parallel and perpendicular to the rolling direction, shocked and unshocked. The tests on 2024-T35 showed little difference between the shocked and unshocked conditions, but did suggest that the corrosion rate for the shocked condition was lower. At higher potentials where pitting originates, the results were consistent with enhanced pitting resistance after laser shocking. The tests on 7075-T651 showed much less effect of shocking. There was an indication that there was an increase in pitting resistance, but not on pit propagation behavior after shocking. Overall, the results indicated that the effect of shocking on stress corrosion cracking resistance should be greater in 2024-T351 than in 7075-T651 [27].

Crack initiation tests were conducted with specimens fixed in a four-point bend jig with outer fiber stress of 60% of the yield, alternately immersed with a cycle of 10 min immersed and 50 min air dry in 3.5% NaCl over a 21-day period. Both shocked and unshocked specimens showed many secondary intergranular cracks, but shocking did have some effect in making the surface more resistant to corrosive attack. However, this was more pronounced in the 7075-T651 than in the 2024-T351, contrary to the polarization results. Concerning time to initiation of stress corrosion cracks, shocking provided no benefit to 2024-T351, cracks appeared about nine days earlier in shocked than in unshocked specimens. However, 7075-T351 did show some benefit. Cracks appeared in two unshocked specimens after 13 days, whereas it took five more days to initiate cracks in shocked specimens. Unfortunately, the crack propagation studies were inconclusive due to limited specimens and experimental difficulties. Overall, the electrochemical and crack initiation experiments did not indicate which alloy was aided more by laser shocking [27].

This program supported the earlier results that the surface of precipitation hardened aluminum alloys in the peak-aged condition did not increase in hardness with laser shocking at the lower power densities usually applied to them. In any case, laser shock strengthening is only effective for thin sections, but can be enhanced by simultaneous, split beam shocking. The very limited corrosion investigation suggested that laser shocking could benefit the 2024 alloy, while the corrosion cracking investigation indicated it could benefit 7075.

Late in the 1970s a research program supported by the Army Research Office investigated the possibility of developing pressure-induced ω phase in titanium-vanadium alloys using laser induced shock waves [21]. To increase the chance for success, it was necessary to increase the laser induced shock pressure on the Ti-V disk specimens. Two approaches were evaluated, one using a high acoustic impedance tungsten backup to a 2.5 mm-thick Ti-V disk to reflect a magnified compressive wave from the back surface of the target, and the other to simultaneously laser shock the front and back surfaces of the Ti-V disk, superimposing the compressive waves at the mid-plane of the disk. Modeling these two scenarios with a quartz overlay at a laser power density of 3 GW/cm^2 predicted peak pressures of 10.2 GPa with the tungsten disk backup compared to 12.5 GPa with simultaneous laser shocks. Unfortunately, no ω phase was detected by either X-ray or microstructural analysis, perhaps because the pressure pulse was too short.

Beginning in 1977, Battelle, sensing commercial potential in laser shock processing, began to fund exploratory research to demonstrate benefits for commercial applications. This required identifying applications where laser shocking could enhance properties of commercial alloys to increase their commercial value. It was suggested by Steve Ford that Battelle consider fretting fatigue around fastener holes in aircraft structures, a concern in the late 1970s. The test specimen is shown in Figure 9a [29]. This specimen paired a tensile specimen and rectangular pad of 7075-T6 aluminum fastened together with a steel aerospace quality aircraft fastener through a hole in the pad and the gauge length of the tensile specimen. The difference in the cross-sectional areas of the pad and tensile specimen caused a 30% load transfer, creating a cyclic fretting strain differential between the two pieces at the fastener hole. Laser shocking was simultaneously applied to both sides of the fastener hole of the fatigue specimen with a 13 mm-diameter spot centered on the hole. The tensile fretting fatigue results are shown in Figure 9b. These very encouraging and welcome results pointed toward a focus on fatigue related properties as a promising path to commercial use for laser shock processing [26].

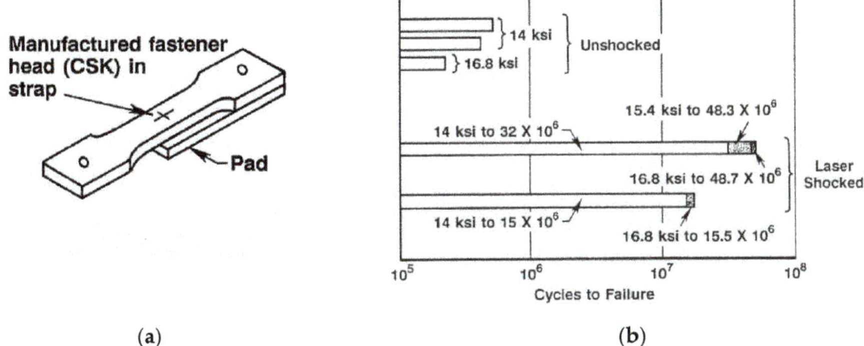

(a) (b)

Figure 9. Fretting fatigue of 7075-T6 laser shocked and unshocked in tension with 30% load transfer, $R = 0.1$. (14 ksi = 96.5 MPa, 15.4 ksi = 106 MPa, 16.8 ksi = 115.7 MPa): (**a**) the test specimen; (**b**) test results. The stresses indicate steps in the applied stress [29].

Post-test examination showed the fretting surface contained short fretting cracks, but no differences due to laser shocking. At the time, the reason for the life improvement was not clear. It was speculated that the fatigue life improvement may have been due to compressive residual stress, but an earlier measurement of residual stress in 7075 showed only about 10 ksi (68.9 MPa) surface compressive stress. This earlier measurement was the first measurement of residual stress in a laser shocked surface and there was no other data to compare it to. This low surface stress can now be attributed to a low power density shot. It was also puzzling that the fretting test was duplicated with a shot peened surface and there was no life increase, although it was expected that the surface compressive residual stress would be much higher than 10 ksi (68.9 MPa). It was only after residual stress measurements were made later, that the cause of the extended fatigue life in the laser shocked specimens was understood to be the deeper compressive stress inhibiting the growth of the short surface fretting cracks deeper into the surface.

It was then decided to do a quick test to determine whether crack propagation could be slowed by laser shocking as would be expected if residual stresses were induced. A 0.5 mm deep notch was machined into each side of the hole in the dog-bone tensile fatigue specimen used for the fretting fatigue tests and laser shocked as in the fretting test. The specimens were tested at 82.7 MPa, somewhat lower than the fretting fatigue tests. After the test, the unshocked specimen had a single crack emanating from the root of each notch, one across the width and the other nearly across the width, failing at 4.3×10^5 cycles. By comparison, the laser shocked specimen did not fail from the notches, instead, repeated failure of the grips necessitated terminating the test at 2.3×10^6 cycles. After the test, several small cracks were observed at the root of each notch with the maximum crack growth being 0.8 mm [26]. This dramatic demonstration of crack growth retardation after laser shocking confirmed significant potential for laser shock processing to enhance fatigue properties; another encouraging early result.

This led to the first study of the effect of laser shocking on fatigue strength. Some interest had been expressed concerning increasing the fatigue strength of welds in aluminum, so welded 5456-H116 aluminum alloy tensile fatigue specimens were tested after laser shocking the weld and heat affected zone. The results of these first fatigue tests on laser shocked specimens are shown in Figure 10. At 25 ksi (172.3 MPa), laser shocked specimens ran out at 5×10^6 cycles, compared to typical runouts below 17 ksi. The fatigue life was improved by more than an order of magnitude [30].

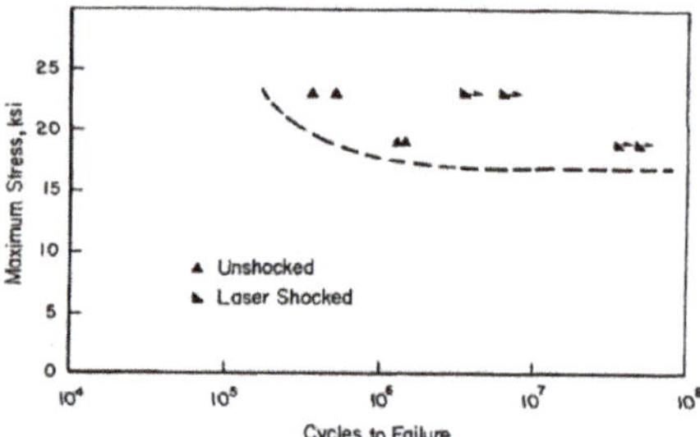

Figure 10. Effect of laser shocking on the fatigue life of welded 5456 aluminum, tension, $R = 0$. The dashed line represents the typical, unshocked tensile fatigue curve for this condition (10 ksi = 68.9 MPa) Reproduced with permission from [30], Springer US, 2019.

Other exploratory tests funded by Battelle included laser shocking ceramics and stainless steel. Laser shocking silicon nitride showed a small hardness increase after laser shocking, indicating it might be possible to develop a compressive surface stress in this ceramic. Additionally, an attempt was made to create a compressive residual stress near the back surface of yttrium stabilized zirconium coupons by driving the tetragonal to monoclinic phase transformation with the reflected tensile wave. This transformation is accompanied by a volume increase and can be activated by a localized tensile stress. It was considered that the toughness of this ceramic could be complimented by a compressive residual stress created by the volume expansion. However, for the limited conditions tried, laser shocking caused only cracking and fracture of the zirconia. Further, to take advantage of the high work hardening behavior of 304 stainless steel, the surface was shock hardened with multiple shots on the same spot. The surface hardness increased steadily with the number of shots, increasing nearly 70% in hardness after 10 shots [30]. Wear and galling tests after laser shocking showed no discernable improvement in wear, but did appear to reduce galling.

Throughout the 1970s, laser shocked microstructures were examined by transmission electron microscopy in aluminum alloys, including weldments, 304 stainless steel, and Ti-V alloys. The dislocation microstructures were those typically observed in shock hardened alloys. They consisted of greatly increased dislocation density, dense dislocation tangles, some evidence of bands of high dislocation density indicating localized high shear strain in 7075. Some twinning was observed in 304 stainless steel. The first transmission electron microscopy micrographs of high pressure laser shocked structures were made by Wilcox [12].

Based on the fretting fatigue results and the non-propagation of cracks from a notch in the fastener hole of the fretting fatigue specimen described above, in 1978 the US Air Force funded a program to investigate laser peening fastener holes in 2024-T3 and 7075-T651 alloys to mitigate crack initiation and propagation from these holes in aircraft structures [31]. The investigations included fatigue tests for large laser spots centered on 3 mm diameter holes in 3 and 6 mm thick sheet, crack initiation and growth with laser spots slightly overlapping each side of the hole, fretting fatigue, and a limited comparison between constant stress amplitude cycling and a flight-by-flight spectrum (variable stress cycling) for fatigue testing. Quartz and black paint overlays were used throughout the program except for limited tests with water and transparent plastic tape overlays on black paint. In retrospect, it is not clear why quartz overlays continued to be used. It was probably because it was desirable to maximize the pressures for the power densities used at the time. The fatigue specimens were large, 457 mm long with a 250 mm × 102 mm gauge section. Two 3.2 mm diameter holes were drilled along the central axis of each gauge section 102 mm apart. Each hole had side notches having a radius of 0.75 mm to facilitate crack initiation. An 11 mm-diameter laser spot was centered on the predrilled hole, providing 3 mm of laser shocked surface surrounding the notches. The crack initiation and propagation specimens had only one hole with an 11 mm spot overlapping the notches on each side of the hole to provide a longer laser shocked path in front of the cracks.

Residual stress measurements on laser shocked specimens were made to confirm the expectation that laser shock induced compressive stresses were the source of the fatigue life improvements previously observed. These surface stress profiles were measured using X-ray diffraction with measurements spaced across the diameter of the laser spot as shown in Figure 11. The measurements were made to determine whether the magnitude of the surface stress depended on drilling the hole before or after laser shocking. The profile before hole drilling shows the maximum compressive stress at mid-radius as confirmed later by others, but the residual stress outside the hole is the same whether the hole is drilled before or after laser shocking. Based on these results, the holes were predrilled during fabrication of the test specimens and the laser spots were centered on the hole. A few tests were made using water and plastic adhesive tape overlays at higher power densities with mixed results.

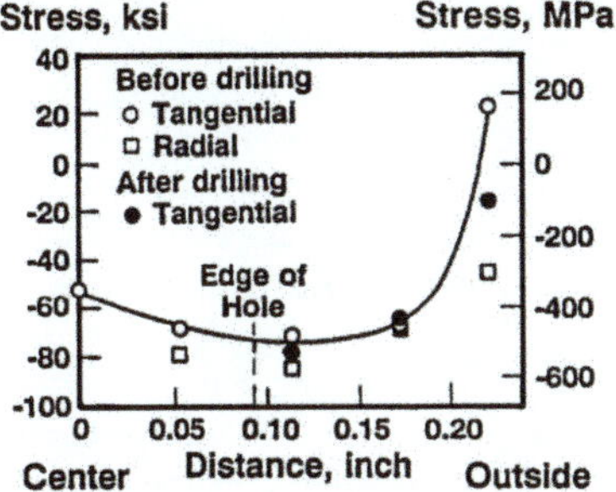

Figure 11. The first residual stress measurements on laser shocked 7075-T651, 5 GW/cm^2, 3 mm quartz and black paint overlays, 11 mm spot diameter, 6 mm-thick specimens (0.1 inches = 2.5 mm) [29].

The fatigue life of 2024-T3 was extended up to an order of magnitude for both the 3 and 6 mm thicknesses after laser shocking around the holes. However, laser shocked 7075-T651 showed an increase in fatigue life only for the 3 mm thickness specimens. In fatigue testing using a flight-by-flight stress spectrum (a cyclic stress profile having varying stress amplitudes that simulates stress variations during service), 7075 showed improvement by laser shocking at the 40 ksi (275.6 MPa) maximum stress, but little or no benefit at 15 ksi (103.4 MPa) or 17 ksi (117.1 MPa) constant stress amplitude tests. This was attributed to the lower average stress level for the flight-by-flight tests.

The crack propagation results for 2024-T351 are shown in Figure 12. For comparison, the top two sets of bars represent fatigue lives of non-precracked specimens shocked with a 13 mm diameter spot centered on the 6 mm hole. The crack propagation specimens were pre-fatigued to grow a 0.5 mm crack from the notches on each side of the hole, then laser shocked with 11 mm spots as shown in Figure 12a. The effect of laser shocking ahead of the pre-existing crack on fatigue life is shown in the lower set of bars in Figure 12b. Laser peening over an existing crack significantly slowed the crack growth rate and produced a fatigue life approaching that of the non-precracked condition.

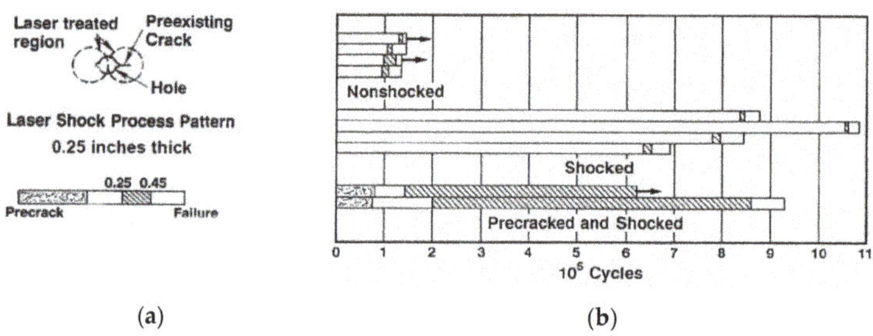

Figure 12. The effect of laser shocking around holes to mitigate crack initiation and propagation in 2024-T351 6 mm thick plate: (**a**) the laser shock pattern around the hole; (**b**) test results. The cross-hatched portions are the cycles for the longest crack to increase from 6 to 11 mm long from the center of the hole (0.25 inches = 6 mm) [29].

Fatigue tests using a flight-by-flight spectrum on precracked specimens of 7075-T651 showed a significant reduction in crack propagation rate by half to a third, probably due to the number of low load levels in the flight spectrum. Low-load-transfer fastener joint fretting tests for 7075-T651 showed a factor of 2–3 improvement in life for lower maximum load flight-by-flight tests, but none for higher maximum load tests. In light of other work on 7075 aluminum before and after this program, it is clear that the higher strength 7075-T651 specimens were not laser shocked with sufficient intensity to achieve better fatigue results [30,31].

At the completion of the program, although some benefits were demonstrated, they were not sufficient to continue the program. Looking back, this outcome can be attributed in a large part to having used lower power densities than are now applied, not applying multiple impacts and not shocking material a larger distance from the edge of the hole. Additionally, in retrospect, over 30 years later, Ivetic et al. demonstrated that drilling the hole after laser peening may well have led to longer fatigue lives in this program by reducing or eliminating the mid-thickness tensile residual stress on the hole surface [32]. In this case, even though it may have extended the fatigue life significantly, it would probably have been difficult to implement in the manufacturing process. At the U.S. Air Force's request, one part of the program developed a design for a pre-prototype laser looking forward to eventual commercialization of laser shock processing. Later, this design provided the starting point for designing and building an industrial pre-prototype demonstration laser at Battelle in the mid-1980s.

Although the results of the program were disappointing, the team gained a great deal of valuable experience. The laser peened area around the holes should extend further from the hole. Multiple shots and higher power densities should be applied to achieve deeper residual stresses and/or cold work.

In addition, applying multiple shots on the inside surface of the hole to inhibit in-hole crack initiation would have given better results. These lessons would be applied in the future.

After the U.S. Air Force program ended in 1979, Battelle funded a program to extend the investigation of laser shocking and fatigue phenomena in an aircraft structural alloy, 2024-T3 aluminum [33]. The work focused on issues associated with fastener holes noted in the preceding Air Force program. There was still no emphasis on using water as the transparent overlay for process development work at this point, so this program relied primarily on quartz overlays to enhance the shock pressures. Acrylic transparent overlays were also used for residual stress comparisons. The acrylic overlay produced residual stress levels and depths comparable to the quartz overlay, but showed scatter that indicated more testing would be necessary to use it with confidence.

The fastener holes were 4.7 mm in diameter. The laser spots were either 11 or 16 mm in diameter and placed concentric to the holes after the holes were drilled. A few tests were made using spring loaded momentum traps placed on the rear surface of a hole to explore processing changes to address instances where there was laser beam access from only one side of a thin section and it was necessary to minimize distortion.

In the Air Force program, it was observed that during fatigue of the laser shocked holes, the crack initiated on the surface of the hole at mid-thickness where the compensating tensile residual stress resided. A comparison of the crack initiation and propagation behavior for unshocked and split beam shocked holes is shown in Figure 13 as maps of the progression of the crack front. In the unshocked condition, the crack opens along the entire height of the hole before propagating away from the hole with a straight front. In the shocked condition the crack initiates on the hole surface at mid-thickness of the sheet, followed by tunneling between the compressive surface stresses until it is beyond the laser shocked area. While tunneling it is not visible on the surface and when the ends of the crack do break through to the surface, the compressive stress clamps it closed, making it very difficult to detect. By the time the crack is detected outside the laser shocked spot, it is already many millimeters long, and rapidly propagates to failure. Not being able to see a propagating crack concerned the Air Force.

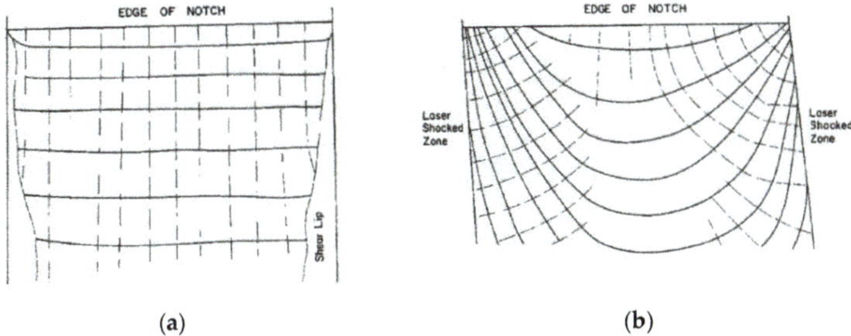

Figure 13. Maps of the crack front progression from the tip of the notch in the side of a hole in 2024-T3: (**a**) not shocked; (**b**) shocked both sides simultaneously with a split beam [33].

To address this problem, the shape of the beam was changed from a solid spot to an annular shape as shown in Figure 14a. This would enable a crack emerging from the hole to be observed at the surface shortly after initiation, but slow its growth when it encountered the compressive stresses from the annular beam. The annular beam was applied concentric to the hole with about 2 mm between the edge of the hole and the inside edge of the annular spot. It turned out that this configuration also created a lower surface compressive stress inside the annulus, in the unshocked region to the edge of the hole. This laser shocking configuration was effective in slowing crack propagation outward from the fastener hole, but not as effective as a full circular spot, as shown in Figure 14b. However, the annular beam would provide some factor of safety for inspection or delaying a repair, by, in this case, about a factor of two.

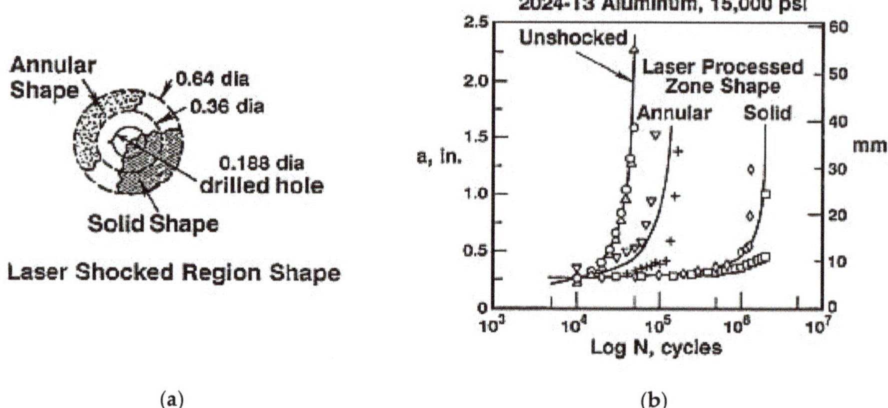

(a)   (b)

Figure 14. Crack propagation comparison for solid spot and annular spot laser shocking. The hole is 4.8 mm diameter. The annular beam has 9.1 mm ID and 16.3 mm OD; (a) annular spot application configuration; (b) crack growth comparison for shocked annular ring (15,000 psi = 103 MPa) [29].

Despite their commercial importance, up to this time no laser shock processing had been tried on steels. In 1980, Battelle funded a small exploratory task on 4340 steel having hardness levels of 42 Rc and 54 Rc. 4340 steel is often used in cyclic fatigue environments. The first fatigue tests used dog-bone shaped sheet specimens 1.5 mm thick and 38 mm wide having side notches 15.2 mm deep with a root radius of 7.6 mm giving a stress concentration of $K_T = 1.3$ at the bottom of the notches. The 7.6 mm of steel bridging the roots between the opposing notches was laser shocked on opposite sides simultaneously with a 10 mm-diameter spot, applying either one or five shots at 8.5 GW/cm^2, 15 ns, using quartz and black paint overlays. The surface compressive stress reached about half the tensile strength of the steel after five shots for each hardness. The depth of the compressive stress was limited by the sheet thickness to about 0.45 mm for both one and five shots.

The fatigue results for the 54 Rc hardness specimens after five shots are shown in Figure 15. The unshocked curve beyond 10^5 cycles is handbook data. Specimens 2, 3, and 4 were step loaded. The fatigue results were very encouraging with the fatigue strength increasing over 70% after laser shocking. These tests demonstrated that a significant increase in fatigue life could be achieved by laser shocking both sides of a thin cross section in the vicinity of a stress riser. This would later be the case for laser shocking the leading edge of airfoils.

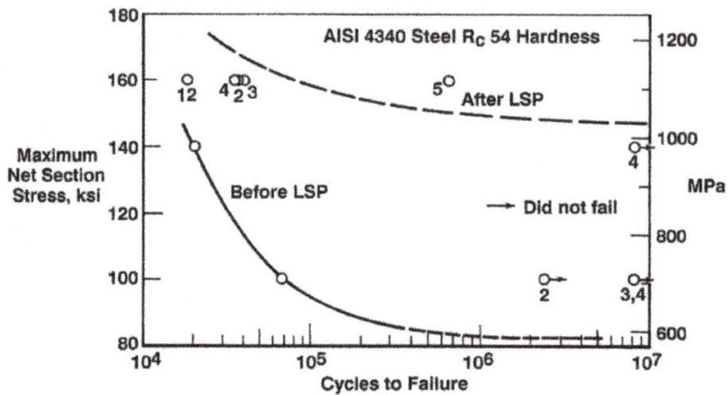

Figure 15. First fatigue life tests on laser shocked steel, 4340 steel at 54 Rc hardness, $R = 0.1$ [29].

Another set of tests using 4340 steel at 54 Rc involved laser shocking directly into the notch of beam specimens loaded in four-point bending. The specimens were 7.5 mm wide by 19 mm high by 204 mm long. The notch in the tensile surface of the beam had a root radius of 4.5 mm and a depth of 1.5 mm. It was laser shocked with multiple shots using a 9 mm diameter spot centered on the notch. The increase in fatigue strength was at least 30% over the notched, unshocked condition. At that load level the beam deformed under the loading rods, preventing testing at higher loads.

These tests demonstrated that a significant increase in fatigue life could also be achieved by laser shocking directly into a stress riser such as a notch or fillet in a thick section, e.g., a change in diameter of a shaft or the fillet at the base of an airfoil.

By 1980, after seven years of research, a basic understanding of the process had been achieved and its potential for increasing the hardness, strength and fatigue properties of metals had been demonstrated, along with some understanding of the effects of part shape and size. However, funding for further investigations of laser shock processing became difficult to obtain. The response for further funding from supporters of the technology was "It is time to go out and find someone interested in developing it commercially for specific applications." You have a "solution looking for a problem". The search for funding was hindered by the current large size of the laser, the slow repetition rate, and probable high costs of building a viable production prototype laser with no identifiable critical need. It was difficult for potential users to look past the current circumstances and envision a viable commercial process.

Finishing up the funded programs in 1981 and 1982, Clauer presented a paper at the Conference on Lasers in Materials Processing in Los Angeles in 1983 [33]. He believed this was the beginning of a long interruption in the development of laser shock processing until another group and organization in a more favorable situation continued the effort. Fortunately, this was not the case.

4. Path to Commercialization

The key event that enabled the development to continue, closely followed Clauer's presentation at the conference. Within a week after returning from the conference, he received a call from the plant manager of Wagner Casting Company, a cast iron foundry for automotive parts in Decatur, Illinois. The plant manager had attended the talk, and upon returning to Decatur, immediately discussed the possibilities of the process with the company management. The discussion concerned the potential of using laser shock processing to upgrade the fatigue properties of iron castings to make them competitive with wrought steel parts at a lower cost. Following a visit to Battelle by Wagner management it was decided Battelle would laser peen and fatigue test several different types of cast iron specimens. A few showed an increase in tensile fatigue strength of 10–15% encouraging further interest.

Wagner Castings was also considering buying a powder metallurgy plant and developed an interest in evaluating the potential of laser shock processing to improve the fatigue properties of ferrous powder metallurgy parts for automotive use. To investigate this possibility, automotive iron-nickel powder was pressed and sintered to 89% density directly into net shape tensile fatigue specimens. The specimens were $100 \times 25 \times 6$ mm having side notches 6 mm deep with a 6 mm radius. These were then laser shocked directly into the notches with an 11 mm diameter spot using water and black paint overlays over a range of pulse energies and 30 ns pulse length. The results are shown in Figure 16. Figure 16a shows the surface residual stress versus cumulative energy on a spot, because different combinations of pulse energy and number of shots were applied. The surface residual stress was unchanged after multiple shots of 30 J and single shots up to 70 J. For single shots of 100 J and multiple shots of 50 J and above there was a steady increase in the compressive stress with increasing cumulative energy. Figure 16b shows the fatigue life versus cumulative energy. It was somewhat unexpected to see that the fatigue life increased steadily with increasing cumulative energy, considering the porous nature of the sintered specimens. The fatigue life increased from 5×10^4 cycles to runouts at nearly 6×10^6 cycles using multiple 70 and 100 J pulses. All future laser peening from this point was performed with black paint and water overlays.

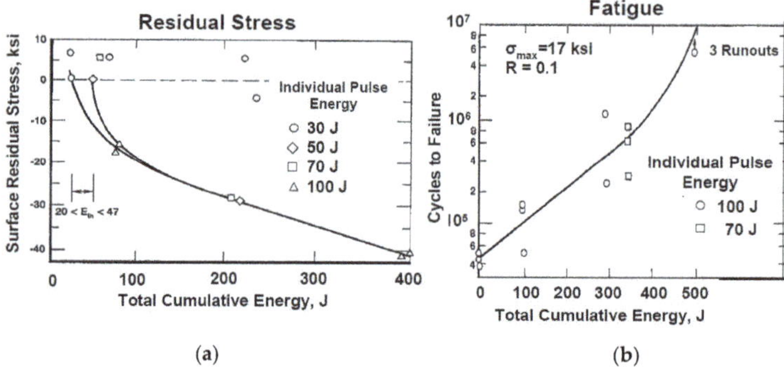

Figure 16. Surface stress and fatigue life of laser shocked, pressed and sintered Fe-2Ni-0.5C powder: (**a**) surface residual stress; (**b**) cyclic fatigue life.

Wagner's interests, along with the two larger programs on laser shock processing for enhancing fatigue properties of aluminum alloys described earlier, were now defining the primary focus of laser shock processing as laser shock peening (LSP). Based on the promise shown by LSP to deliver deep compressive stresses and improve fatigue properties of metals beyond that attainable by shot peening, along with other considerations, Wagner Castings purchased the exclusive worldwide license for LSP from Battelle.

To successfully commercialize LSP, it was necessary to design and build a high energy pulsed laser system that would demonstrate that LSP was capable of operating in a manufacturing environment. The system had to have a reasonably small foot print, a pulse frequency and energy demonstrating a capability to produce a reasonable throughput of product and meet environmental and safety requirements: all at a reasonable cost per shot. In 1984, Wagner Castings funded the design and construction of a first-generation prototype laser to demonstrate commercial viability and to process candidate commercial parts for potential users. Harold Epstein designed the laser and Jeff Dulaney oversaw the fabrication and testing of the prototype. In addition, also instrumental in the success of the prototype system were Battelle colleagues Mark O'Loughlin and Steven Toller. The prototype laser, shown in Figure 17, became operational in 1986, producing two beams of 50 J each, 20 ns at 0.5 Hz. In the Figure a He-Ne laser beam defines the beam path of the Nd-glass laser beam. The two cabinets behind the operator contained the electronic control system and capacitors. This was a major step away from the large system shown in Figure 1.

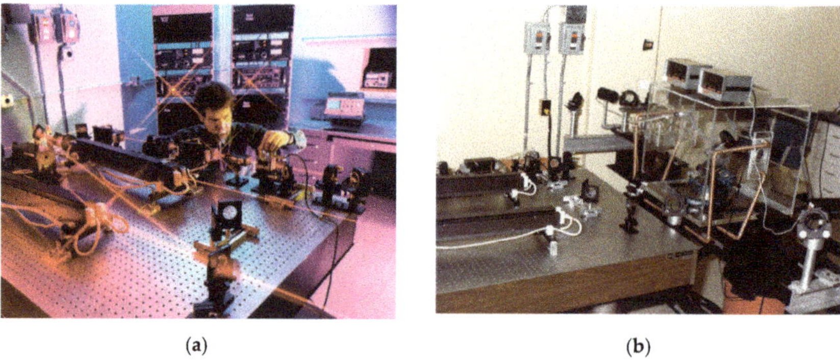

(a)   (b)

Figure 17. Laser designed and built at Battelle to demonstrate the ability to build the small, high energy pulsed lasers necessary to commercialize laser shock peening; (**a**) laser; (**b**) work cell.

After the new laser became operational, it was used for all subsequent laser shock peening using a Plexiglas box containing a three-axis stage and flowing water as the peening cell. During break-in testing of the equipment, it was found that the residual stress profiles from the prototype laser were unacceptable compared to the profiles generated by the CGE laser. The CGE laser used an aluminum blow-off foil to provide isolation between the oscillator and the amplifier chain, but this element was left out of the new prototype laser. In comparing the laser pulse temporal profiles generated by the two lasers, it was obvious that the aluminum blow-off foil of the CGE laser produced a sharp rise time on the leading edge of the pulse. The CGE laser pulse shown in Figure 4 clearly has a rise time of less than 5 ns. Now recognizing that a laser pulse having a steep rise time was a key element for better results, especially at higher power densities, a laser pulse rise-time modifying device was incorporated into the prototype laser. Rise-time modifying devices (e.g., Pockels cells, SBS cells, etc.) are now used in all production laser peening systems that do not naturally produce a rise time less than 5 ns.

In 1984, a marketing effort for laser peening was begun. The market was the equipment and parts producers, but very few, if any, people in industry had ever heard of laser peening. John Koucky, Vice President of Engineering for Wagner Castings, led this effort assisted by Clauer. Over the next nine years, they made many calls and presentations to companies throughout the aerospace, automotive, medical and other industries. With time, awareness and interest in laser peening slowly began to build. A major selling point of laser peening compared to other existing and developing surface treatment technologies such as shot peening and its variations and water jet peening, was that the residual compressive stress extends much deeper below the surface. The compressive stress introduced by the latter technologies is nominally 0.1–0.5 mm deep. By comparison, laser peening extends from nominally 1–1.5 mm in most applications. From this effort, there was a steady flow of parts and laboratory specimens to laser peen for fatigue testing and residual stress analysis, many of which had a problem of premature failure or a need to extend life and reliability without redesign. Although there was much interest in view of the fatigue and compressive residual stress benefits, there was a reluctance to implement the process commercially. The most significant obstacle was the absence of an immediate capability to provide laser peening services or production-ready systems. Another was that without an exceptionally dramatic improvement in properties or a compelling need to avert a crisis in product performance, the inertia stemming from the need to perform qualification testing, modify specifications, change product flow and possibly introduce a new process onto the factory floor was too great to overcome. The comment was made that it is much easier to introduce a new alloy into a part than a new process, even though the end result in product improvement might be greater. Fortunately, the crisis needed to pull LSP into an industrial process was coming.

In October and December of 1990, in-flight engine shut downs occurred during two B-1B bomber flights due to severe engine damage resulting from a fan blade being ingested into the engine [34]. These two incidents led to grounding for more than 50 days of all B-1B bombers not on nuclear flight status. The cause was traced to foreign object damage (FOD) on the first stage fan blades in the F101 engines. During flight, fatigue cracks initiating from the FOD propagated across the airfoil, causing it to separate from the blade and pass through the engine. To avoid these events, preflight checks of the leading edge of all the first stage fan blades were required, since a very small dent in the leading edge of only 0.25 mm could be a problem. Before each flight a thread or thumbnail would be passed along the leading edge of each first stage fan blade. If a defect snagged the thread or fingernail, the blade would be replaced before the next flight. These measures dramatically increased the maintenance costs to keep the planes flying and an element of risk was still present.

Serendipitously, early in 1991, Koucky and Clauer made a presentation at the Air Force Aeronautical Research Laboratories (ARL) within Wright-Patterson Air Force Base. Immediately after the meeting, William Cowey of ARL set up a meeting with the manufacturer of the engines, General Electric Aircraft Engines (GEAE) in Evendale, Ohio. Out of this meeting began the relationship between LSP Technologies and GEAE that became the genesis for solving the FOD problem with the F101 engine and the first commercial application of LSP. The action item from this first meeting was to laser peen a GEAE test coupon. The laser peened coupon demonstrated to GEAE the potential effectiveness of LSP.

Battelle/Wagner was then asked to laser peen four first stage F-101 engine fan blades for GEAE to test and evaluate. Battelle was asked to laser peen the leading edge using just one peening condition. When the laser peened blades were returned to GEAE, they gave them a rigorous fatigue test. Instead of placing a small damage site equivalent to the ones requiring a blade to be replaced in an engine, a 6 mm-deep notch was hacksawed into the leading edge in the fatigue sensitive location. When tested, the blades displayed the fatigue life of an undamaged blade, much to the disbelief of everyone concerned. After a few more test sequences, in 1995 the Air Force performed their own independent evaluation of the laser shocked blades [35]. They compared the fatigue properties of blades given several different surface treatments on the leading edge and types of simulated damage. The treated conditions included undamaged blades, blades shot peened to the manufacturing specification, high intensity shot peening to achieve greater compressive stress depth, and LSP. After laser peening, simulated foreign object damage was introduced into the leading edge of the surface treated blades, either a 6 mm-deep v-notch pressed into the edge by a chisel, or a 3 mm-deep electrical discharge machined notch. The former had a highly deformed, work hardened surface, whereas the latter had a recast surface layer which probably contained a fine network of shrinkage cracks. The fatigue test results are shown in Figure 18 [34]. The testing consisted of vibrating the airfoils at high frequency using a high velocity air jet, beginning at a maximum stress amplitude of 138 MPa, testing for 10^6 cycles, then raising the stress amplitude by 69 MPa and again testing to 10^6 cycles. This sequence was repeated until the airfoil failed. The results validated the original test results. As shown in Figure 18, damage with no pretreatment caused a large degradation in fatigue strength. Both shot peening treatments improved the fatigue strength somewhat compared to the untreated condition. The laser peened blades amazingly retained the fatigue strength of the undamaged blades. This wholly unexpected result occurred because the compressive residual stress extended through the thickness of the thin leading edge of the blade. The compensating tensile residual stress was positioned behind the laser peened strip, not at mid-thickness up to the leading edge.

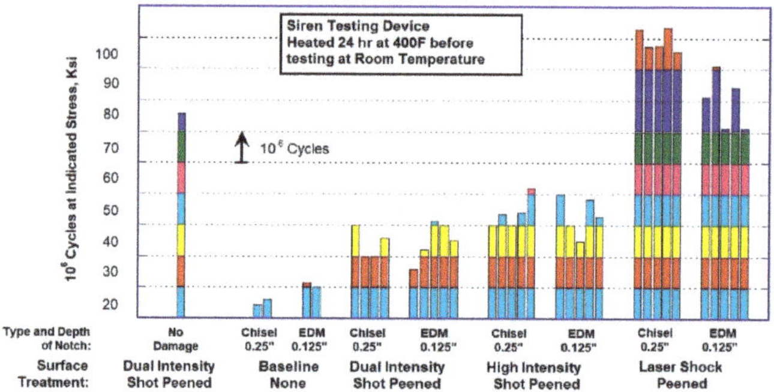

Figure 18. Fatigue strength of F101 fan blades comparing the influence of various surface treatments for protecting against the loss of fatigue strength from foreign object damage.

While the necessary further processing and testing of the laser peened fan blades was underway, by 1994, Wagner Casting had become a subsidiary of another company with a different business focus and they returned the license to Battelle. Dulaney, who had been leading the laser physics team, took advantage of this opportunity. He left Battelle in late 1994 to start up LSP Technologies (LSPT). His goal was to take LSP to the market. In 1995, he acquired an exclusive worldwide license from Battelle. Clauer retired from Battelle and joined LSP Technologies in 1995 to continue his quest to help commercialize LSP. Meanwhile, after a thorough testing program, GEAE and the US Air Force decided that laser peening had to be applied to the F101 1st stage fan blades to remove their vulnerability to FOD.

The most immediate need was to obtain a laser system capable of production laser peening. There were no commercial lasers available and all laser manufacturers were deemed to be high-risk suppliers of a high-energy, high-reliability laser system that met GEAE's specifications. With no existing commercial alternative, GEAE contracted with LSP Technologies to design and fabricate three production-capable lasers, under the strict guidance and control of GE's Corporate Research and Development center in Schenectady, New York. Todd Rockstroh and Seetharamaiah Mannava were the technical contacts within GEAE. These systems were delivered to GEAE in Evendale, Ohio throughout the late-1990s and were the first commercial lasers sold specifically for laser shock peening. With this equipment GEAE began production laser peening F101 1st stage fan blades in 1997 and became the first industrial user of LSP. After laser peening was applied to in-service blades reinstalled in the engines, they were no longer vulnerable to normal FOD and the preflight fan blade inspections were terminated.

In 1996, following LSP Technologies' delivery of the first production laser to GEAE, LSPT began a program funded by the US Air Force, to design, fabricate, and demonstrate a second-generation production laser for in-house use. This laser was completed and successfully demonstrated the following year and served many years as a production laser. In 1999, the Air Force awarded LSPT a multi-million dollar joint program with GEAE and Pratt and Whitney to design and build a production laser peening system as a manufacturing system prototype, expand the use of laser peening within gas turbine engines and develop non-aerospace commercial laser peening opportunities. In 1999, LSPT began working with Pratt and Whitney to apply laser peening to the airfoils of an integrally bladed rotor (IBR) for the F119 engine used on the F-22 fighter aircraft. LSPT began production laser peening on the IBRs in March 2003, becoming the first commercial provider of laser peening services. LSPT has grown and expanded production to other turbine airfoils and non-turbine components since then.

Through the late 1990s and early 2000s the financial support of the U.S. Air Force Aeronautical Research Laboratory at Wright-Patterson Air Force Base was critical to developing laser shock peening as an industrial process. In addition to the financial support, the technical support and advocacy of the Air Force Man Tech project engineers played a crucial role.

In the late-1990s, a third company became interested in laser peening technology. The Metal Improvement Company (MIC) began working with the Lawrence Livermore National Laboratory to design and build a slab laser for laser peening. They began production laser peening of fan blades for the Rolls Royce Trent 800 commercial gas turbine engines in 2003, and have expanded production to other blades and components since then. They have made significant innovations in laser peening systems and have extended laser peening to a production laser peen forming process forming wing skins for the Boeing 747-8 aircraft. With their entry into the field as an independent supplier of laser peening services, laser peening overcame a significant barrier for growth. Without the backup of an alternative provider of laser peening services, manufacturers have some reluctance to commit to using laser peening only to find that at some point they may be losing the only service provider.

5. Global Expansion of Laser Shock Peening

While process and business developments were being vigorously pursued in the United States, strong, productive research programs were being pursued elsewhere. In the mid-1980s, interest in laser shock peening took hold in France, an extension of their years of previous work with high-energy pulsed lasers. A short time before 1986, Jean Fournier began the first investigation into laser shock processing in France as his doctoral dissertation under Professor Remy Fabbro at the Ecole Polytechnique in Paris. In 1986, Fournier and Fabbro visited Battelle for a mutually beneficial extended discussion of laser peening with Clauer. Their first paper on laser shock waves using a confined plasma concerned determining the impulse imparted to copper specimens [36]. This was followed by a publication describing a model for pressure pulse generated by a confined plasma in 1990 [37]. In the years since then, Fabbro, Patrice Peyre and colleagues conducted an extensive, broadly-based program in laser peening centered around their high energy pulsed laser facilities. Their investigations have covered many aspects of laser peening, including material property effects such as fatigue, stress corrosion, wear, laser beam interactions with water overlays and target materials, pressure measurements for a variety of laser pulse widths, wavelengths and power densities, and the pressure spatial distribution within the laser spot. In addition, a significant amount of modeling of the pressure pulse and in-material shock wave behavior was pursued. This effort has supported a large number of doctoral dissertations and research programs, particularly in the late 1990s and has been sustained for over 30 years in their universities and laboratories. The French programs have made an important and significant contribution to advancing the understanding of laser shock processing to where it stands today.

In China, interest in laser induced shocks using a confined plasma first appeared in 1996 from the University of Science and Technology in Hefei, where Zhiyong Li and colleagues studied the attenuation of laser-induced shock waves in copper using an acrylic transparent overlay [38]. At that same time, Yongkang Zhang initiated a laser shock processing program in Nanjing University [39]. Later, Yongkang Zhang and Jianzhong Zhou directed laser shock processing programs at Jiangsu University in fatigue, modeling and laser shock metal forming [40]. Recently Zhang moved to Guangdong University of Technology to set up another laser shock laboratory. In addition to these university programs there are several other laboratories doing research in laser processing in China.

There are a number of other productive university and national laboratory programs in laser shock processing other countries, but unfortunately, there is not enough space to acknowledge their efforts.

The globalization of laser peening is also evident in the growth of the number of patents related to laser shock processing, most of them about laser peening. Due to the possibility that laser shock processing could become a commercial process, early on there was an awareness that it was necessary to patent important aspects of the process and equipment developments being discovered. The resulting growth in the number of patents related to laser shock processing is shown in Figure 19. The growth rate accelerated dramatically with the beginning of production at GEAE and the startup of LSP Technologies in the mid-90s. Later, around 2000, the numbers of World, European and Asian patents accelerated. In 2012 the total patents reached 380, representing 15 countries, with the major regions being the United States, Europe, Japan, and China. The number of technical papers related to laser peening follows a similar trajectory, growing slowly until the mid-90s, then rapidly accelerating. At this time the number of technical papers concerning laser shock processing is approaching 800–1000, representing the research of tens of laboratories.

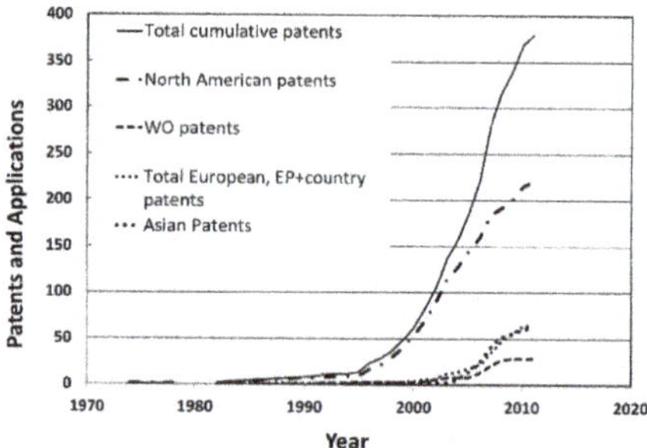

Figure 19. Growth of the number of patents associated with laser shock peening (LSP).

A Modified Laser Peening Process is Developed

In 1993, Yuji Sano at Toshiba in Japan, unaware of the laser peening work elsewhere, independently began developing a laser peening system entirely different from the high energy pulsed lasers being used in the USA and Europe at the time [41]. His system was specifically developed to treat structural features inside water-filled boiling water reactors (BWR) and pressurized water reactors (PWR) to mitigate stress corrosion cracking problems. The prevailing laser peening approach taken by the United

States and France at that time used Nd-glass lasers with 1054 nm wavelength, pulse energies of 10–40 J, pulse widths of 10–30 ns, and spot diameters $\geq$2 mm, with five or more spots per cm^2 applied at 1–5 Hz on surfaces covered with a protective opaque overlay of tape or paint. The modified approach developed by Sano at Toshiba used Nd-YAG lasers frequency doubled to produce a beam of 532 nm wavelength, pulse energies of $\leq$0.1 J, pulse widths of $\leq$10 ns, spots diameters $\leq$1 mm, with thousands of spots per cm^2 applied under water with no protective overlay. The 532 nm wavelength decreases the absorption of the beam while passing through the water. This process is referred to Laser Peening without Coating (LPwC). Although the bare surface being peened initially experiences tensile stress on the surface due to melting and lesser thermal effects, as the density of spots applied increases, the subsurface compressive stress increases and "bleeds" through to the surface, reversing the initial tensile stress. Although the small spot size precludes achieving compressive stress much deeper than 1 mm, the magnitude and depth of the compressive stresses are comparable to those achieved by the historical peening conditions using higher energy, larger spots.

In 1994, Sano achieved the first demonstration of compressive surface stress for LPwC and in 1999 was able to make the first application to a BWR shroud. This was followed by the first application using fiber-delivery of the laser beam in 2001 and by the first application to nozzles in a PWR in 2002. In 2006 the development of an ultra-compact portable system was completed and first used in applications in 2012. Through the early 2000s Sano decreased the size of the laser, increased the repetition rate and developed a portable laser peening system, paving the way for this process to be used in air outside reactor vessels. LPwC began to be applied to nuclear steam turbine blades in 2010. Recently, a hand-held laser has been developed.

In the late 1990s, Professor José Ocaña and colleagues at the Universidad Politécnica de Madrid, began developing a comprehensive model for laser shock processing [42]. In the early 2000s they initiated an experimental program using an approach similar to Sano's, but with a slightly larger spot size of 1.5 mm diameter and higher pulse energy of 2 J applied at 10 Hz. Ocaña and his colleagues have contributed significantly to the understanding of laser processing technology over the last 20 years. Their investigations have been wide ranging, including theory, overlay effects on residual stresses and surface roughness with and without black paint overlay, fatigue [43], hardness, and wear on a number of metal alloys. They have also demonstrated the potential of laser shock processing for micro-metal forming for Micro-Electro-Mechanical Systems (MEMS) applications [44]. In addition, his laboratory's laser has been available for others to use to pursue their research.

6. Present Status of Laser Shock Processing

Laser shock peening is now firmly entrenched as a mature commercial process to mitigate fatigue problems and for highly controlled bending or forming of aluminum plate into complex contours. Publicly available industrial specifications exist for laser peening (AMS2546) and the commercial providers are AS9100 certified. There are currently two laser peening companies, LSP Technologies located in Dublin, Ohio and the Metal Improvement Company (MIC) located in Livermore, California, giving customers the opportunity to choose the best fit for their needs. These peening companies have expanded their customer base worldwide; LSP Technologies from the United States into Germany and China, and MIC from the United States into Great Britain. GEAE does laser shock peening in-house for its own parts only, and is not a commercial supplier of laser peening. Business partnerships and alliances have begun to develop around this technology. In 2010, LSP Technologies and General Electric entered into an intellectual property cross-licensing agreement, allowing each access to the other's patents and intellectual property. In August 2012, the cross-license was expanded to allow sub-licensing of each other's patents.

Growth has been slow, but steady, as with most new industrial processes. The years following laser peening's entry into the market have provided potential users the opportunity to evaluate its commercial viability and reliability, its adaptability to new applications and manufacturing environments, its decreasing cost trajectory and its versatility. As this scenario has been unfolding, unanticipated opportunities for applying the technology, each with their own challenges, are appearing. Meeting these challenges to implement laser peening in new ways makes this an exciting time in the growth of the technology and the marketplace.

Fortunately, the two laser peening providers have taken a different approach to the market place, giving customers a choice. LSP Technologies provides both laser peening systems and laser peening services in-house. It has recently developed the Procudo Laser Peening System, the first commercially available laser developed specifically for laser peening, shown in Figure 20. The Procudo Laser Peening Systems requested by customers include the Procudo Laser and custom peening systems designed and engineered for the user's particular needs. The Procudo Laser produces pulses up to 10 J at 1–20 Hz. Considering the range of pulse frequencies available for laser peening, it is desirable that the effect on the target is independent of frequency for the same spot size. Figure 21 shows the consistency in the residual stresses produced from 1 to 20 Hz on Ti-6Al-4V. The processing conditions were water and black tape overlays, 2.5 mm diameter spots, 9 GW/cm^2, five layers with same spot pattern. The residual stress was determined by the slitting technique.

Figure 20. The next generation laser peening laser, the Procudo Laser Peening System by LSP Technologies, Inc.

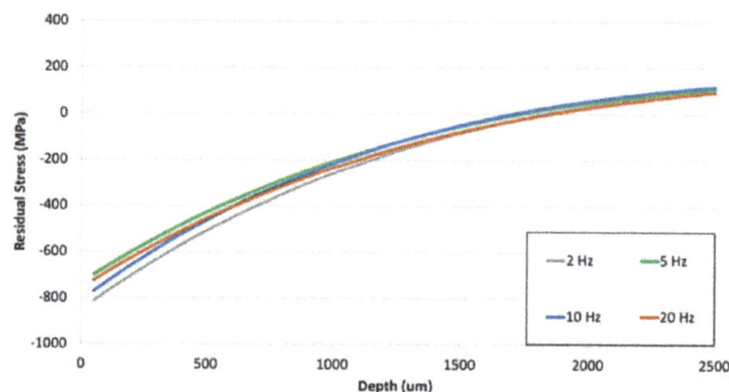

Figure 21. Residual stress profiles in 19 mm-thick Ti-6Al-4V coupons produced by the Procudo Laser Peening System applied at 1, 5, 10, and 20 Hz.

MIC provides laser peening services both on-site and off-site. It appears that off-site laser peening is either done in the customer's facility through a business agreement, or uses MIC's truck-transported lasers and processing systems set up in the customer's facility. MIC uses their own custom designed lasers.

The second industrial application of laser peening, controlled bending or curvature of metal wing skins for aircraft has been implemented for contoured wing skins by MIC on Boeing aircraft in Boeing's facility. This process has also been demonstrated by LSP Technologies on large aluminum plates in collaboration with Navy projects, forming desired contours with great accuracy.

A recent third application is now mitigation of stress corrosion cracking of stainless steel casks for holding spent nuclear fuel in storage by MIC [45].

One of the important tools now available to decrease the cost and time to move a part benefitting from laser peening into production is finite element modeling. In the 1990s, it was realized that getting a part approved for production entailed extensive exploratory processing and testing to ascertain the best processing conditions for maximum benefit to the peened part, followed by further processing and testing to qualify the process and the part for production. The substantial expense and time involved in this endeavor was a significant negative factor when considering new applications. It was clear that to minimize the informed, empirical approach to selecting the initial processing conditions, a modeling approach to preselect the most promising exploratory processing conditions was needed. The first 1D codes developed at Battelle in the 1970s discussed earlier, were limited to predicting surface pressures and included only hydrodynamic attenuation of the shock wave in the material. In 1990, Fabro et al. published their extended 1D model to elucidate the various physical processes occurring in confined, laser induced plasmas, providing an incentive for laser peening modeling efforts ever since [37]. Subsequently the French teams modeled the in-material shockwave behavior extensively with their SHYLAC code [46]. To initiate modeling development for laser peening in the United States, the Air Force supported a joint LSP Technologies-Ohio State University (OSU) program as a dissertation study in 1998 [47]. Abaqus finite element software was used to model in 2D and limited 3D with explicit and implicit steps to predict the magnitude and gradient of the compressive residual stress. The intent was to eliminate modeling the laser-material interaction step and instead to apply just the pressure pulse to the surface of the model over the area of a laser spot. The peak pressure of the pulse for a selected power density was taken from peak pressure vs. power density plots as shown in Figure 6. Models of different thickness, single and split beam applications, the Johnson-Cook constitutive equation, various yield criteria, and wave damping methods were among the aspects investigated. The predicted results were compared to experimental residual stress measurements for Ti-6Al-4V as shown in Figure 22. Very good agreement was obtained for single shots at two power densities, one of which was predicted before making the measurements. For multiple shots, the model predicted higher compressive stresses, but was not checked experimentally. The results for the split beam application was not even close. This was attributed to not having the ability to handle the extensive reversed plasticity occurring in this case. Unfortunately, an extensive publication of this work did not occur. The only publication of this work was by Clauer et al. [48,49]. Braisted and Brockman's model was similar to the OSU model [50]. In 2003, Peyre et al. in France [51,52], and in 2004, O'caña et al. in Spain [42,53], also used finite element models for 2D and 3D modeling of shock waves and residual stresses. In 2012, Brockman, et al. did an extensive modeling analysis of the non-uniformities in the residual stress field in a laser peened volume, demonstrating that care must be taken to ensure that these uniformities do not jeopardize the reliability of the processing [54]. Fortunately, the process is robust enough that these concerns can be alleviated to a degree by spot overlap in practice. These and many other modeling efforts have greatly increased the understanding of the process. However, expanding this modelling approach to larger areas containing many spots, multiple layers and nonplanar geometries typical of most laser peened parts, required substantial computer time, model tweaking and further development.

Fortunately, in the early 2000s a much different modeling approach to representing residual stresses in laser peened parts appeared. This approach could be applied to a finite element model of any desired size or shape and did not require simulating the passage of a shock wave to develop the stress field. Instead, a residual stress gradient is created in a finite element model by inserting an appropriate eigenstrain distribution into it [55–57]. The eigenstrains are derived from actual residual stress magnitudes and gradients measured on coupons of an alloy of interest after laser peening with different conditions. The eigenstrains are inserted into a finite element model of the part geometry under the anticipated laser peened area by either inserting a distribution of different thermal expansion coefficients at element nodes and raising the temperature one degree, or by maintaining a constant thermal expansion coefficient in the model and imposing a temperature on the surface to develop a thermal gradient into the model. This eigenstrain approach makes it relatively easy to explore how the residual stress field over the laser peened area adapts to different geometries. By this means, the extent and shape of the area to be processed and the appropriate range of processing intensities can now be determined relatively quickly before processing the first test parts. This approach now reduces the time and expense for developing new applications and increasing the odds of a successful implementation of the process.

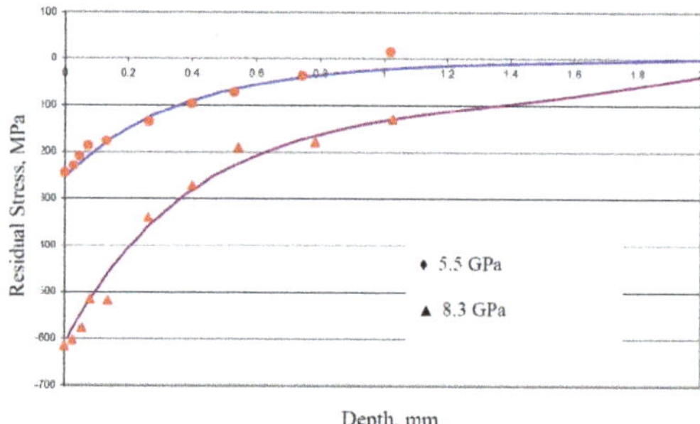

Figure 22. Comparison of experimental and predicted residual stress gradients in Ti-6Al-4V [48,49].

Over the last 25 years, it has been established that laser peening is a robust and versatile process. Deep compressive stresses can be obtained with a broad range of laser capabilities and laser peening conditions. One way of illustrating this is to categorize laser shock processing systems into roughly three categories according to the energy range of the pulses generated. The original lasers were high energy pulsed lasers having Nd-glass rods producing a beam with a wavelength of 1054 nm. These lasers output pulses of 10–40 J, 10–30 ns long, operating at 1–5 Hz. The working spot sizes are 3 mm–10 mm in diameter. Larger spot sizes have the advantage producing deeper compressive stresses in thicker sections. In the mid-1990s, it was demonstrated that acceptable results could also be obtained with what we may describe as intermediate energy and low energy lasers. The intermediate energy lasers have Nd-YAG rods producing a beam with a wavelength of 1064 nm. These lasers output pulses of 1–10 J, nominally 10 ns long, operating at 1–20 Hz. The working spot size is 1–2 mm diameter. The low energy pulsed lasers also have Nd-YAG rods. These lasers output pulses of ≤1 J, ≤10 ns long, operating at 40–100+ Hz. The working spot sizes are ≤1 mm in diameter. In general, decreasing the laser's energy per pulse is a trade-off enabling pulsing the laser at higher frequencies. The laser footprint of these lasers is nominally in the range of 1–6 m^2 with the low energy lasers at the small end. This range of options enable the size of laser peening systems to be scaled in size, cost and capability to fit the needs of the manufacturer and product. The trend now appears to favor the intermediate energy pulsed lasers. Higher beam repetition rates favor the use of a tape opaque overlay or processing the bare metal surface.

While the forgoing described the present state of laser peening, some of the activities pursuing the use of laser-induced shock waves in a broader context are nearing commercialization to meet specific industry needs. For example, the Laser Bond Inspection (LBI) system developed by LSP Technologies, Inc. to evaluate the strength of adhesive bonds in composite bonded structures is used to evaluate the strength at the bond interface between composite layers. The major aircraft manufacturers are working towards implementing this technology as a method to evaluate the bond strength integrity during the manufacture of adhesively bonded structures [58,59].

Another example is the LASer Adhesion Test (LASAT) developed by the French National Center for Research. This is a method of measuring and testing the adhesion of thin films to metal and ceramic substrates. There are variations of this test for different situations, and tests may be performed with a confined plasma in air or unconfined in a vacuum.

7. The Future

Predicting what lies ahead is always risky, but a few comments will be ventured. More laser peening facilities will be equipped with the intermediate size lasers due to their small footprint and lower operating costs. These lasers will be located in, and operated by, the companies and incorporated within the normal flow of their production lines.

The low energy, high frequency laser systems will find their niche, perhaps first as their small size and portability will enable laser peening of critical locations in large structures to benefit from laser peening.

Eventually, applications will be found for other aspects of laser shock processing. Several have been explored in laboratories and have shown promise technically. However, they have to address a real need to justify the costs of further development. Some of these are metal forming of small objects such as MEMs components, welding of dissimilar metals for small assemblies, surface imprinting of shape memory alloys, laser peening additively manufactured parts and using thermomechanical processing, i.e., laser peening at elevated temperatures, where it contributes to higher compressive residual stress and strength.

One last comment: it has been a wonderful gift to have had the opportunity to be part of the birth and maturation of laser shock peening over the past 46 years.

Funding: The organizations funding the research described here were identified within the review text.

Acknowledgments: The author would like to acknowledge again those people named herein and those who are not, who have made the many valuable contributions leading to the successful development of laser shock peening.

Conflicts of Interest: The author declares no conflicts of interest.

References

1. Askaryan, C.A.; Moroz, E. Pressure on evaporation of matter in a radiation beam. *J. Exp. Theor. Phys.* **1962**, *43*, 2319–2320.
2. Neuman, F. Momentum transfer and cratering effects produced by giant laser pulses. *Appl. Phys. Lett.* **1964**, *4*, 167–169. [CrossRef]
3. Gregg, D.W.; Thomas, S.J. Momentum transfer produced by focused laser giant pulses. *J. Appl. Phys.* **1966**, *37*, 2787–2789. [CrossRef]
4. Braginskii, V.B.; Minakova, I.I.; Rudenko, V.N. Mechanical effects in the interaction between pulsed electromagnetic radiation and a metal. *Sov. Phys. Tech. Phys.* **1967**, *112*, 753–757.
5. Afanasev, Y.V.; Krokhin, O.N. Vaporization of matter exposed to laser emission. *Sov. Phys. JETP* **1967**, *25*, 639–645.
6. Skeen, C.H.; York, C.M. Laser-induced "blow-off" phenomena. *Appl. Phys. Lett.* **1968**, *12*, 369–371. [CrossRef]
7. O'Keefe, J.D.; Skeen, C.H. Laser-induced stress-wave and impulse augmentation. *Appl. Phys. Lett.* **1972**, *21*, 464–466. [CrossRef]
8. Anderholm, N.C. Laser-generated stress waves. *Appl. Phys. Lett.* **1970**, *16*, 113–115. [CrossRef]
9. Anderholm, N.C. Paper Bk9. *APS Bull.* **1968**, *13*, 388.
10. Mirkin, L. Plastic deformation of metals caused by a 10-8 sec laser pulse. *Sov. Phys. Dokl.* **1970**, *14*, 207–208.
11. Metz, S.A.; Smidt, F.A. Production of vacancies by laser bombardment. *Appl. Phys. Lett.* **1971**, *19*, 207–208. [CrossRef]
12. Fairand, B.P.; Wilcox, B.A.; Gallagher, W.J.; Williams, D.N. Laser shock-induced microstructural and mechanical property changes in 7075 aluminum. *J. Appl. Phys.* **1972**, *43*, 3893–3895. [CrossRef]
13. Mallozzi, P.; Fairand, B. Altering Material Properties. US3850698A, 26 November 1974.
14. Hahn, G.T.; Mincer, P.N.; Rosenfield, A.R. The Fe−3Si steel etching technique for local strain measurement. *Exp. Mech.* **1971**, *11*, 248–253. [CrossRef]

15. Clauer, A.H.; Fairand, B.P.; Wilcox, B.A. Pulsed laser induced deformation in an Fe-3 wt pct Si alloy. *Met. Trans. A* **1977**, *8A*, 119–125. [CrossRef]

16. Fairand, B.P.; Clauer, A.H.; Jung, R.G.; Wilcox, B.A. Quantitative assessment of laser-induced stress waves generated at confined surfaces. *Appl. Phys. Lett.* **1974**, *25*, 431–433. [CrossRef]

17. AFWAL-TR-66-48. *The Solid/Vapor Equation of State and Hydrodynamics Routine Are Based On*; U.S. Airforce Wright Aeronautical Laboratory: Wright-Patterson Air Force Base, OH, USA, 1966.

18. Fox, J.A. Effect of water and paint coatings on laser-irradiated targets. *Appl. Phys. Lett.* **1974**, *24*, 461–464. [CrossRef]

19. Fairand, B.P.; Clauer, A.H. Use of Laser Generated Shocks to Improve the Properties of Metals and Alloys. In Proceedings of the SPIE 0086, Industrial Applications of High Power Laser Technology, San Diego, CA, USA, 30 December 1976; Volume 86, pp. 112–119.

20. Fairand, B.P.; Clauer, A.H. Effect of water and paint coatings on the magnitude of laser-generated shocks. *Opt. Commun.* **1976**, *18*, 588–591. [CrossRef]

21. Fairand, B.P.; Clauer, A.H. Laser generation of high-amplitude stress waves in materials. *J. Appl. Phys.* **1979**, *50*, 1497–1502. [CrossRef]

22. Fairand, B.P.; Clauer, A.H. Laser generated stress waves: Their characteristics and their effects to materials. *AIP Conf. Proc.* **1979**, *50*, 27–42.

23. O'Keefe, J.D.; Skeen, C.H.; York, C.M. Laser-induced deformation modes in thin metal targets. *J. Appl. Phys.* **1973**, *44*, 4622–4626. [CrossRef]

24. Yang, L.C. Stress waves generated in thin metallic films by a Q-switched ruby laser. *J. Appl. Phys.* **1974**, *45*, 2601–2608. [CrossRef]

25. Clauer, A.H.; Fairand, B.P.; Wilcox, B.A. Laser shock hardening of weld zones in aluminum alloys. *Metall. Trans.* **1977**, *8*, 1871–1876. [CrossRef]

26. Clauer, A.H.; Fairand, B.P. Interaction of laser-induced stress waves with metals. In *Applications of Lasers in Materials Processing*; Metzbower, E.A., Ed.; American Society for Metals: Metals Park, OH, USA, 1979; pp. 1–22.

27. Clauer, A.H.; Fairand, B.P.; Slater, J.E. *Laser Shocking of 2024 and 7075 Aluminum Alloys*; National Aeronautical and Space Agency: Wright-Patterson Air Force Base, OH, USA, 1977.

28. Herring, G.; Olson, R.B. *The Effect of Aging Time on Spallation of 2024-T6 Aluminum*; U.S. Army Materials and Mechanics Research Center: Watertown, MA, USA, 1971.

29. Clauer, A.H.; Dulaney, J.L.; Rice, R.C.; Koucky, J.R. Laser Shock Processing for Treating Fastener Holes in Aging Aircraft. In *Durability of Metal Aircraft Structures*; Atluri, S.N., Harris, C.E., Hoggard, A., Miller, N., Sampath, S.G., Eds.; Atlanta Technology Publications: Atlanta, GA, USA, 1992; pp. 350–361.

30. Clauer, A.H.; Holbrook, J.H.; Fairand, B.P. Effects of Laser Induced Shock Waves on Metals. In *Shock Waves and High-Strain-Rate Phenomena in Metals: Concepts and Applications*; Meyers, M.A., Murr, L.E., Eds.; Springer: Boston, MA, USA, 1981; pp. 675–702.

31. Ford, S.C.; Clauer, A.H.; Fairand, B.P.; Galliher, R.D. *Investigation of Laser Shock Processing*; U.S. Air Force Wright Aeronautical Laboratories: Wright-Patterson Air Force Base, OH, USA, 1980; Volume 2.

32. Ivetic, G.; Meneghin, I.; Troiani, E.; Molinari, G.; Ocaña, J.; Morales, M.; Porro, J.; Lanciotti, A.; Ristori, V.; Polese, C.; et al. Fatigue in laser shock peened open-hole thin aluminium specimens. *Mater. Sci. Eng. A* **2012**, *534*, 573–579.

33. Clauer, A.H.; Walters, C.T.; Ford, S.C. The Effects of Laser Shock Processing on the Fatigue Properties of 2024-T3 Aluminum. In *Lasers in Materials Processing*; ASM International: Metals Park, OH, USA, 1983; pp. 7–22.

34. Rockwell B-1 Lancer: Operational History: Strategic Air Command. Available online: http://en.wikipedia.org/wiki/Rockwell_B-1_Lancer (accessed on 9 May 2019).

35. Thompson, S.D.; See, D.W.; Lykins, C.D.; Sampson, P.G. Laser Shock Peening vs Shot Peening—A Damage Tolerance Investigation. In *Surface Performance of Titanium*; Gregory, J.K., Rack, H.J., Eylon, D., Eds.; TMS: Warrendale, PA, USA, 1996; pp. 239–252.

36. Fabbro, R.; Fournier, J.; Fabre, E.; Leberichel, E.; Hannau, T.; Corbet, C. Experimental Study of Metallurgical Evolutions in Metallic Alloys Induced by Laser Generated High Pressure Shocks. *Laser Processing: Fundamentals, Applications, and Systems Engineering* **1986**, *SPIE 668*, 320–324.

37. Fabbro, R.; Fournier, J.; Ballard, P.; Devaux, D.; Virmont, J. Physical study of laser-produced plasma in confined geometry. *J. Appl. Phys.* **1990**, *68*, 775–784. [CrossRef]

38. Li, Z.; Zhu, W.; Cheng, J.; Zhou, G.; Guo, D. Experimental study on attenuation of high power laser-induced shock waves in copper. *Chin. Sci. Bull.* **1996**, *41*, 1694–1696.

39. Zhang, Y.K.; Zhang, S.; Yu, C.; Tang, Y.; Zhang, H. Laser shock processing for fatigue and fracture resistance. *Sci. Chin.* **1997**, *40*, 170–177. [CrossRef]

40. Zhou, J.Z.; Yang, J.C.; Zhang, Y.K.; Zhou, M. A study on super-speed forming of metal sheet by laser shock waves. *J. Mater. Process. Technol.* **2002**, *129*, 241–244. [CrossRef]

41. Sano, Y.; Mukai, N.; Aoki, N.; Konagai, C. Laser Processing to Improve Residual Surface Stress of Metal Components. In Proceedings of the Digest IEEE/Leos 1996 Summer Topical Meeting, Keystone, CO, USA, 5–9 August 1996; pp. 30–31.

42. Ocana, J.; Molpares, C.; Morales, M. A Model for the Coupled Predictive Assessment of Plasma Expansion and Material Compression in Laser Shock Processing Applications. In *High Power Ablation II*; Phipps, M., Niino, C.R., Eds.; International Society for Optics and Photonics: Bellingham, WA, USA, 2000; Volume 3885.

43. Rubio-González, C.; Ocaña, J.L.; Gomez-Rosas, G.; Molpeceres, C.; Paredes, M.; Banderas, A.; Porro, J.; Morales, M. Effect of laser shock processing on fatigue crack growth and fracture toughness of 6061-T6 aluminum alloy. *Mate. Sci. Eng. A* **2004**, *386*, 291–295. [CrossRef]

44. Ocaña, J.L.; Morales, M.; Porro, J.A.; García-Ballesteros, J.J.; Correa, C. Laser shock microforming of thin metal sheets with Ns lasers. *Phys. Proc.* **2011**, *12*, 201–206. [CrossRef]

45. Curtis, W. Laser peening by CWST contributes to successful nuclear canister storage program. Available online: https://www.shotpeener.com/library/pdf/2018017.pdf (accessed on 9 May 2019).

46. Peyre, P.; Berthe, L.; Scherpereel, X.; Fabbro, R.; Bartnicki, E. Experimental study of laser-driven shock waves in stainless steels. *J. Appl. Phys.* **1998**, *84*, 5985–5992. [CrossRef]

47. Nam, T. Finite Element Analysis of Residual Stress Field Induced by Laser Shock Peening. Ph.D. Thesis, Ohio State University, Columbus, OH, USA, 2002.

48. Clauer, A.; Brockman, R.; Braistead, W.; Noll, S.; Lee, J.; Gilat, A. Modeling Residual Stresses for Laser Shock Peening. In Proceedings of the 5th National Turbine Engine High Cycle Fatigue Conference, Chandler, AZ, USA, 7–9 March 2000.

49. Clauer, A.; Lahrman, D. Laser shock processing as a surface enhancement process. *Key Eng. Mater.* **2001**, *197*, 121–144. [CrossRef]

50. Braisted, W.; Brockman, R. Finite Element simulation of laser shock peening. *Int. J. Fatigue* **1999**, *21*, 719–724. [CrossRef]

51. Peyre, P.; Sollier, A.; Chaieb, I.; Berthe, L.; Bartnicki, E.; Braham, C.; Fabbro, R. FEM simulation of residual stresses induced by laser peening. *Eur. Phys. J. Appl. Phys.* **2003**, *23*, 83–88. [CrossRef]

52. Peyre, P.; Chaieb, I.; Braham, C. FEM calculation of residual stresses induced by laser shock processing in stainless steels. *Modell. Simul. Mater. Sci. Eng.* **2007**, *15*, 205–221. [CrossRef]

53. Ocaña, J.L.; Morales, M.; Molpeceres, C.; Torres, J. Numerical simulation of surface deformation and residual stresses fields in laser shock processing experiments. *Appl. Surf. Sci.* **2004**, *238*, 242–248. [CrossRef]

54. Brockman, R.A.; Braisted, W.R.; Olson, S.E.; Tenaglia, R.D.; Clauer, A.H.; Langer, K.; Shepard, M.J. Prediction and Characterization of residual stresses from laser shock peening. *Int. J. Fatigue* **2012**, *36*, 96–108. [CrossRef]

55. Korsunsky, A.M. The modelling of residual stresses due to surface peening using eigenstrain distributions. *J. Strain Anal. Eng. Des.* **2005**, *40*, 817–824. [CrossRef]

56. DeWald, A.T.; Hill, M.R. Eigenstrain-based model for prediction of laser peening residual stresses in arbitrary three-dimensional bodies Part 1: Model description. *J. Strain Anal. Eng. Des.* **2009**, *44*, 1–11. [CrossRef]

57. DeWald, A.T.; Hill, M.R. Eigenstrain-based model for prediction of laser peening residual stresses in arbitrary three-dimensional bodies Part 2: Model verification. *J. Strain Anal. Eng. Des.* **2009**, *44*, 13–27. [CrossRef]

58. Bossi, R. Introduction to Technical Focus Issue on NDT of Adhesive Bonds. *Mater. Eval.* **2009**, *67*, 818–827.

59. Bossi, R.; Housen, K.; Walters, C. Laser bond inspection device for composites: Has the holy grail been found? *NTIAC Newslett.* **2005**, *30*, 1–7.

Article

Experimental Determination of Electronic Density and Temperature in Water-Confined Plasmas Generated by Laser Shock Processing

Cristóbal Colón [1],*, María Isabel de Andrés-García [1], Cristina Moreno-Díaz [1], Aurelia Alonso-Medina [1], Juan Antonio Porro [2], Ignacio Angulo [2] and José Luis Ocaña [2]

[1] Department of Applied Physics, ETSIDI, Universidad Politécnica de Madrid, Ronda de Valencia 3, 28012 Madrid, Spain
[2] Department of Applied Physics, ETSII, Universidad Politécnica de Madrid, José Gutiérrez Abascal 2, 28006 Madrid, Spain
* Correspondence: cristobal.colon@upm.es; Tel.: +34-910677607

Received: 17 June 2019; Accepted: 19 July 2019; Published: 22 July 2019

Abstract: In this work, diagnoses of laser-induced plasmas were performed in several Laser Shock Processing (LSP) experiments using the Balmer Hα-line (656.27 nm) and several Mg II spectral lines. A Q-switched laser of Nd:YAG was focused on aluminum samples (Al2024-T351) in LSP experiments. Two methods were used to diagnose the plasma. The first method, which required two different experiments, was the standard for establishing the electronic temperature through the use of a Boltzmann Plot with spectral lines of Mg II and self-absorption correction. The Stark width of the Balmer Hα-line was used to determine the electron density in each of the cases studied. The second method had lower accuracy, but only required an experimental determination. Two parameters, the electronic temperature and the electron density, were obtained with the aid of the Hα-line in a single data acquisition process. The order of magnitude of the temperature obtained from this last method was sufficiently close to the value obtained by the standard method (within a factor lower than 2.0), which is considered to be important in order to allow for its possible use in industrial conditions.

Keywords: laser shock processing; plasma diagnosis; electron density

---

1. Introduction

Laser Shock Processing (LSP) is based on focusing a pulse of a high energy laser (I > 10^9 W/cm^2, τ < 50 ns) over a piece of metal. It results in the instantaneous vaporization of the piece's surface and the generation of a high temperature and density plasma composed of the different ionized species of elements present in the piece and in the atmosphere. The high pressure of the plasma generates a shock wave that propagates to the piece, which affects its mechanical characteristics. Today, LSP is a consolidated alternative for improving the surface properties of metal alloys. This technique has been studied since the 1960s, when Askar'yan and Moroz [1] discovered that a high-energy laser pulse produces backpressures in the surface of the metal material on which the laser is focused (target). In Fairand et al. [2] it was found that laser shocking induced a tangled dislocation substructure similar to explosively shocked aluminum. The stress waves were studied using the piezoelectric response of X-cut quartz-crystal disks by Yang [3]. Since then, numerous works have aimed at finding the best conditions in which the technique should be applied. In 1990, Fabbro et al. [4] quantitatively described the evolution of the plasma in a LSP experiment under confined geometry using different characteristic plasma-dynamics stages, including plasma pressure buildup, plasma development under the laser irradiation, and final plasma expansion until the pressure decreases and it is too low to cause plastic deformation in the material. An explanation of this phenomenon was given by Berthe et al. [5].

In 1997, Sano et al. [6] confirmed that the underwater shock processing with YAG was feasible to improve residual stress in metal. There are different studies about several methods to confine the plasma (see, e.g., Morales et al. 2009 [7]), indicating the crucial role of a plasma-confining layer in withstanding both the laser irradiation (with minimum internal breakdown) and the plasma pressure needed to transmit to the shocked target. The use of glass or quartz plates as confining layers are theoretically very suitable, but are generally only valid for a few laser shots and in measurements made in the laboratory, which are far from the needs imposed by industrial application of the technique, so purified water is normally used as the LSP confining medium.

The observation of the plasma shock wave through the confining medium has always been a subject of much interest in order to characterize the actual impulsion transmitted to the shocked material, a key point regarding the systematic monitoring and control of LSP processes in an industrial environment (see, e.g., Berthe et al. [5] and Ocaña et al [8]).

In 2009, Martí-Lopez et al. [9] performed interferometric measurements in order to record hemispherical shock fronts, cylindrical shock fronts, plane shock fronts, cavitation bubbles, and phase disturbance tracks. A summary of the different results obtained in the LSP experiments are presented by Ocaña et al. [8] along with some conclusions about the LSP technology as a profitable industrial method.

Recently, works on the effect of the LSP process have also been presented by some of the authors of this study. A comparison between experimental values and numerical predictions of the inside propagation of the residual stresses in plates of Al2024-T351 of 2 mm thickness was presented in 2015 by Ocaña et al. [10]. Also in 2015, a study on the influence of the randomness of the superposition of pulses on the final anisotropy of the residual stresses induced in the treated material was completed by Correa et al. [11].

Several theoretical works try to model the LSP process with the purpose of finding the practical criteria for its optimization. A self-closed thermal model for LSP was described by Wu and Shin [12]. Unlike most of the existing laser peening models, there are no free parameters in this model, and all the variables are calculated based on related physics theories. Later, a model-based systematization of process-optimization criteria and a practical assessment on the real possibilities of the technique along with practical results at a laboratory scale on the application of LSP to characteristic high-elastic limit metallic alloys were presented in 2008 by Morales et al. [13]. A calculation model conceived for the analysis of the problem of laser shock wave generation and propagation was presented in 2013 by Ocaña et al. [14]. However, these works were not able to predict the best conditions of application of the LSP technique with changes in the target materials. This is due in part to the large number of physical processes involved. As an example, although these models provide, as indicated, good predictions about the shock wave on the surface of the sample, their forecasts on the parameters of the plasma are too far from the experimental values, even when the plasma is confined in air. There is no experimental information on these properties in the case of plasma that is confined by water in flow.

The presence of hydrogen in the plasma is proof that chemical attacks on the surface of the sample can occur during the process. The interaction between the laser, the plasma, and the surface of the sample goes beyond the mechanical impact of the shock wave on the target. The relative presence of different ions in the plasma and its chemical interaction with the surface cannot be established without the knowledge of the plasma electron density number and its electronic temperature.

The purpose of this work was to experimentally estimate, by means of a practical (ideally industrially applicable) procedure, the electron density and the electronic temperature of the plasma in water-flow conditions to study chemical interactions between plasma and the surface of the sample and improve the theoretical treatments of the LSP process.

These studies of electron density number and plasma temperature determination have been performed for years using the technique known as Laser Induced Breakdown Spectroscopy (LIBS). The LIBS technique is based on the study of the emission of the different atomic and ionic species present in plasma. An excellent compilation of experiments and applications of the LIBS technique can be found in the literature (see, e.g., Musazzi and Perini [15]). This technique has been used in several scientific applications by the authors signing this work. In 2006, Colón and Alonso-Medina [16] measured Stark broadening of several Pb II spectral lines, and in 2011, Alonso-Medina [17] measured the broadening Stark of Pb III spectral lines. The LIBS technique also stands out as an analytical technique in different industrial applications (see, e.g., Noll et al. [18]).

In 2006, El Sherbini et al. [19] used the Hα-line to measure electron density in an experiment of the laser-induced breakdown of plasma in air. In 2010, Parigger and Oks [20] presented a review about plasma diagnostics based on Stark broadening of hydrogen Balmer lines in laser-induced breakdown of plasma. Later, in 2012, the Hα-line was used again by El Sherbini et al. [21] to determine the electron density in aluminum plasma and to correct the self-absorption in the Mg I and Mg II spectral lines present in this plasma. In 2004, De Giacomo et al. [22] completed several experiments of laser induced breakdown spectroscopy in aqueous solution. Nath and Khare [23] studied laser-induced breakdown experiments in water in 2010 using the emission bands of different molecular species.

As mentioned in a previous work by Moreno-Díaz et al. [24], the flow of water in LSP conditions weakens the emission of all species present in the plasma, with some exceptions. In the present work, in addition to the emission of hydrogen, already mentioned above, a weak emission of Mg II in the 279.5 nm zone was targeted. As in the previous work, the electron density was measured using the Stark width of the Hα-line. Now, the temperature can be estimated directly under LSP conditions using the emission of the spectral lines of Mg II.

The current work used the Stark width of the Hα-line on a sample of aluminum alloy (Al2024-T351) in LSP conditions (with water as the confining environment). The novelty of this study is in the measurement of the electronic temperature, which was performed directly in the LSP experiments using a Boltzmann plot with measurements of the intensities of the spectral lines of species present in the plasma in the water flow, not in air as was used in the earlier work of Moreno-Diaz et al. [24]. Measurements were taken with different delay times from the laser pulse (2–5 μs), while the plasma cooled adiabatically, allowing us to obtain the width and shift of the Hα-line in all cases. In order to obtain the best conditions for measurements, different gate times of 100 ns, 200 ns, 300 ns, 500 ns, and 1000 ns were used.

In order to verify the suitability of the proposed one-determination diagnosis procedure, a comparison of the obtained temperatures with those estimated from the shift of the Hα-line (two determinations needed) was successfully performed. This diagnosis with a single experiment would allow for the much-needed real-time monitoring of plasma behavior as described in Ocaña et al. [25] and Takata et al. [26].

In this paper, in Section 2 (Materials and Methods) the equipment and the experimental procedures are described. In Section 3 (Results) the analysis of the results is presented by comparing the values of the plasma temperature achieved by two different methods. Finally, in the Sections 4 and 5 (Discussions and Conclusions, respectively) we present the possibilities of plasma diagnosis through a single experiment and through a second experiment without using the Mg II lines.

2. Materials and Methods

The experimental setup used in this work is the same as that used by the authors in a previous work (Moreno-Diaz et al. [24]). A schematic diagram of the experimental set-up is shown in Figure 1a,b.

A laser pulse of a Q-switched Nd:YAG laser (of 10 ns, 1.06 μm of wavelength, and 2.5 J per pulse) was focused using a 20 cm focal lens on a sample of certified aluminum alloy (Al2024-T351). The composition of the sample included in addition to aluminum around 4% Cu, 1.5% Mg, and 0.6% Mn. As can be observed in Figure 1, a constant water flux was supplied in this experiment. A typical crater, see Figure 2, of ~2 mm diameter was produced in the sample surface after the laser pulse. The image was obtained with a confocal microscope (LEICA DCM 3D´Leica Microsistemas S.L.U., L'Hospitalet de Llobregat, Spain). The irradiance of laser was approximately 10 GW/cm^2.

The plasma light was collected by an optical fiber and directed to the spectrograph provided with a diffraction grating of 1800 grooves/mm. The placement of a lens to collect more plasma light may be an alternative in air, but the lens has been shown to be badly damaged by water splashes in the laboratory and probably would be even more so in an industrial environment.

Emission spectra were acquired using a spectrograph (Horiba Jobin Ybon FHR1000, HORIBA UK Limited, Northampton, United Kingdom) equipped with an ICCD camera (Andor, model iStar 334T, Oxford Instruments, Concord, MN, USA).

Measurements were taken 2 mm from the target surface (where the signal/noise ratio was the best). As we mentioned in our previous work, in spite of our attempts, the quality of the signal prevented us from making the Abel inversion, and so our data are spatially integrated and present the temporary integration of the measurement gate time.

A low-pressure Ne lamp (Oriel 6032, Newport, Irvine, CA, USA) was used to calibrate the wavelength scale. The instrumental profiles in the interest range were measured with a He-Ne laser, checking its full width at half maximum. The instrumental bandwidths in this range were found on average to be 0.18 ± 0.01 Å. In this bandwidth, it was found by numerical adjustment that the Lorentzian contribution was practically null.

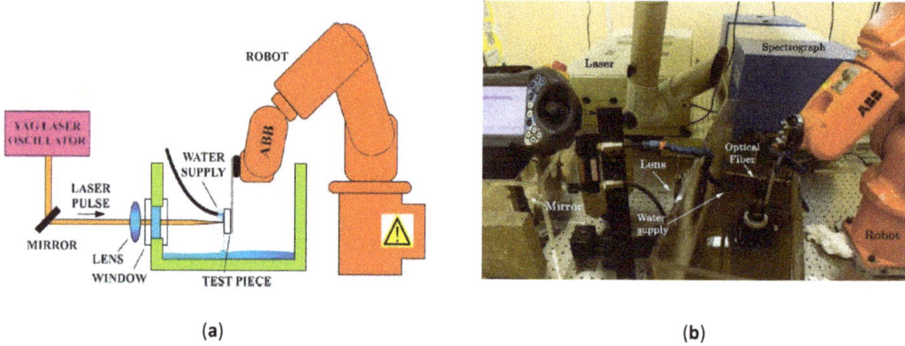

(a)   (b)

Figure 1. Experimental setup used in the reported Laser Shock Processing (LSP) experiments: (**a**) Scheme; (**b**) Photo.

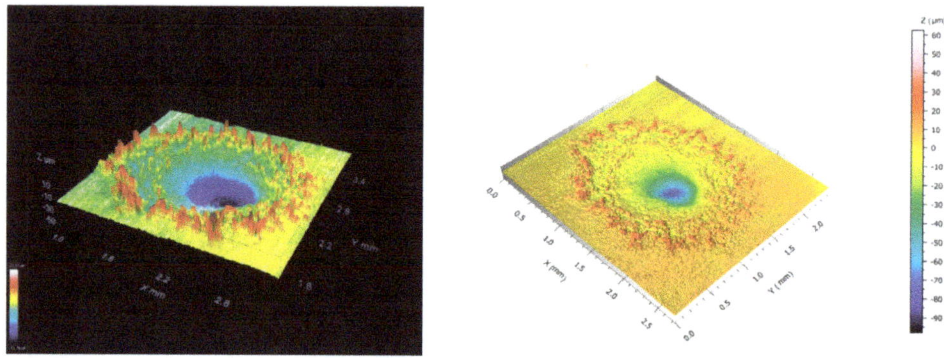

Figure 2. A typical surface change after an LSP experiment. Images obtained via scanning confocal microscopy.

Relevant LSP experiments were performed both under air and water confinement. Under air-confinement conditions, the composition of the plasma reflected the composition of the sample, obtaining spectral lines of Al I and Al II, Mg I and Mg II, Cu I and Cu II, and Mn I and Mn II, in addition to the Hα-line of the hydrogen, which can only be due to very weak traces of hydrogen in the plasma and a possible reaction between the ionized aluminum and the residual humidity of the air [21]. The presence of the water flow weakened the emissions of spectral lines differently than it did those of the Hα-line, which had a similar profile in air and in water flow in all cases. This effect can be observed in Figure 3.

As the profiles and the displacement of the Hα-line were practically identical under air and water flow confinement (LSP condition), the hypothesis of the similarity of the plasma behavior and properties under both conditions (already recognized in the previous work [24]) is considered sound.

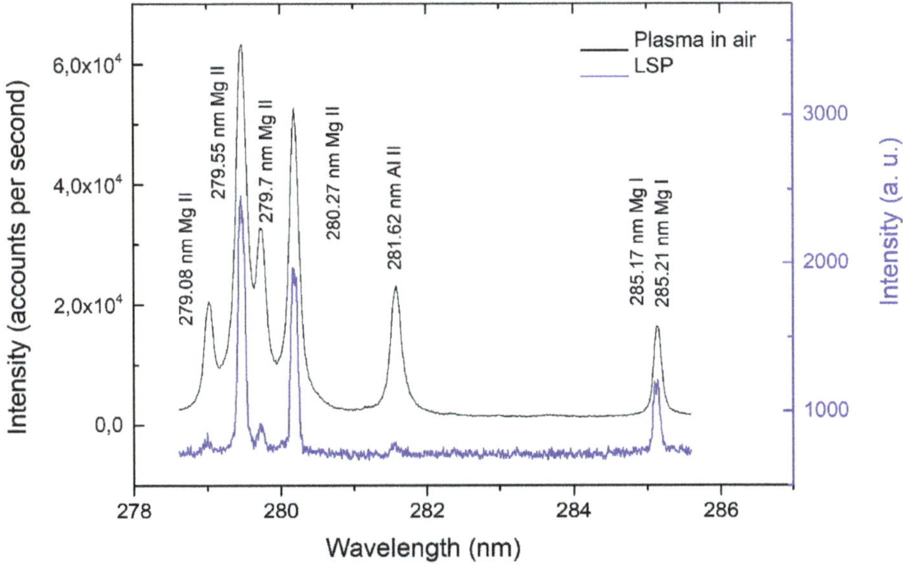

Figure 3. Mg II emission lines in air and LSP conditions. Delay time: 2 μs. Gate time: 1000 ns.

In Figure 4a–c, images of the plasma trace at 2.5, 4, and 5 μs after laser pulse (gate time of 300 ns in all cases) and the corresponding spectra can be observed. As previously indicated, the presence of neutral hydrogen, although important for the obtained results, is residual, and confirms the practical non-absorption of the Hα line. The hostile conditions in which these measures must be carried out, with water in flow and splashing, makes it very difficult to experimentally contrast this claim. Nevertheless, the Hα lines did not show any self-absorption signs. In other words, there is full symmetry and no anomalous dip in the center. These facts have already been mentioned by El Sherbini et al. [19,21] in the electron density range of these kinds of experiments (described in Moreno-Diaz et al. [24]). Another relevant feature was the appearance of a red shift of the peak of the Hα-line which, in turn, was observed to decrease with the delay time.

In order to perform the analysis of the spectral lines, the obtained experimental profiles were adjusted to Voigt profiles generated numerically. To obtain the Lorentz experimental broadening of the line, the contribution of the instrumental profile was discounted.

Unlike the Lorentzian contribution, which was practically null, the Gaussian contribution of the instrumental profile of the used experimental device was important in relation to the Gaussian contribution found in the Voigt profiles of the Hα-line analyzed in this work. This fact advises against the use of this contribution to estimate the temperature of thermal agitation of hydrogen, which only in conditions of full thermodynamic equilibrium would coincide with the sought electronic temperature.

Figure 5 shows four examples of the fitting from a Voigt profile to experimental line emissions of the plasma. The Hα-lines corresponding to 4 μs delay time and gate times (emission integration times) of 100 ns, 300 ns, 500 ns, and 1000 ns are presented.

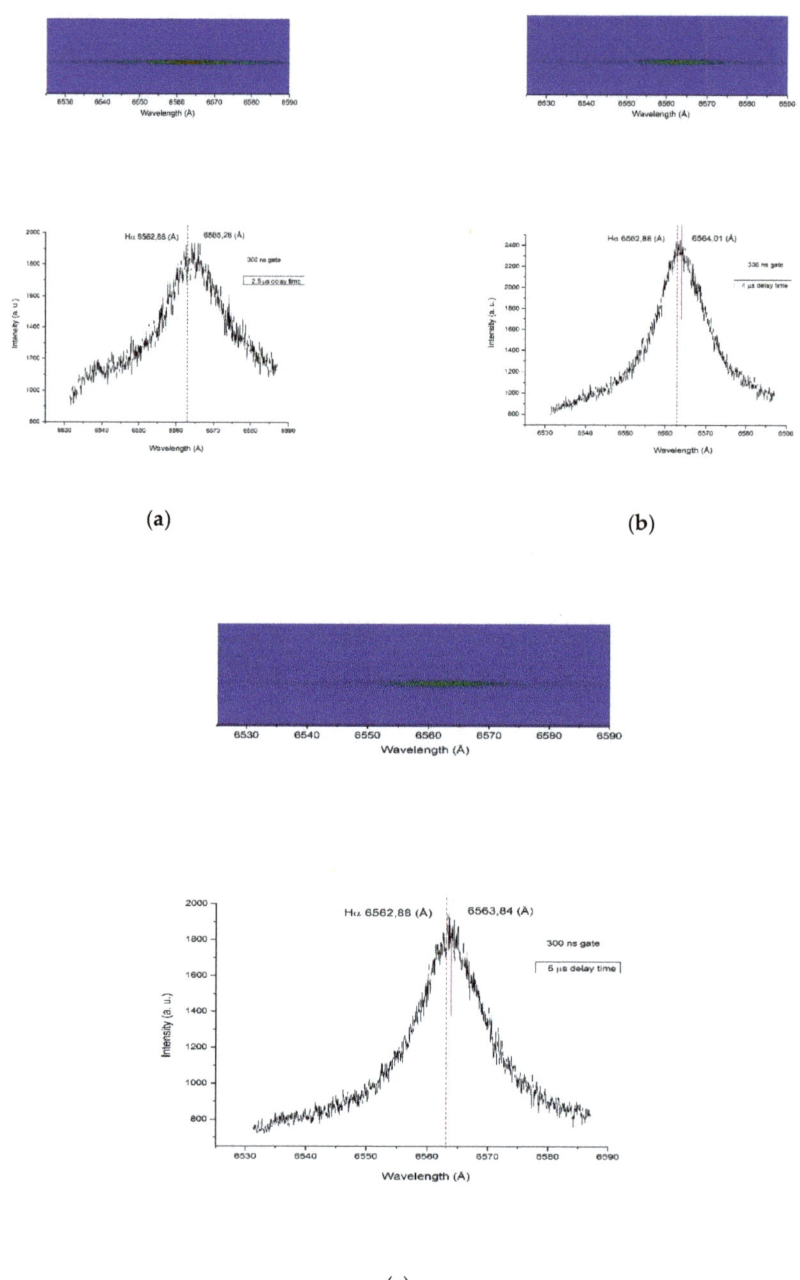

(a)

(b)

(c)

Figure 4. Spectrally resolved images of the plasma (with a gate time of 300 ns) at (**a**) 2.5 μs, (**b**) 4 μs, and (**c**) 5 μs after laser pulse with a wavelength range from 6530 to 6590 Å and spectra obtained from these images of the Hα-line in LSP conditions.

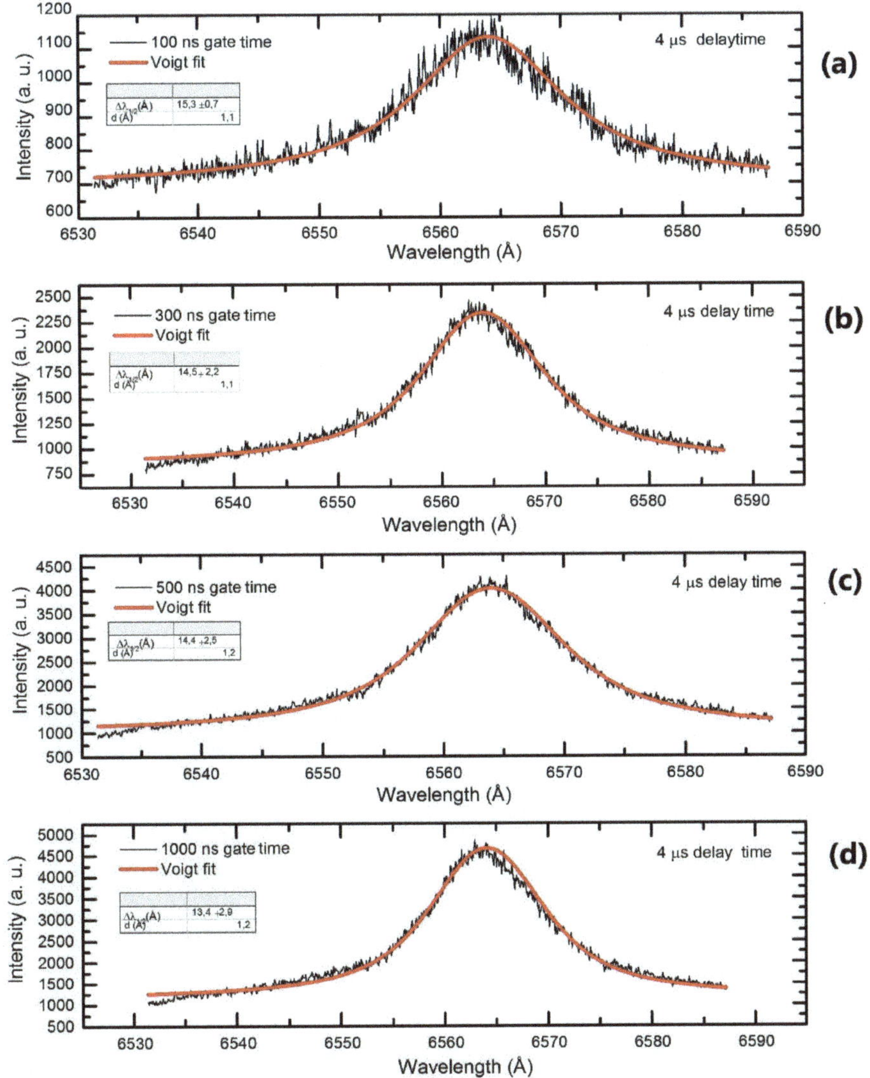

Figure 5. Fitting from a Voigt profile to experimental line emissions of the plasma with gate times of (**a**) 100 ns, (**b**) 300 ns, (**c**) 500 ns, and (**d**) 1000 ns at 4 µs delay time with a wavelength range from 6530 to 6590 Å.

The corresponding Stark broadening and red shift appear in the left side of every spectrum. As can be observed, the best line profiles are at 300 and 500 ns. At 100 ns and 1000 ns there are slight asymmetries that are attributed to the fact that signal integration times are not adequate. At 100 ns, the signal integration time is too short and at 1000 ns, it is too long. It is also observed that despite these disadvantages for the 100 ns and 1000 ns gate times, the Stark widths and shifts obtained are the same in all gates within the experimental error margins.

3. Results

Plasma emission spectra in the range of 652.0–659.0 nm at different times after the laser pulse (2–5 µs) and with several gate times (100, 200, 300, 500, and 1000 ns) were obtained. The spectra were measured in LSP conditions (flow water over the sample). The main feature of the spectrum was the appearance of the Hα-line emission. In contrast to the spectrum obtained in air, in this case and as was observed in our previous work (Moreno-Diaz et al., [24]), the second order line of 327.39 nm Cu I does not appear.

The results obtained about Stark width and red shift after the fitting processes of the experimental data are presented in Table 1. In the first column the delays are presented, and in the successive columns, for each gate time, the corresponding Stark width and shift are shown. In these data, $\Delta\lambda_{1/2}$ (Å) and d (Å) represent the Lorentzian component (in Å) of the Voigt profile (once the instrumental contribution had been discounted) and the Stark red shift, respectively.

Table 1. Experimental Stark broadening ($\Delta\lambda_{1/2}$ (Å)) and red Stark shift (d (Å)) of the Hα-line obtained in the LSP experiments of this work.

Time Delay (µs)	1000 ns Gate		500 ns Gate		300 ns Gate		200 ns Gate		100 ns Gate	
	$\Delta\lambda_{1/2}$ (Å)	d (Å)	$\Delta\lambda_{1/2}$ (Å)	d (Å)	$\Delta\lambda_{1/2}$ (Å)	d (Å)	$\Delta\lambda_{1/2}$ (Å)	d (Å)	$\Delta\lambda_{1/2}$ (Å)	d (Å)
5.0	13.0 ± 0.5	1.0	13.8 ± 0.4	1.0	13.3 ± 0.4	1.0	11.5 ± 0.9	(1.1)	11.8 ± 3.1	(1.0)
4.5	13.0 ± 1.4	1.1	13.9 ± 0.1	1.0	14.5 ± 0.2	0.9	12.3 ± 2.7	(1.0)	13.9 ± 0.2	(1.0)
4.0	13.4 ± 2.9	1.2	14.4 ± 2.5	1.2	14.5 ± 2.2	1.1	14.7 ± 1.9	(1.2)	15.3 ± 0.7	(1.1)
3.5	14.6 ± 0.1	1.2	15.6 ± 2.4	1.4	15.9 ± 1.8	1.5	16.8 ± 1.9	(1.6)	17.1 ± 3.9	(1.7)
3.0	17.2 ± 0.6	1.5	15.7 ± 1.9	1.3	17.2 ± 0.8	1.9	17.7 ± 0.8	(2.0)	16.9 ± 1.9	(1.7)
2.5	17.5 ± 2.5	1.8	20.2 ± 1.4	2.2	19.4 ± 3.4	2.4	19.6 ± 3.0	(2.5)	20.4 ± 5.3	(2.3)
2.0	18.7 ± 2.0	2.1	22.5 ± 0.4	2.6	22.5 ± 7.4	2.8	18.7 ± 2.3	(3.0)	21.0 ± 4.5	(-)

The experimental red shifts obtained in this work were affected by a general uncertainty of around 15%, except for those that appear in parentheses (with time gates of 100 ns and 200 ns and for which the statistics of the data are poor, and the uncertainty may be higher). However, they are included in the table since they seem to indicate the experimental trend indicated by Griem [27]. The 100 ns and 200 ns time gates appear to have light-integration times that are too short. These low statistics are the reason why the use of short gate times for measurements in industrial conditions are discouraged.

An example of these results can be observed in Figure 6, where the displacement of the maximum emission of the Hα-line can be clearly appreciated.

3.1. First Method

3.1.1. Electron Densities Determination

The following expression (given by Ashkenazy et al. [28]) was used to obtain the electron number density, N_e, from the width of the Hα-line:

$$N_e = 8.02 \times 10^{12} \, (\Delta\lambda_{1/2}/\alpha_{1/2})^{3/2} \ \text{cm}^{-3} \tag{1}$$

where $\alpha_{1/2}$ is half the width of the reduced Stark profiles in Å. Values of $\alpha_{1/2}$ for the Balmer series can be found in Griem [29] and also in Kepple and Griem [30].

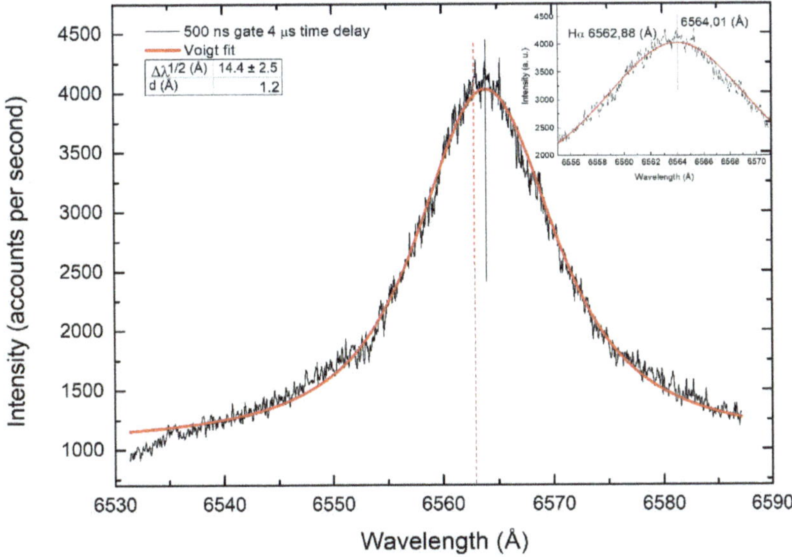

Figure 6. Voigt profile fitting to experimental Hα-line emission of the plasma with a gate time of 500 ns at 4 μs delay time with a wavelength range from 6530 to 6590 Å.

In these references [29,30], it can be observed that the parameter $\alpha_{1/2}$ depends on the electron density and that it also varies slowly with the temperature, T (around 6% at 10,000 K). It is well known that for plasmas in air, in experimental conditions analog to those in this work, temperature will be between 10,000 and 15,000 K (El Sherbini et al. [19]; De Giacomo et al. [22]). In this work, as the profile of the Hα-line in air and in the LSP conditions were similar we used the parameter given by Griem at a temperature of 15,000 K. This assumption is justified in the next paragraph.

The values of the electron density obtained by this procedure are shown in Table 2 and were obtained using a value of $\alpha_{1/2}$ for an electron density of 10^{17} cm^{-3}. Subsequently, an iterative method was used until the used parameters corresponded to the electron densities obtained. The uncertainties of the central values included the deviations in the Stark broadening, shown in Table 1, plus 6% due to temperature uncertainty.

Table 2. Electron densities deduced from experimental values of Stark broadening of the Hα-line obtained in the LSP experiments of this work.

Time Delay (μs)	1000 ns Gate N_e (10^{17} cm^{-3})	500 ns Gate N_e (10^{17} cm^{-3})	300 ns Gate N_e (10^{17} cm^{-3})	200 ns Gate N_e (10^{17} cm^{-3})	100 ns Gate N_e (10^{17} cm^{-3})
5.0	1.5 ± 0.1	1.6 ± 0.1	1.5 ± 0.1	1.2 ± 0.1	1.3 ± 0.3
4.5	1.5 ± 0.2	1.7 ± 0.1	1.8 ± 0.1	1.4 ± 0.3	1.7 ± 0.1
4.0	1.6 ± 0.4	1.8 ± 0.3	1.8 ± 0.3	1.8 ± 0.2	2.0 ± 0.1
3.5	1.8 ± 0.1	2.0 ± 0.3	2.1 ± 0.2	2.1 ± 0.3	2.4 ± 0.6
3.0	2.3 ± 0.1	2.0 ± 0.2	2.4 ± 0.1	2.4 ± 0.1	2.4 ± 0.3
2.5	2.3 ± 0.3	2.9 ± 0.2	2.8 ± 0.1	2.8 ± 0.4	3.1 ± 0.8
2.0	2.6 ± 0.3	3.4 ± 0.1	3.5 ± 1.1	2.7 ± 0.3	3.2 ± 0.7

3.1.2. Electron Temperatures

As in the previous work (Moreno-Diaz et al. [24]), the electron temperature can be estimated by a Boltzmann plot under the assumption that the plasma is in local thermodynamic equilibrium (LTE) (Griem [27]; Griem [29]) using:

$$I_{ij}{}^\lambda = (A_{ij}g_i/U(T))N \cdot \exp(-E_i/kT)$$
$$\ln(I_{ij}{}^\lambda/A_{ij}g_i) = \ln(N/U(T)) - (E_i/kT) \tag{2}$$

For a transition from a higher state i to a lower state j, I_{ij}^λ is the integrated measured integral line intensity in counts per second, A_{ij} is the transition probability, λ is the wavelength of the transition, E_i is the excited level energy, g_i is the energy and statistical weight of level i, $U(T)$ is the atomic species partition function, N is the total density of emitting atoms, k is the Boltzmann constant, and T is the temperature in Kelvin. If the dependency of $\ln(I_{ij}^\lambda/A_{ij}g_i)$ vs. E_i is plotted for lines of known transition probability (Boltzmann plot) the resulting straight line would have a slope of $-1/kT$, and, therefore, the temperature can be obtained without any previous knowledge of the total density of atoms or the atomic species partition function.

In several of the described cases, the temperature can be determined directly through this procedure because, as already mentioned, the emission of Mg II close to 279.5 nm, although weak, it was sufficient for the desired temperature estimation. Spectral lines used in the Boltzmann plot along with the transition probabilities and the energies corresponding to different starting levels were presented in Table 2 of the previous work (Moreno-Diaz et al. [24]). In addition, that table shows the ωref (impact broadening factors) measured by different authors that is necessary for the line self-absorption corrections.

In order to correct for the above mentioned self-absorption of Mg II lines, a procedure similar to that previously used and described in Moreno-Diaz et al. [24] was followed. Voigt profiles were adjusted to Mg II lines to determine their Lorentz broadening and apparent electron densities ($N_e{}^*$). These values were then compared to the real electron densities (N_e), estimated from the H_α-line, in order to obtain the corresponding self-absorption (SA) coefficients in every line. Finally, with these coefficients, the relative intensity in the limit of null self-absorption (negligible self-absorption) was estimated.

An example of the results obtained by means of this procedure is displayed in Figure 7. In this Figure, for times up to 2 μs after laser pulse (500 ns temporal gate), a temperature of 18,700 ± 2000 K can be determined before correcting the line's self-absorption. Once the self-absorption has been corrected, temperatures of 15,200 ± 2500 K are estimated. In the calculations, the uncertainty includes the statistical error from fitting of the temperature value and the statistical uncertainty in the intensities (~10%). These temperature values are maintained within the experimental range for temporal gates of 1000 ns and 2000 ns.

This process was also used for a 3 μs delay from the laser pulse and with those gate times where the intensity of the Mg II lines in the LSP conditions allowed for their discrimination from the background noise (500 ns and 1000 ns). The result obtained in this case was 14,000 ± 2500 K (using the self-absorption corrected values).

For delays equal to or greater than 3.5 ns, the estimation of the temperature by means of this method was been possible because of the lack of clear discrimination of the emission of the Mg II levels from the noise.

In a previous work of the authors (Moreno-Diaz et al. [24]) temperatures of about 10,900 K were estimated by means of an indirect procedure at a delay of 5 μs from the laser pulse. The values obtained here are in-line with that result and with the values that can be found in the bibliography without the presence of water (i.e., with ambient air).

The final results obtained in this work with this first procedure for electron densities and temperatures are shown in Figure 8. As it can be observed, the electron density and temperature decrease with an increase in the delay time. These values are similar to those obtained by other authors in similar experiments without water, as expected from the results obtained in previous work.

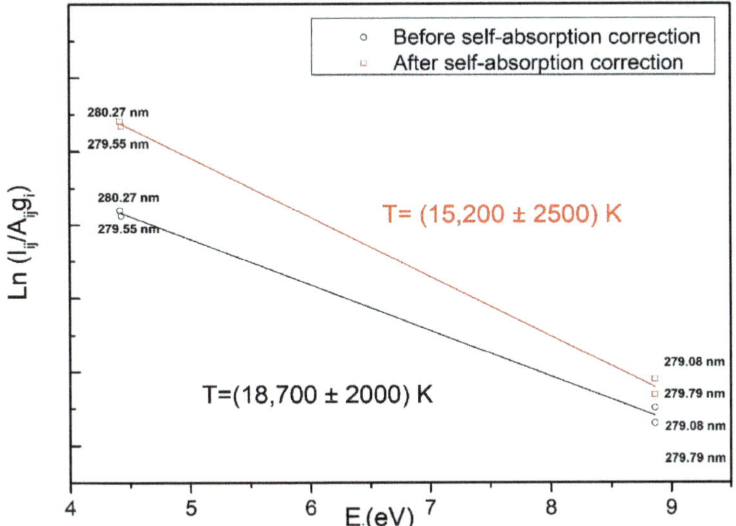

Figure 7. Boltzmann plot for Mg II spectral lines from Laser Induced Plasma (Al 2024 target) in LSP conditions. The spectrum was recorded at a 2 µs delay time from the laser pulse and a gate time of 500 ns.

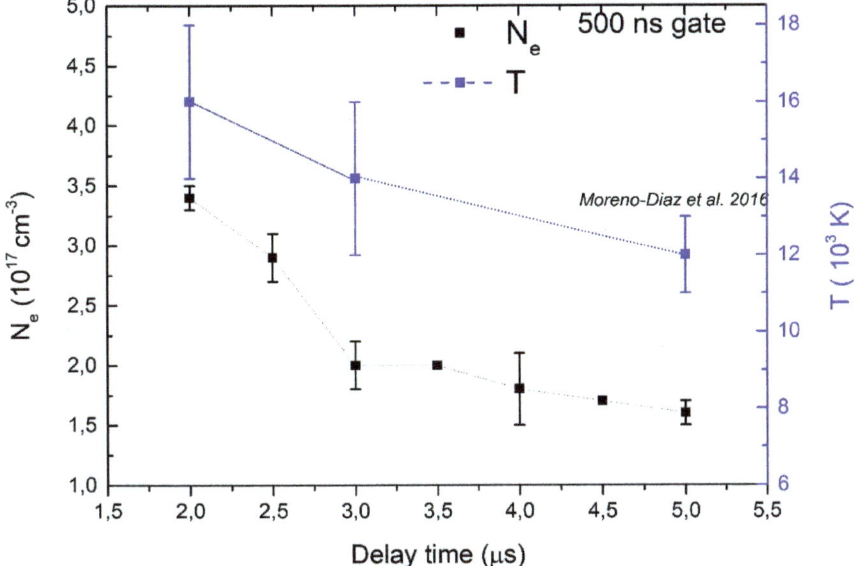

Figure 8. Electron density and electronic temperature versus delay time obtained with a gate time of 500 ns.

3.2. Second Method

In order to diagnose the plasma with a single experiment, the Stark width and the Stark shift of the Hα-line were used together for the first time in a combined way on the basis of the theoretical values of Griem [30] in the Equation (1). The experimental data presented in Table 1 were cross-checked with all the parameters provided by Griem and those situations that were coherent were conserved. As an example, for a 500 ns measurement window and a delay time of 4 μs, the experimental value obtained from $\Delta\lambda_{1/2}$ = 14.4 $\pm$ 2.5 Å (see Figure 5), only provides a coherent value of electron density for the parameter $\alpha_{1/2}$ at the electron density of $(1.78 \pm 0.3) \times 10^{17}$ cm^{-3} and at the temperature of 14,000 $\pm$ 2000 K. With a parameter $\alpha_{1/2}$ corresponding to any other pair of values of electron density and temperature, the numerical results obtained would not be consistent. If $N_e = 10^{17}$ cm^{-3} and T = 10,000 K are chosen, the tabulated value of $\alpha_{1/2}$ would be 0.0186, which would give an electron density of 1.729×10^{17} cm^{-3} (about 70% different to the initial value of 10^{17} cm^{-3}).

By using this method, the values for N_e and T presented in Table 3 were obtained. To confirm these results, the red shift values of the measured Hα-line were compared to the theoretical values provided by different authors [20,29] for the electron densities and the temperatures estimated in this work. An example of this comparison is presented in Table 4. In most cases the obtained value was very close to the theoretical values of these authors within the acknowledged experimental uncertainty levels, which allows us to state that the estimated values for electron density and temperature were consistent with the experimental results for the widths and shifts obtained in this work.

Table 3. Electron densities and temperatures deduced from experimental values of Stark broadening of the Hα-line obtained in the LSP experiments of this work. N_e is 10^{17} cm^{-3} units and T is 10^3 K units.

Delay (μs)	1000 ns Gate		500 ns Gate		300 ns Gate		200 ns Gate		100 ns Gate	
	N_e	T	N_e	T	N_e	T	N_e	T	N_e	T
5.0	1.5 $\pm$ 0.1	13	1.6 $\pm$ 0.1	13	1.5 $\pm$ 0.1	10	1.2 $\pm$ 0.1	10	1.3 $\pm$ 0.3	13
4.5	1.5 $\pm$ 0.2	13	1.7 $\pm$ 0.1	13	1.8 $\pm$ 0.1	17	1.4 $\pm$ 0.3	14	1.7 $\pm$ 0.1	18
4.0	1.6 $\pm$ 0.4	14	1.8 $\pm$ 0.3	14	1.8 $\pm$ 0.3	21	1.8 $\pm$ 0.2	15	2.0 $\pm$ 0.1	22
3.5	1.8 $\pm$ 0.1	16	2.0 $\pm$ 0.3	15	2.1 $\pm$ 0.2	23	2.1 $\pm$ 0.3	18	2.4 $\pm$ 0.6	27
3.0	2.3 $\pm$ 0.1	18	2.0 $\pm$ 0.2	19	2.4 $\pm$ 0.1	24	2.4 $\pm$ 0.1	20	2.4 $\pm$ 0.3	28
2.5	2.4 $\pm$ 0.3	20	2.9 $\pm$ 0.2	21	2.8 $\pm$ 0.1	25	2.8 $\pm$ 0.4	23	3.1 $\pm$ 0.8	28
2.0	2.7 $\pm$ 0.3	26	3.4 $\pm$ 0.1	24	3.5 $\pm$ 1.1	27	2.7 $\pm$ 0.3	27	3.2 $\pm$ 0.7	28

Table 4. Experimental Stark shifts of the Hα-line obtained in this work compared to the theoretical values found in the literature.

Delay (μs)	1000 ns Gate			500 ns Gate			300 ns Gate		
	d (Å) This Work	d (Å) Griem	d (Å) Oks	d (Å) This Work	d (Å) Griem	d (Å) Oks	d (Å) This Work	d (Å) Griem	d (Å) Oks
5.0	1.0	0.9	0.6	1.0	0.9	0.6	1.0	0.9	0.6
4.5	1.1	0.9	0.6	1.0	1.0	0.7	0.9	1.2	0.8
4.0	1.2	1.0	0.7	1.2	1.1	0.7	1.1	1.3	0.9
3.5	1.2	1.2	0.8	1.4	1.3	0.8	1.5	1.5	1.0
3.0	1.5	1.6	1.1	1.3	1.4	0.9	1.9	1.8	1.3
2.5	1.8	1.7	1.2	2.2	2.1	1.4	2.4	2.2	1.5
2.0	2.1	2.2	1.5	2.6	2.6	1.8	2.8	2.8	1.9

4. Discussion

According to the obtained results, both presented methods provide about the same values of electronic densities, but the second method leads to temperature values higher (around a 1.7 factor) than those obtained by the first procedure, as can be seen in Table 5. In this table, values of 5, 3, and 2 μs for the delay times and 500 ns for the time gate (with the best time of integration of the light) were used.

The authors consider that this deviation (that must not be considered as dramatic provided the general uncertainties considered in the kind of plasma diagnosis envisaged in the paper) may be attributed to the inherent inaccuracy of the procedure used to estimate the Hα-line shifts. Finer adjustments of the profile of the line (beyond a Voigt profile and considering asymmetries) are expected to yield more precise values of the temperatures using the described combinations of Stark widths and Stark shifts as described in the second method. In view of the promising character of this method for a practical determination of plasma parameters in one single step, these kinds of improvements are under way and are the subject of upcoming research.

The most important feature of the presented development and its comparison to the standard procedures is that within the experimental uncertainties, the authors consider that the described second method is sufficiently precise and practical to be implemented at an industrial level of LSP applications.

Table 5. Electronic temperatures obtained from Boltzmann plot compared with the temperatures estimated from red shifts of the Hα-line obtained in this work at 5, 3, and 2 μs delay times and 500 ns of time gate.

N_e (10^{17} cm^{-3})	T (10^3 K) from Boltzmann plot	d (Å) Griem [a]	d (Å) This Work	T (10^3 K) from Red Shifts
1.6 ± 0.1	11	0.8	1.0	13
2.0 ± 0.2	14	1.1	1.3	18
3.4 ± 0.1	15	1.8	2.6	26

[a] Griem [30].

In view of the results obtained with the 500 ns window, we think that this second method based on both the Stark broadening and Stark shift of the Hα-line allows for the diagnosis of the plasma in industrial conditions, with good precision for the electron density and an acceptable uncertainty for temperature

5. Conclusions

In the present paper, the problem of direct spectroscopic monitoring of the laser induced plasma in water confined LSP processes is discussed as a feasible procedure aimed at answering the existing need for the development of process monitoring and control in industrial-scope applications. The authors present a procedure of diagnosis in relevant LSP conditions (confining water flow) using the spectroscopic possibilities offered by the Stark broadening and shift of the Hα-line (656.27 nm) to estimate electron densities and temperatures. The procedure has been demonstrated for the case of Al2024-T351 alloy using the luminescence of Mg II lines emitted by the traces of this element present in the alloy.

A Q-switched laser of Nd:YAG (2.5 J per pulse, 10 ns of pulse duration) was focused on an aluminum sample (Al2024-T351) used as the demonstrator. The Stark width and shift of the Balmer Hα-line (656.27 nm) was obtained with different delay times after the pulse of the laser (2–5 μs) and with several gate times (100, 200, 300, 500, and 1000 ns). The electron densities and electronic temperatures were estimated from this experimental data, the procedure being reproducible for any alloy containing Mg traces without lack of generality.

The measurement of the electronic temperature was performed directly under realistic LSP conditions. In the relevant experiments, the electron density and the temperature ranged between 1.2 × 10^{17} and 3.5 × 10^{17} cm^{-3} and between 10,000 and 16,000 K, respectively. These values are similar to those obtained by other authors in similar experiments without water. The best time of integration of the light was found to be about 500 ns. The rest of measured windows present different types of problems due to a lack of statistics or because they exceed the plasma evolution times (in the case of the 1000 ns gate time), this parameter being adjustable for each particular case of interest.

The novelty of the presented work stems from its proved capability to estimate the mentioned parameters under realistic water-confined LSP conditions using an improved procedure over previous developments, allowing single, direct determinations and shorter delay times from the laser pulse incidence. This procedure to estimate both electronic density and temperature density by a single experimental determination is considered an improvement on the existing plasma monitoring of LSP processes and opens the door for the much needed real-time monitoring of plasma behavior.

Author Contributions: Conceptualization: C.C., A.A.-M., J.L.O., M.I.d.A.-G.; Methodology: C.C., A.A.-M., J.L.O., J.A.P., M.I.d.A.-G., C.M.-D.; Investigation: C.C., A.A.-M., J.L.O., J.A.P., M.I.d.A.-G., C.M.-D., I.A.; Writing—review and editing: C.C., M.I.d.A.-G.

Funding: This work was financially supported by the Spanish DGI project MAT2015- 63974-C4-2-R.

Conflicts of Interest: The authors declare no conflict of interest.

References

1. Askar'yan, G.A.; Moroz, E.M. Pressure on evaporation of matter in a radiation beam. *JETP* **1963**, *16*, 1638–1644.
2. Fairand, B.P.; Wilcox, B.A.; Gallagher, W.J.; Willians, D.N. Laser Shock-Induced microstructural and mechanical property changes in 7075 Aluminum. *J. Appl. Phys.* **1972**, *43*, 3893–3895. [CrossRef]
3. Yang, L.C. Stress waves generated in thin metallic films by a Q-switched ruby laser. *J. Appl. Phys.* **1974**, *45*, 2601–2607. [CrossRef]
4. Fabbro, R.; Foumier, J.; Ballard, P.; Devaux, D.; Virmont, J. Physical study of laser-produced plasma in confined geometry. *J. Appl. Phys.* **1990**, *68*, 775–784. [CrossRef]
5. Berthe, L.; Fabbro, R.; Peyre, P.; Tollier, L.; Bartnicki, E. Shock waves from a water-confined laser-generated plasma. *J. Appl. Phys.* **1997**, *82*, 2826–2833. [CrossRef]
6. Sano, Y.; Mukai, N.; Okazaki, K.; Obata, M. Residual stress improvement in metal surface by underwater laser irradiation. *Nucl. Instrum. Methods Phys. Res. B* **1997**, *121*, 432–436. [CrossRef]
7. Morales, M.; Porro, J.A.; Blasco, M.; Molpeceres, C.; Ocaña, J.L. Numerical simulation of plasma dynamics in laser shock processing experiments. *Appl. Surf. Sci.* **2009**, *255*, 5181–5185. [CrossRef]
8. Ocaña, J.L.; Molpeceres, C.; Porro, J.A.; Gómez, G.; Morales, M. Experimental assessment of the influence of irradiation parameters on surface deformation and residual stresses in laser shock processed metallic alloys. *Appl. Surf. Sci.* **2004**, *238*, 501–505. [CrossRef]
9. Martí-López, L.; Ocaña, R.; Porro, J.A.; Morales, M.; Ocaña, J.L. Optical observation of shock waves and cavitation bubbles in high intensity laser-induced shock processes. *Appl. Opt.* **2009**, *48*, 3671–3680. [CrossRef]
10. Ocaña, J.L.; Correa, C.; Porro, J.A.; Díaz, M.; Ruiz de Lara, L.; Peral, D. Induction of through-thickness compressive residual stress fields in thin Al2024-T351 plates by laser shock processing. *IJMSI* **2015**, *6*, 725–736. [CrossRef]
11. Correa, C.; Peral, D.; Porro, J.A.; Díaz, M.; Ruiz de Lara, L.; García-Beltrán, A.; Ocaña, J.L. Random-type scanning patterns in laser shock peening without absorbing coating in 2024-T351 Al alloy: A solution to reduce residual stress anisotropy. *Opt. Laser Technol.* **2015**, *73*, 179–187. [CrossRef]
12. Wu, B.; Shin, C. A self-closed thermal model for laser shock peening under the water confinement regime configuration and comparisons to experiments. *J. Appl. Phys.* **2005**, *97*, 113517-1–113517-11. [CrossRef]
13. Morales, M.; Ocaña, J.L.; Molpeceres, C.; Porro, J.A.; García-Beltrán, A. Model based optimization criteria for the generation of deep compressive residual stress fields in high elastic limit metallic alloys by ns-laser shock processing. *Surf. Coat. Technol.* **2008**, *202*, 2257–2262. [CrossRef]
14. Ocaña, J.L.; Porro, J.A.; Morales, M.; Iordachescu, D.; Díaz, M.; Ruiz de Lara, L.; Correa, C.; Gil Santos, A. Laser Shock Processing: An emerging technique for the enhancement of surface properties and fatigue life of high-strength metal alloys. *IJMMP* **2013**, *8*, 38–52. [CrossRef]
15. Musazzi, S.; Perini, U. *Laser Induced Breakdown Spectroscopy Theory and Applications*; Springer: Berlin, Germany, 2014.
16. Colón, C.; Alonso-Medina, A. Application of a laser produced plasma. Experimental Stark widths of single ionized lead lines. *Spectrochim. Acta B* **2006**, *61*, 856–863. [CrossRef]
17. Alonso-Medina, A. Measured Stark widths of several spectral lines of Pb III. *Spectrochim. Acta B* **2011**, *66*, 439–443. [CrossRef]

18. Noll, R.; Sturm, V.; Aydin, Ü.; Eilers, D.; Gehlen, C.; Höhne, M.; Lamott, A.; Makowe, J.; Vrenegor, J. Laser-induced breakdown spectroscopy—From research to industry, new frontiers for process control. *Spectrochim. Acta B* **2008**, *63*, 1159–1166. [CrossRef]

19. El Sherbini, A.M.; Heagzy, H.; El Sherbini, T.M. Measurement of electron density utilizing the H_α-line from laser produced plasma in air. *Spectrochim. Acta B* **2006**, *61*, 532–539. [CrossRef]

20. Parigger, C.G.; Oks, E. Hydrogen Balmer series spectroscopy in laser-induced breakdown plasmas International. *IRAMP* **2010**, *1*, 13–23.

21. El Sherbini, A.M.; Al Amer, A.A.S.; Hassan, A.T.; El Sherbini, T.M. Measurements of plasma electron temperature utilizing magnesium lines appeared in laser produced aluminum plasma in air. *OPJ* **2012**, *2*, 278–285. [CrossRef]

22. De Giacomo, A.; Dell´aglio, M.; De Pascale, O. Single Pulse-Laser Induced Breakdown Spectroscopy in aqueous solution. *Appl. Phys. A* **2004**, *79*, 1035–1038. [CrossRef]

23. Nath, A.; Khare, A. Spectroscopic Investigations on Laser Induced Breakdown in Water. *J. Phys. Conf. Ser.* **2010**, *208*, 012090-1–012090-4. [CrossRef]

24. Moreno-Díaz, C.; Alonso-Medina, A.; Colón, C.; Porro, J.A.; Ocaña, J.L. Measurement of plasma electron density generated in an experiment of Laser Shock Processing, utilizing the Hα-line. *J. Mater. Process. Technol.* **2016**, *232*, 9–18. [CrossRef]

25. Ocaña, J.L.; Molpeceres, C.; Morales, M.; Porro, J.A. Application of plasma monitoring methods to the optimized design of laser shock processing applications, High-Power Laser Ablation VI. *Proc. SPIE* **2006**, *6261*, 626124.

26. Takata, T.; Enok, M.; Chivavibul, P.; Matsui, A.; Kobayashi, Y. Acoustic Emission Monitoring of Laser Shock Peening by Detection of Underwater Acoustic Wave. *Mater. Trans.* **2016**, *57*, 674–680. [CrossRef]

27. Griem, H.R. Shift of hydrogen lines from electron collisions in dense plasmas. *Phys. Rev. A* **1983**, *28*, 1596–1601. [CrossRef]

28. Ashkenazy, J.; Kipper, R.; Carner, M. Spectroscopic measurements of electron density of capillary plasma based on Stark broadening of hydrogen lines. *Phys. Rev. A* **1991**, *43*, 5568–5574. [CrossRef]

29. Griem, H.R. *Spectral Line Broadening by Plasmas*; Elsevier: New York, NY, USA, 1974.

30. Kepple, P.; Griem, H.R. Improved Stark profile calculations for the hydrogen lines H_α, H_β, H_γ and H_δ. *Phys. Rev.* **1968**, *173*, 317–325. [CrossRef]

Article

Simulation and Experimental Comparison of Laser Impact Welding with a Plasma Pressure Model

Sepehr Sadeh, Glenn H. Gleason, Mohammad I. Hatamleh, Sumair F. Sunny, Haoliang Yu, Arif S. Malik * and Dong Qian

Department of Mechanical Engineering, Erik Jonsson School of Engineering and Computer Science, The University of Texas at Dallas, 800 W. Campbell Rd., Richardson, TX 75080, USA; sepehr.sadeh@utdallas.edu (S.S.); glenn.gleason@utdallas.edu (G.H.G.); mohammad.hatamleh@utdallas.ed (M.I.H.); sumair.sunny@utdallas.edu (S.F.S.); haoliang.yu@utdallas.edu (H.Y.); dong.qian@utdallas.edu (D.Q.)
* Correspondence: arif.malik@utdallas.edu; Tel.: +1-972-883-4550

Received: 7 October 2019; Accepted: 29 October 2019; Published: 7 November 2019

Abstract: In this study, spatial and temporal profiles of an Nd-YAG laser beam pressure pulse are experimentally characterized and fully captured for use in numerical simulations of laser impact welding (LIW). Both axisymmetric, Arbitrary Lagrangian-Eulerian (ALE) and Eulerian dynamic explicit numerical simulations of the collision and deformation of the flyer and target foils are created. The effect of the standoff distance between the foils on impact angle, velocity distribution, springback, the overall shape of the deformed foils, and the weld strength in lap shear tests are investigated. In addition, the jetting phenomenon (separation and ejection of particles at very high velocities due to high-impact collision) and interlocking of the foils along the weld interface are simulated. Simulation results are compared to experiments, which exhibit very similar deformation and impact behaviors. In contrast to previous numerical studies that assume a pre-defined deformed flyer foil shape with uniform initial velocity, the research in this work shows that incorporation of the actual spatial and temporal profiles of the laser beam and modeling of the corresponding pressure pulse based on a laser shock peening approach provides a more realistic prediction of the LIW process mechanism.

Keywords: high-velocity impact welding; laser impact welding; finite element simulation; experimental analysis

1. Introduction

High-velocity impact welding (HVIW) methods are rapidly gaining popularity in joining of similar and dissimilar metals. Compared to conventional fusion-based welding techniques, impact welding is very rapid and could be applied to a wide variety of metals. The primary advantage of impact welding is that metals are joined due to the high impact, thus achieving melting temperatures is not required. Therefore, metals with widely differing melting temperatures and mechanical properties can be welded by impact alone, hence, avoiding the formation of brittle intermetallic compounds. During the short impact welding process, and depending on materials, their surface morphologies, and impact angle and velocity, material flow 'waves' that lead to interlocking and bonding may be observed at the impact interface. Depending on the end-use application and dimensions of the impacting metals, various means have been used for impact generation. From larger to smaller scale, the four main impact-driven joining methods are Explosive Welding (EXW), Magnetic Pulse Welding (MPW), Vaporizing Foil Actuator Welding (VFAW), and LIW, named according to their source of impact upon launching a flyer plate towards a target (base) plate. The high-impact pressure loads are generated in EXW, MPW, VFAW, and LIW, respectively, due to explosive detonation, large magnetic field, high voltage rapid vaporization of a metal foil, and surface ablation of a metal by a focused laser beam.

In the past few decades, both experimental and numerical investigations have been conducted into the aforementioned processes. In previous numerical studies on HVIW of both similar and dissimilar metals, an initial flyer velocity, impact angle, deformed flyer shape (at impact), and corresponding simplified boundary conditions have been assumed. These numerical studies have focused on mimicking the material jetting and morphology of the weld interface. In addition, only two-dimensional simulations of small sections of the entire model have been considered even though the actual dimensions of the flyer and target plates, as well as their boundary conditions, are important to accurately capture the deformed shapes prior to and during impact.

In consideration of the above, we briefly introduce the various HVIW methods and a discussion of related experimental and numerical studies reported in the literature.

1.1. Explosive Welding

In EXW, a controlled detonation provides the source of the high-velocity impact. The explosion accelerates a flyer plate towards a base (target) plate, leading to a weld due to high-velocity collision. Since its discovery in 1957 [1], EXW has been studied by many researchers. Szecket and Mayseless [2] reported on methods of introducing an 'artificial disturbance' to stabilize the wave formation at the weld interface when joining similar metals (copper to copper, and mild steel to mild steel). Mousavi and Al-Hassani [3] designed a physical experiment to mimic the conditions of explosive welding using a pneumatic gun. They performed oblique welding of 3 mm thick flyers of copper, stainless steel, titanium, and zirconium to 30 mm thick base plates of mild steel. Assuming a wide range of constant initial impact angles and flyer velocities, they performed 2D Eulerian simulations of the process to show spallation and jetting of material at the collision interface. Aside from Eulerian simulation methods, meshfree methods such as the Material Point Method (MPM) and Smooth Particle Hydrodynamics (SPH) have also been used to simulate EXW. For example, Wang et al. [4] utilized MPM to model the welding of a 10 mm thick copper flyer plate to a 50 mm thick steel base plate driven by the explosion of ammonium nitrate. Their method did not simulate the formation of a wavy interface but did capture the 'flight' shape of the flyer. Wang et al. [5] simulated EXW of aluminum, steel, and copper flyers against titanium base plates using SPH, but with constant initial flyer velocities assumed. They observed waves having larger wavelengths and greater amplitudes in the collision of stainless steel and titanium compared to aluminum and titanium. Zhang et al. [6] implemented a density-adaptive SPH technique to model EXW, including the detonation of the explosive. They reported that the attainment of requisite impact angles in a parallel flyer/target setup is more challenging compared to oblique orientations.

1.2. Magnetic Pulse Welding

In MPW, the sudden discharge of a capacitor generates very high electric currents through a coil, leading to a very high magnetic field pressure that accelerates toward one another the parts to be welded. Welding of a wide range of similar and dissimilar materials in different MPW configurations have been reported. For example, Okagawa and Aizawa [7] proposed a parallel MPW configuration that led to seam welds between 1 mm thick aluminum sheets. They experimentally studied the seam weld shearing strength and observed its dependence on the kinetic energy of the sheets before the collision and magnetic pressure after the collision. Watanabe et al. [8] obtained lap joints for aluminum flyer plates and iron, nickel, and copper base plates. They reported that exceeding a certain level of discharge energy resulted in a decrease in amplitude and wavelength of material interflow at the weld interface. Lee et al. [9] produced lap joints between 1.2 mm thick aluminum flyer plates and 1 mm thick low-carbon steel plates using MPW. They observed a slight flattening of the grains of aluminum plates in the vicinity of the weld after impact, but no change was apparent in the grain structure of the steel plates. Their nano-indentation hardness tests revealed increased hardness in the intermediate layer between the plates. Ben-Artzy et al. [10] showed that, in tubular MPW joints, reflected shock waves generate interference waves along the weld interface, and that the wavelength of interference waves is proportional to the free path of the shock wave propagation (the first path between front face and back surface) on the interior side of the weld. Göbel et al. [11] made a comparison between the physics involved in EXW and MPW. They reported that in MPW the impact velocity and angle changes along the weld, and that wave creation is possible at lower velocities in MPW compared to EXW. This allows for different welding windows (in terms of impact angle and velocity) in MPW and EXW. Raoelison et al. [12] performed an Eulerian simulation of MPW assuming a linear flyer velocity distribution having a mean value of 600 m/s for aluminum workpieces. They predicted thermomechanical material flow in the form of particle jetting. Cui et al. [13] proposed a method for MPW of 1.4 mm thick carbon-fiber-reinforced plastic and 1 mm thick aluminum tubes. They numerically imitated the MPW process using a coupled electromagnetic and mechanical simulation with constant stress elements.

1.3. Vaporizing Foil Actuator Welding

In VFAW, a sudden capacitor-discharge generates a very high current through a thin conductor foil, vaporizing it almost instantaneously. This creates a very high-pressure plasma, which accelerates the flyer plate towards the target plate. The technique was introduced by Vivek et al. [14] in 2013. They successfully welded copper to titanium, copper to steel, aluminum to copper, aluminum to magnesium, and titanium to steel. Moreover, they quadrupled the weld strength of the titanium/steel combination by introducing a thin nickel interlayer. Hahn et al. [15] attempted an experimental comparison of VFAW and MPW and achieved success only for VFAW. Using the same charging energies of the pulse generator, with VFAW they reached velocities three times those for MPW. Vivek et al. [16] demonstrated two VFAW experiments to join copper sheets with 3 mm thick zirconium-based bulk metallic glass (BMG). In one experiment, the 0.5 mm thick copper sheet was directly launched towards the BMG, resulting in a straight welded interface. In the other experiment, a 0.5 mm thick titanium sheet was first launched towards a 0.25 mm thick copper sheet, consequently launching it towards the BMG, which resulted in a wavy weld interface and no devitrification of the BMG. Nassiri and Kinsey [17] created 2D simulations for VFAW of 2 mm thick flyer and 3 mm thick base plates of aluminum using ALE and SPH methods. In both methods, an initial constant flyer velocity and impact angle were assumed. While the SPH method was able to simulate material jetting, it was found to be less accurate compared to ALE. Nassiri et al. [18] conducted a similar comparison study for VFAW of 0.5 mm thick titanium flyer and 1 mm thick copper plates. They validated the ALE and SPH simulation results through VFAW experiments. ALE was deemed incapable of mimicking material vorticities. Chen et al. [19] machined slanted grooves in 6.3 mm thick steel plates at different angles ranging from 8 to 28 degrees and joined them to 0.508 mm thick aluminum flyers using VFAW. They used this technique to control the impact angle, and thus obtained various morphologies at the weld interface. In a separate study [20] that implemented this procedure, aluminum flyer plates of the same thickness were welded to 1.905 mm thick titanium plates. Gupta et al. [21] applied the same technique and similar foil thicknesses in VFAW of copper to titanium, and vice versa. They created an explicit thermomechanical Eulerian simulation with no slip in contact of Eulerian parts but again assuming a constant initial flyer velocity. It was shown that for capture of interfacial waves, sufficient mesh refinement must be implemented. Groche et al. [22] designed an experiment for process window acquisition in HVIW. They used 2 mm thick aluminum workpieces and reached total normal impact velocities up to 262 m/s (131 m/s for each workpiece). It was shown that in cases where the collision point velocity exceeds the speed of sound in aluminum, jetting would not occur, resulting in no bonding.

1.4. Laser Impact Welding

LIW was patented by Daehn and Lippold [23] in 2011. Similar to laser shock peening, a focused laser beam is used to ablate a sacrificial layer, which is placed on the surface of a metal flyer foil. Rapid vaporization of this ablative layer creates a high-pressure plasma. By using a transparent overlay, the plasma is confined to further increase its pressure. The plasma generates shock waves and accelerates the flyer towards the target metal. Upon collision, jetting and interlocking of the foils occur along a weld interface. Since the deformation of the flyer in LIW is a direct consequence of the shock pressure load generated by the intense laser pulse, there likely exist large velocity gradients amongst regions of the flyer foil upon laser incidence, depending on the spatial profiles of the laser beam and associated pressure pulse. Furthermore, the temporal profiles of the laser pulse and corresponding pressure load further determine the nature of and time to impact in LIW. Thus, assuming a single-valued initial flyer velocity in the simulation of LIW neglects the effects of spatial and temporal profiles of the laser pulse and its plasma pressure load. On that premise, a review of the experimental and numerical studies of LIW follows.

As reported in the literature, successful welds of similar and dissimilar metal foils have been achieved by LIW. Wang et al. [24] used aluminum flyers to investigate the effect of transparent overlay material and laser spot size on the weld quality. They welded nickel flyer (50 μm thick) and base plates using flat and corrugated surfaces on the base. Corrugated base plates provided greater surface area for welding, but large impact angles resulted in no bonding. In a separate experiment using Photonic Doppler Velocimetry (PDV), they revealed that the maximum flyer velocity was achieved within 0.2 μs after impact and in less than 30 μm displacement from the initial position. In addition, they welded 25.4 μm thick aluminum flyer foils to 75 μm thick titanium base foils [25]. It was found that separation of black paint in the ablated area renders it a more suitable ablative layer than black tape. They further studied the effects of laser spot size, the gap between the aluminum flyer and titanium target foils, flyer thickness (from 50 to 250 μm), and transparent overlay material on the weld strength and area [26]. The weld strength and area were evaluated using peel tests and voltage drop measurements [27] across the welds, respectively. It was shown that increasing laser fluence increases the impact velocity (measured by PDV) until the ablative layer reaches its energy absorption limit. It was revealed that by increasing the laser spot size, and thus decreasing the laser fluence, achieving a successful weld becomes more difficult. However, welds achieved by larger spot sizes resulted in greater weld strength and area compared to those achieved by smaller spot sizes. In addition, smaller spot sizes resulted in a greater number of waves with higher amplitude and shorter wavelength on the weld interface. Wang et al. [28] studied the effect of laser fluence on weld interface morphology during oblique LIW of 0.1 mm thick aluminum flyer foils to aluminum and copper base foils of the same thickness. While the authors have reported the laser pulse energy values, it is vital to note that the laser-induced effects heavily depend on the laser spot size among other parameters such as laser wavelength, pulse duration, and the irradiated material [29]. For instance, using the same laser pulse energy, laser impact might generate ultrasounds or shock waves depending on the laser spot size, pulse duration, etc. Superior results (for laser spot size of 6 mm and impact angle of 20 degrees) were obtained at laser fluences of 13.44, 14.15, and 14.85 J/cm^2 when they welded 0.05 mm thick sheets of aluminum and copper flyers to 0.1 mm thick sheets of aluminum base foils [30]. It was found that doubling and tripling the standoff distance between the foils increased the weld diameter by 50% and 83%, respectively. Furthermore, they numerically simulated LIW using the SPH technique in a 2D domain. A constant initial velocity and arc shape in the flyer plate was assumed and showed that increasing the standoff distance resulted in a narrower but taller springback region. The boundary condition applied to the base plate in the numerical model was different from the experimental conditions. In the experiment, the base plate was placed on a back-support while it was left free at the bottom in the numerical simulation. Using the same numerical method and boundary conditions, they simulated LIW of 0.03 mm thick aluminum and titanium flyer plates to 0.1 mm thick copper base plates [31]. Springback of the foils, spallation in the base plate, the jetting, and the wavy interface was simulated and found to be similar to the experimental results. Numerical simulation also revealed that the jetting was mainly produced from a very thin layer of the flyer plate. The numerical simulation was also used to compare LIW of 0.05 mm thick aluminum flyer plates to steel base plates of the same thickness and vice versa [32]. It was shown that using aluminum as the flyer and steel as the target resulted in more successful welds compared to the case where the roles were interchanged. More importantly, swapping the materials changed the direction of wave formation with reference to the impact weld direction. In performing LIW of 0.03 mm thick aluminum flyer plates to 0.08 mm thick brass target plates [33], the authors showed that by increasing the laser fluence, the melting and thus formation of intermetallic compounds was increased, but so did the amplitude and wavelength of the weld interface waves, and the effect of the latter overcame that of the former. Therefore, overall the bond strength was increased with increasing laser fluence. In another experimental work, the authors achieved LIW of 0.03 mm thick crystalline copper flyers to 0.028 mm thick Fe-based metallic glass targets [34]. Copper foil was annealed and attained a finer grain structure prior to LIW. Increased nanoindentation hardness was observed in the copper, the weld interface, and the metallic glass after impact. No crystallization occurred due to LIW and the metallic glass retained an amorphous microstructure after bonding (tested and confirmed only for 1.5 mm laser spot diameter and 47.25 J/cm^2 laser fluence). Liu et al. [35] proposed a preforming technique in LIW of 0.03 mm thick titanium flyers to 0.1 mm thick copper targets. In this method, a preformed local hump replaced the gap between the foils normally used in LIW. The jetting, springback and wavy interface were all observed similar to other LIW experiments (the preforming method was not explained, but the overall dimensions of the flyer after preforming were provided). In addition, they studied the effect of adding a 0.02 mm thick aluminum foil as an interlayer between the copper flyer and base foils having 0.03 and 0.05 mm thicknesses, respectively [36]. It was reported that the nanoindentation hardness levels increased more in the flyer compared to the target. Lap shearing tests resulted in failure in the upper weld interface (between the flyer and the interlayer).

In following the above review of prior research into general HVIW methods, and LIW in particular, the remainder of this paper is organized as follows: Section 2 describes in detail the experimental setup and materials and methods used to study the effects of incorporating measured spatial and temporal profiles of the laser pulse to better simulate the LIW process. Section 3 contains a comprehensive description of the LIW numerical modeling procedures. The results of modeling and comparisons to the experiments are discussed in Section 4.

2. Experimental Procedures

In the experimental setup, 76.2 mm × 25.4 mm × 3.175 mm borosilicate glass samples, 5 mm × 5 mm × 0.05 mm aluminum alloy 1100 foils, and 10 mm × 10 mm × 0.05 mm 304 stainless steel foils were used as transparent overlay, flyer, and base (target) plates, respectively. A very thin layer of black Rust-Oleum enamel aerosol paint was sprayed on one side of the flyer to serve as a sacrificial ablative layer, preventing the top surface of the flyer from melting. A small thin piece of a clear double-sided tape (slightly larger than the laser spot size) affixed the flyer (from its painted side) to the transparent overlay. Target foils were attached to a fixed metal specimen using clear tape along the opposite edges, as shown in Figure 1. To create a gap as the standoff distance between the metal foils, thin glass coverslips were used as spacers. Standoff distance was controlled by the number of glass coverslips placed between the transparent overlay and the fixed metal specimen. The transparent overlay and the glass spacers were attached to the fixed metal specimen using black tape wrapped around the entire sample on both ends, as shown in Figure 2c. The surface of the metal specimen was covered with black tape to prevent it from bonding with the base foil.

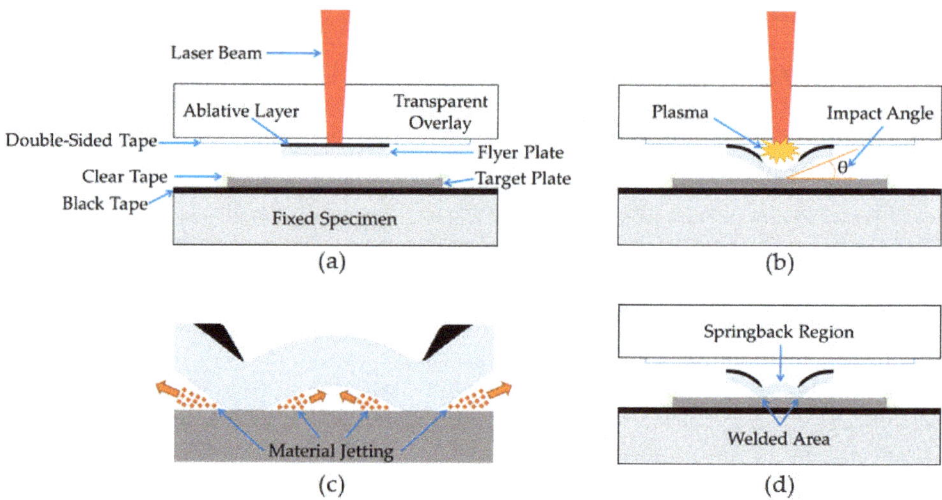

Figure 1. Schematic of the LIW setup and different stages of the experiment: (**a**) LIW components; (**b**) Laser impact; (**c**) Material jetting in the impact region; (**d**) Springback region and welded area. Note: Spacers are not shown.

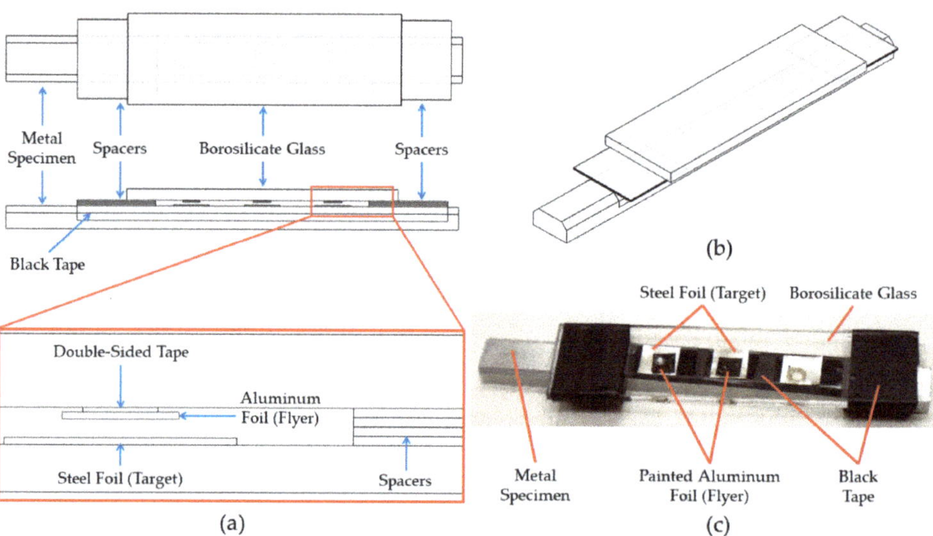

Figure 2. Structure of a LIW specimen: (**a**) Schematic of the LIW specimen (top and side views); (**b**) Schematic of the LIW specimen (isometric view); (**c**) An actual LIW specimen after laser impact.

A Spectra-Physics Quanta Ray Pro-350 infrared laser system (technical specifications listed in Table 1), was utilized for LIW experiments. The laser beam pulse was characterized as depicted in Figure 3. The spatial profile of the laser pulse was measured using an Ophir SP928 high-speed camera utilizing the BeamGage® software. The temporal profile of the laser pulse was measured using an Ophir FPS-1 fast photodetector in conjunction with a Teledyne LeCroy Waverunner 204Xi DSO high-resolution oscilloscope. Measured spatial and temporal profiles of the laser pulse are shown in Figures 4 and 5, respectively.

Table 1. Laser parameters [37,38], with permission from ASME, 2019.

Laser Parameter (Units)	Type/Value(s)
Pulse Type	Q-Switched
Laser Wavelength (nm)	1064
FWHM Pulse Width (ns)	17
Average Pulse Energy (J)	2.5–3 (±2%)
Laser Spot Size (mm)	3.2 ± 0.1
Incident Peak Power Density (GW/cm^2)	1.8–2.2

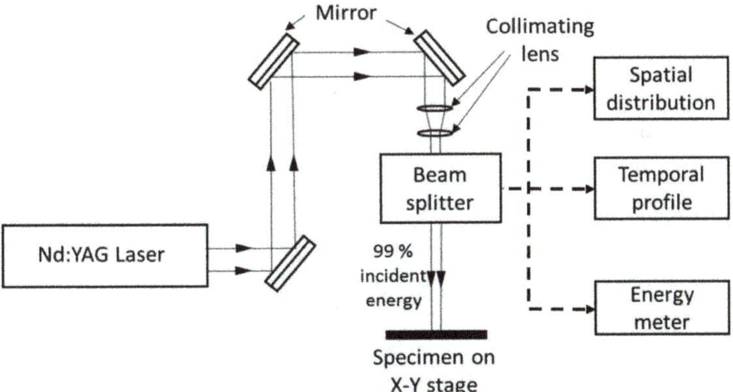

Figure 3. LIW and laser characterization setup. reproduced from [37], with permission from ASME, 2019.

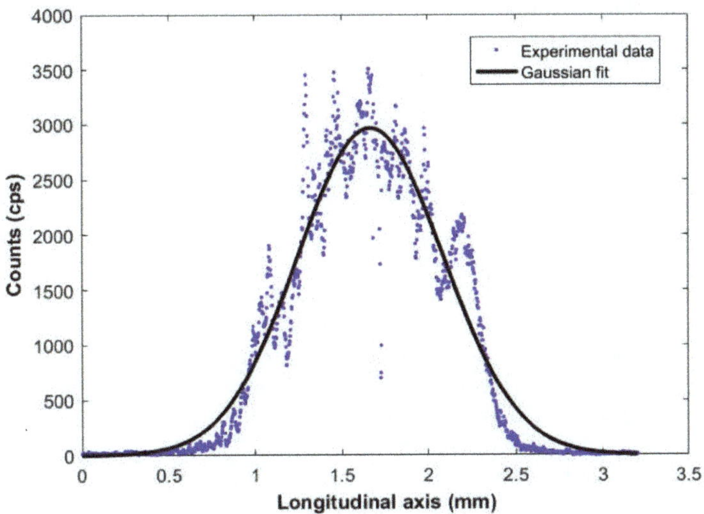

Figure 4. Measured laser pulse spatial profile (assumed axisymmetric) [37,39], with permission from ASME, 2019.

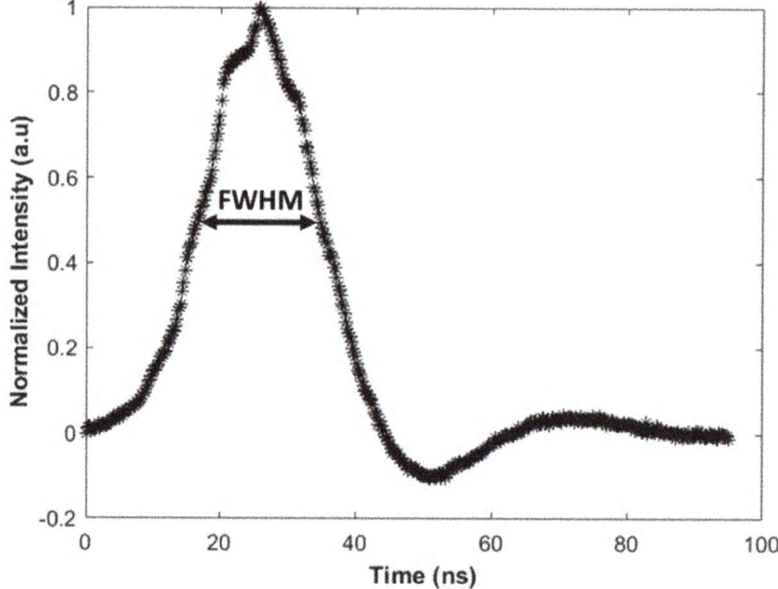

Figure 5. Measured laser pulse temporal profile [37,39], with permission from ASME, 2019.

As can be seen in the earlier Figure 1a, the laser beam passes through the transparent overlay and ablates the sacrificial layer. Since a large amount of energy (2.5–3 J) is released in a very short amount of time (~17 ns) over a small area (0.08 cm^2), the ablative layer reaches extremely high temperatures (above 10,000 °C [40]) and vaporizes instantaneously creating the plasma depicted in Figure 1b. The plasma continues to absorb the energy of the laser and further expands while entrapped in the confinement layer. This leads to the creation of a high-amplitude pressure load penetrating the flyer plate in the form of shock waves. Consequently, the thin flyer plate is launched with high-velocity distribution towards the target plate. In some cases, as reported in the literature [24,30–36], upon collision, bonding is not achieved near the center of the ablated spot since the impact angle is zero and the impact velocity is too great. Instead, the flyer plate (and sometimes also base plate depending on the boundary conditions) springs back. In addition, a jet of metal particles that separates from flyer and base plates is ejected at very high velocities (several thousand m/s), as depicted in Figure 1c. Away from the center point, the impact angle is gradually increased until it reaches the minimum required for a successful weld. Consequently, the welding process initiates and depending on the experimental conditions, flat and/or wavy interfaces can be observed along the weld interface.

After the LIW experiments were performed as described, the welded samples were cut along the centerline of the welds (see Figure 6) and both optical and scanning electron microscope images of the cross-section of the weld were obtained. Prior to cutting, the welded foils were buried in epoxy (poured), wherein the cured epoxy prevented unwanted damage and deformation during the cutting process. After cutting, to reduce the surface roughness of the cut cross-sections, the samples were polished using Allied MetPrep 3^{TM} grinding machine. 6 μm and 1 μm polycrystalline diamond suspensions were used on Gold Label and DiaMat polishing cloths respectively. This was followed by using 0.04 μm colloidal silica suspension on a Final A polishing cloth.

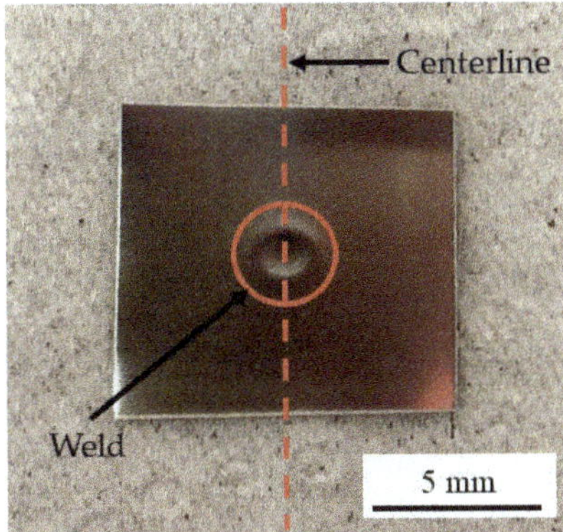

Figure 6. The centerline of the weld.

To measure the strength of the welds, lap shear tests were performed at speed of 0.5 mm/min using a 500 N load cell on an Instron 5969 universal testing system. A schematic of the lap shear test setup is shown in Figure 7. The experimental results are discussed together with the results of the simulations, described next.

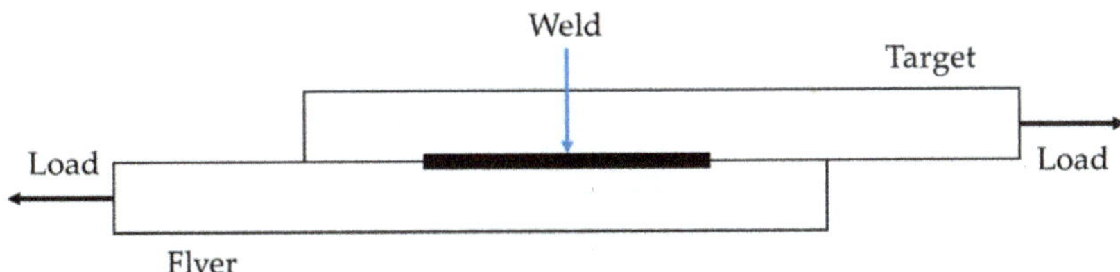

Figure 7. Schematic of the lap shear test.

3. Numerical Simulations

The LIW process, incorporating the spatial and temporal profiles of the laser pulse and its corresponding pressure boundary load, was numerically simulated using two different methods—ALE and Eulerian formulations. As will be shown, these simulations together with the plasma pressure boundary load, provide a more realistic prediction of the LIW process compared to simulations that simply assume an initial constant flyer velocity and flyer shape. In all numerical simulations herein, solution convergence was conducted to determine suitable mesh/grid size. A complete description of the models and techniques used for numerical simulation of the LIW process here is given.

3.1. Plasma Pressure Model

For more realistic simulation of LIW, it is hypothesized that incorporating the transient plasma pressure generated due to laser ablation will provide for improved modeling. Therefore, an axisymmetric Gaussian spatial laser pulse profile (fitted to the measured spatial profile), as well as the measured temporal profile of the laser beam pulse (Figures 4 and 5), were applied in conjunction with the 1D hydrodynamic plasma pressure model by Fabbro et al., which was developed for use in laser shock peening simulation [41]. In this model, during the heating phase, the transient plasma pressure, $P(t)$, is a function of shock impedance Z, plasma thickness $L(t)$, and laser intensity $I(t)$, as shown in Equations (1)–(3).

$$\frac{2}{Z} = \frac{1}{Z_1} + \frac{1}{Z_2}, \tag{1}$$

$$\frac{dL(t)}{dt} = \frac{2P(t)}{Z}, \tag{2}$$

$$I(t) = P(t)\frac{dL(t)}{dt} + \frac{3}{2\alpha}\frac{d[P(t)L(t)]}{dt}, \tag{3}$$

where Z_1 and Z_2 are shock impedances of glass and aluminum, respectively, and α is the fraction of internal energy that dissipates into heat. Coupling Equations (2) and (3) gives Equation (4) [37].

$$\frac{d^2L(t)}{dt^2} = \frac{I(t)}{c_1 L(t)} - \frac{c_1 + c_2}{c_1}\left(\frac{dL(t)}{d(t)}\right)^2 \frac{1}{L(t)}, \tag{4}$$

where $c_1 = \frac{3Z}{4\alpha}$ and $c_2 = \frac{Z}{2}$.

As soon as the laser is switched off (at time t = τ = full width, half maximum (FWHM) of the laser pulse), an adiabatic cooling phase is initiated during which plasma pressure decreases with time as shown in Equation (5).

$$P(t) = P(\tau)\left(\frac{L(\tau)}{L(t)}\right)^{\gamma}, \tag{5}$$

where γ is the adiabatic constant. Parametric constants for modeling of the plasma pressure are listed in Table 2.

Table 2. Parameter values used in plasma pressure modeling [37,42], with permission from ASME, 2019.

Parameter	Magnitude (Units)
Energy ratio (α)	0.25
Adiabatic constant (γ)	1.4
Glass impedance (Z_1)	1.14 (10^6 g/cm^{2}s)
Aluminum impedance (Z_2)	2.75 (10^6 g/cm^{2}s)

Using a substitution, the second-order ordinary differential Equation (4) is rewritten as a system of first-order differential equations and solved numerically using the ode45 solver function in MATLAB® software. Using this solution (in terms of plasma thickness) in Equation (5) and incorporating the measured laser pulse temporal distribution (see Figure 5), the variation of peak plasma pressure over time for an axisymmetric Gaussian spatial pressure was obtained, as shown in Figure 8.

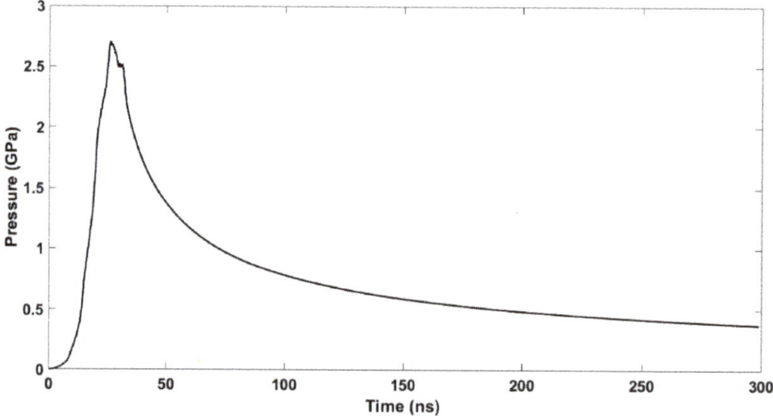

Figure 8. Variation of peak plasma pressure in time.

3.2. Johnson-Cook Model

As a result of very large magnitudes of the pressure load (several GPa) and a very short laser pulse duration (a few ns), extremely high strain rates (over 10^6 s^{-1}) are reached in LIW process, similar to those in laser shock peening. Therefore, the rate-dependent Johnson-Cook (J-C) [43] material constitutive model was implemented to incorporate the effects of strain rate on material behavior. The J-C flow stress (σ_{eq}) is given in Equation (6):

$$\sigma_{eq} = \left(A + B\varepsilon_{eq}^n\right)\left[1 + C\, ln\left(\frac{\dot{\varepsilon}}{\dot{\varepsilon}_0}\right)\right]\left[1 - \left(\frac{T - T_0}{T_0 - T_{melt}}\right)^m\right], \tag{6}$$

where parameters A, B, n, C, and m are quasi-static yield stress, hardening constant, hardening exponent, strain rate constant, and thermal softening exponent, respectively. These material constants are determined empirically. The rest of the parameters in the J-C model are equivalent plastic strain (ε_{eq}), plastic strain rate ($\dot{\varepsilon}$), reference strain rate ($\dot{\varepsilon}_0$), testing temperature (T), reference temperature (T_0), and melting temperature (T_{melt}). J-C parameters of the materials used are listed in Table 3.

Table 3. J-C parameters [32], with permission from Elsevier, 2019.

Material	A (MPa)	B (MPa)	n	C	m	T_0 (K)	T_m (K)
1100 Aluminum	148.4	345.5	0.183	0.001	0.895	293	916
304 Stainless Steel	110	1500	0.36	0.014	1	293	1673

3.3. Mie-Grüneisen Formulation

To govern the hydrodynamic behavior of the materials, an equation of state (eos) was used, which defines the pressure (of a solid at a given temperature) as a function of the density and the internal energy [44]. The Mie–Grüneisen (M–G) eos was applied to model the materials' volumetric strength at the high pressures present in the LIW process. The M–G eos is a function of energy, relating the shock velocity and the particle velocity. Equations (7)–(14) are found in [44]. The M–G eos is given in Equation (7):

$$P - P_H = \Gamma\rho(E_m - E_H), \tag{7}$$

where P, P_H, Γ, ρ, E_m, and E_H are pressure, Hugoniot pressure, Grüneisen ratio, current density, internal energy per unit mass, and Hugoniot specific energy per unit mass, respectively. The Grüneisen ratio is given in Equation (8):

$$\Gamma = \Gamma_0\frac{\rho_0}{\rho}, \tag{8}$$

where Γ_0 is a material constant and ρ_0 is the density at the reference point. The Hugoniot parameters are functions of density only and are related by Equation (9):

$$E_H = \frac{P_H\eta}{2\rho_0}, \tag{9}$$

where η is the nominal compressive volumetric strain and is given in Equation (10):

$$\eta = 1 - \frac{\rho_0}{\rho}, \tag{10}$$

Replacing the terms in Equation (7) with their definitions in Equations (8)–(10) gives Equation (11):

$$P = P_H\left(1 - \frac{\Gamma_0\eta}{2}\right) + \Gamma_0\rho_0 E_m, \tag{11}$$

The Hugoniot pressure is defined through curve fitting to the experimental data and is given in Equation (12):

$$P_H = \frac{\rho_0 c_0^2 \eta}{(1 - s\eta)^2},\tag{12}$$

where c_0 and s linearly relate the shock velocity (U_s) and particle velocity (U_p) by Equation (13):

$$U_s = c_0 + s U_p,\tag{13}$$

Replacing P_H in Equation (11) with its definition in Equation (12) gives the linear Hugoniot form of the M–G eos in Equation (14):

$$P = \frac{\rho_0 c_0^2 \eta}{(1 - s\eta)^2}\left(1 - \frac{\Gamma_0 \eta}{2}\right) + \Gamma_0 \rho_0 E_m,\tag{14}$$

The M–G eos parameters of the materials used in the simulations are listed in Table 4.

Table 4. M–G eos parameters [31,45], with permission from Elsevier, 2018.

Material	ρ (kg/m^3)	c_0 (m/s)	s	Γ_0
1100 Aluminum	2710	5380	1.337	2.1
304 Stainless Steel	7905	4570	1.490	2.0

The different numerical techniques employed for simulation of the LIW process are described next.

3.4. Axisymmetric LIW Simulation

Due to axisymmetric loading conditions of the laser pulse pressure, an axisymmetric numerical analysis was deemed reasonable for simulation of the LIW process (performed using Abaqus/Explicit software version 16.4-4). The LIW simulation was divided into two steps; during the first step, a pressure load was applied to the flyer as inherited from the measured spatial and temporal profiles of the laser beam pulse. The measured spatial profile was inputted as an analytical field to define the spatial distribution of the pressure load. The measured temporal profile was inputted as the amplitude of the pressure load. This pressure load was applied to the top surface of the flyer and launched it towards the target plate until contact was initiated. In the second step, the flyer collided with the target at a very high-speed profile in the center of the laser impact area, resulting in large deformations. Since large deformations in purely Lagrangian finite elements result in excessive distortion, an ALE domain was utilized in the second step. A meshing frequency of 1 increment, with 50 remeshing sweeps per increment and 50 initial remeshing sweeps were used. About 80,000 4-node axisymmetric, quadrilateral elements of 2.5-micron edge length with bilinear displacement, reduced integration, and hourglass control (CAX4R) were used. The target foil was placed on a fixed analytical rigid body representing the fixed specimen in the LIW setup. In addition, another fixed analytical rigid body was included on top of the flyer foil to act as the fixed transparent overlay. The kinematic contact method was implemented between the foils while the penalty contact method was applied between each foil-rigid body pair (flyer-transparent overlay and target-fixed specimen). The contact interaction property in both the kinematic and the penalty methods was "hard" contact allowing separation after contact. A schematic of the LIW setup in the axisymmetric simulation is shown in Figure 9a. The velocities in different regions of the foils as well as the deformed foil shapes after collision were simulated. Details on the axisymmetric simulation results are provided in Section 4.

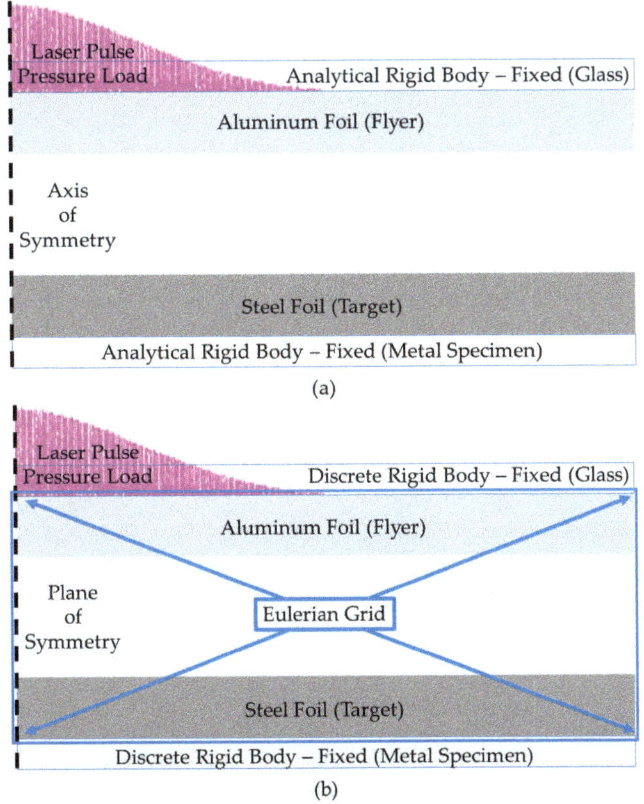

(a)

(b)

Figure 9. Schematic of the LIW simulations: (**a**) Axisymmetric; (**b**) Eulerian.

3.5. Eulerian LIW Simulation

In order to mimic the material jetting phenomenon, an Eulerian model of the LIW process was run in Abaqus/Explicit. In this model, the material's Eulerian volume fraction (EVF) was computed within each element and its movement was tracked as it flowed through the fixed mesh. A material EVF of one means the element is completely filled with that material while a material EVF of zero shows that the material is not at all present in the element. More than one material may be present in Eulerian elements at the same time. If after adding all the material EVFs in an element, the sum is less than one, the software fills the remainder of the element's volume with "void" material which has zero mass and strength [44]. Since Eulerian models in true 2D space are not available in the particular software, a 3D model was created with a thickness of 0.02 mm (four elements thick) in the out-of-plane (depth) direction. However, motion in the depth direction was constrained by a symmetry condition at all nodes, thus rendering the model an equivalent 2D plane strain simulation (due to plate thicknesses being much smaller than other dimensions). An Eulerian grid (meshed control volume) encompassed the entire model in order to track material movement during LIW. Exploiting symmetry, only half of the full geometry was modeled. Approximately 300,000 hexahedral 8-node Eulerian brick elements of 5-micron edge length with reduced integration and hourglass control (EC3D8R) were utilized, which in turn employed predefined volume fractions at the locations of the flyer and target plates. Two fixed discrete rigid bodies were used as the transparent overlay and the metal specimen. Consistent with an Eulerian model description, Abaqus/Explicit allows for load application only at fixed nodes in the meshed control volume. The pressure load described earlier was applied to the top surface of the flyer based on measured spatial and temporal profiles of the laser pulse. Implementation of the measured spatial and temporal profiles as the distribution and amplitude of the load was similar to that of axisymmetric simulation. The only difference was that the load was apportioned among nodal layers of the initial flyer position in the out-of-plane direction, accelerating the flyer towards the target, resulting in jetting and interlocking of the foils along the weld interface upon collision. The general contact method was implemented between all surface pairs using a "hard" contact interaction property allowing separation after contact. A schematic of the LIW in the Eulerian simulation is shown in Figure 9b. Details on the Eulerian simulation results are provided in Section 4.

4. Results and Discussion

Both experimental and numerical results of this study are presented next. The experimental results are included in form of optical and scanning electron microscope images of the cross-sections of the cut LIW samples. Axisymmetric and Eulerian simulation results are discussed, including advantages and disadvantages of the ALE and Eulerian numerical techniques. The results obtained from numerical simulations are compared with the experimental results, and similarities and discrepancies between them are investigated.

4.1. Experimental Results

Different standoff distances and laser fluence values were used in experiments to acquire the LIW process window. These parameters are listed in Table 5. It was seen that regardless of the tested laser fluence values, standoff distances of 0.12 and 0.54 mm were too small and too large respectively. Successful welds were obtained using standoff distances of 0.26 and 0.40 mm, irrespective of the laser fluence values tested in experiments.

Table 5. List of experimental parameters.

Laser Spot Diameter (mm)	Laser Pulse Energy (J)	Laser Fluence (J/cm^2)	Standoff Distance (mm)	Successful Weld?
3.2	2.5 3.0	31.08 37.30	0.12	No
	2.5 3.0	31.08 37.30	0.26	Yes
	2.5 3.0	31.08 37.30	0.40	Yes
	2.5 3.0	31.08 37.30	0.54	No

Using Leica microscopes, optical images of the weld cross-sections were obtained. Sample images for the case with a laser fluence of 31.08 J/cm^2 and a standoff distance of 0.26 mm are shown in Figures 10 and 11. As mentioned in Section 2, in most LIW experimental results reported in the literature, due to very high flyer impact velocities and very low impact angles, springback occurs and welding is not achieved in the center of the laser-ablated region [24,30–36]. This is an unwanted event which decreases the strength and integrity of the weld. Interestingly, in the experimental results here, the springback phenomenon was not observed in the center of the weld region. As seen in Figure 10, the foils are indeed welded together in the center of impact region, while gaps are observed on the outside of the weld. Figure 11 shows the same weld at two higher magnifications. A combination of flat and wavy interface patterns is observed along the weld interface. Elimination of the springback phenomenon in these experiments could be attributed to the polymeric black tape placed between the target foil and the fixed metal specimen. This tape was used to prevent the target foil from bonding with the metal specimen after laser impact. It is hypothesized that direct placement of the target foil on the metal specimen leads to metal foils (flyer in particular) bouncing off of the specimen after impact resulting in springback at the center of the impact region. It is possible that the addition of a polymer interlayer between the target foil and metal specimen elongates the collision time of the foils, preventing the flyer foil from rebounding (which could lead to springback). The validity of this hypothesis needs to be investigated through further experiments as part of future research.

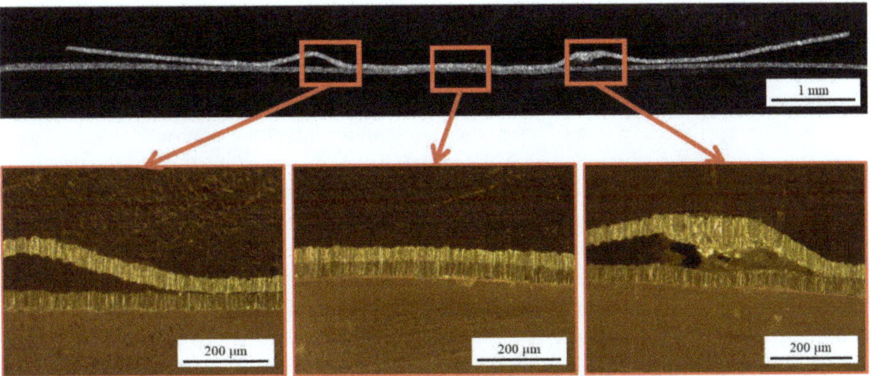

Figure 10. Optical microscope images of different regions of a LIW sample.

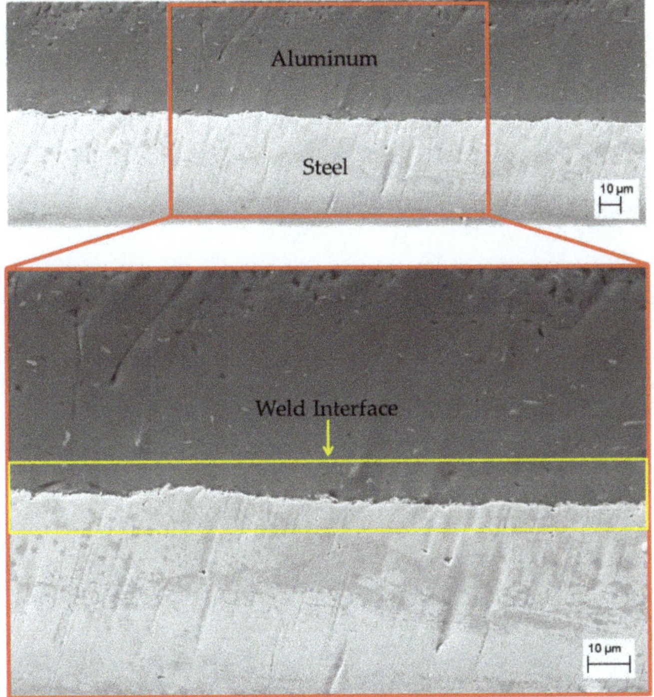

Figure 11. SEM images of a LIW sample at two different magnifications.

Another important observation in these experimental results relates to the overall shape of the welded foils. As seen in Figure 10, while the foils are welded in the central region, there is a gap present on each side of the weld. Since the loading conditions are axisymmetric, this means that the gap has an annular shape. Considering that the experiments include an initial standoff distance between the two foils, and because of the 3D Gaussian profile of the ablation pulse pressure, the generation of a gap that gradually increases with position away from the center of the weld might be expected. However, as seen in Figure 10, starting from the center of the weld, and moving away in the radial direction, the gap gradually increases to a maximum and then starts decreasing, resulting in a shape very similar to the springback region reported in the literature [24,30–36]. Although similar in shape, these are two different phenomena and they happen for different reasons; the annular gap observed in these experiments is attributed to the Gaussian profile of the laser-induced plasma pulse pressure, and the size of the double-sided tape attached to the flyer's laser-ablated area. The double-sided tape is used to enhance the plasma confinement in the glass overlay by eliminating any gap between the flyer foil and the transparent overlay. Due to the Gaussian profile of the pressure load, moving away from the center of the laser-ablated spot in the radial direction, the magnitude of the pressure load decreases significantly. Therefore, if the size of the double-sided tape is considerably larger than that of the laser spot, the laser impact cannot separate the flyer foil from the transparent overlay. In the experiments conducted here, after some trial and error, it was found that a double-sided tape slightly larger than the laser spot resulted in successful plasma confinement and separation of the flyer from the transparent overlay. In the center of the impact area, the flyer detaches from the transparent overlay and is launched towards the target. Away from the center, the resistance of the double-sided tape against separation becomes gradually more significant, together with the decrease in the magnitude of the laser beam pulse pressure. As a result, the size of the gap starts increasing until the edge of the double-sided tape is reached, the flyer is completely detached, and the resistance is no longer present. Subsequently, the gap starts decreasing and then increasing again as the flyer accelerates towards the target.

Lap shear tests were performed on samples with different laser fluences and standoff distances. These results are shown in Figure 12. It was observed that the force linearly increased with displacement until it reached a maximum. Then the weld failed (on the aluminum flyer side) resulting in a sudden drop in force values. Further increasing the displacement resulted in increased tearing of the flyer and thus sliding of the foils. Therefore, the force values gradually decreased until they reached an almost constant value with little change with respect to displacement. Higher fluences resulted in higher maximum force values while increasing the standoff distance decreased the maximum forces.

Numerical results are presented next. Prior to the viewing of numerical results, it is important to note that all experiments were conducted in air at atmospheric pressure. Since the flyer has a supersonic flight, shock waves are generated upon collision of the foils resulting in a strong air-cushion effect. However, the air and its effect on the collision were neglected in the numerical simulations. Therefore, an investigation into the air versus vacuum conditions in the LIW process is a potential area of research as part of future work.

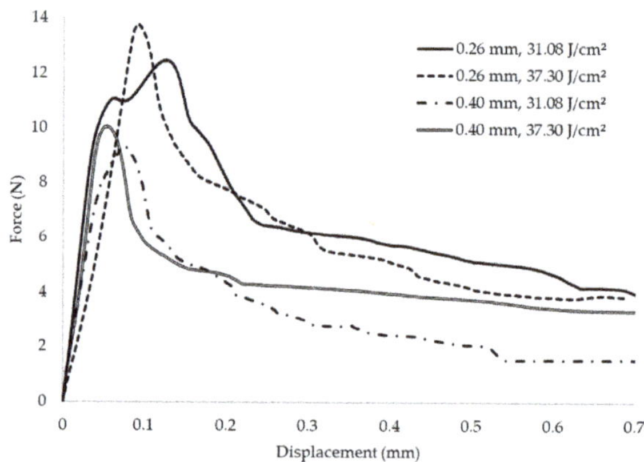

Figure 12. Force-displacement curves obtained from lap shear tests.

4.2. Numerical Results

Axisymmetric ALE and Eulerian numerical methods incorporating the spatial and temporal profiles of the laser beam pressure pulse were implemented in the simulation of the LIW process. In the axisymmetric ALE simulation, the effect of the standoff distance between the foils on the impact angles, nodal velocities, springback, and overall shape of the deformed flyer foil were investigated. Using the Eulerian technique, the jetting phenomenon and the interlocking of the foils along the weld interface were simulated. In addition, the hypothesis that simulation results assuming a constant initial flyer velocity and shape are not as accurate is tested.

The axisymmetric ALE simulation result for the same sample case (laser fluence of 31.08 J/cm^2, and standoff distance of 0.26 mm) is compared to the experimental result, as shown in Figure 13. The diameter of the welded region obtained from experimental and numerical results is 2.5 mm and 2.2 mm, respectively (12% difference). The overall shape of the deformed foils was successfully simulated, showing good agreement with experimental results as seen in Figure 13. Since the effect of the double-sided tape (discussed in experimental results) was neglected, the annular gap observed in the experiments was not fully captured. Therefore, an investigation into the effect of double-sided tape can be pursued in the numerical simulations of the LIW process as part of future research.

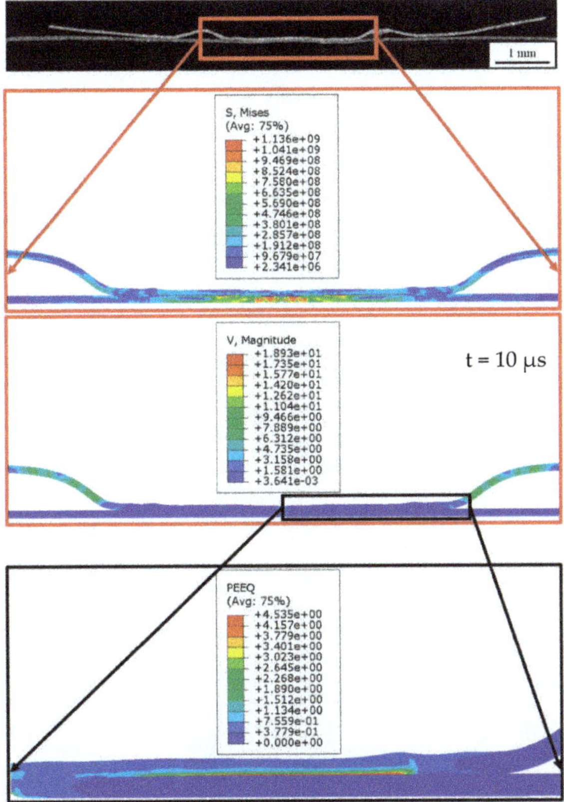

Figure 13. Axisymmetric simulation results: deformed shapes and comparison to experimental results; Contour plots include S: von Mises stress (Pa), V: velocity (m/s), and PEEQ: equivalent plastic strain.

To investigate the importance of the temporal profile of the laser beam pressure pulse, an axisymmetric ALE simulation for a case with a larger standoff distance (0.40 mm) and higher laser fluence (36.06 J/cm^2) was created. The simulation results for this case at different times during the second step of the LIW process are shown in Figures 14 and 15. As seen in these figures, a significant amount of springback is observed. This is due to the fact that the plasma pressure pulse duration (~300 ns, see Figure 8) is significantly shorter than the collision start time (366 ns, see Figures 14 and 15). This means that the pressure load application ends before the flyer collides with the target. Since there is no force to resist the bouncing of the foils, the springback occurs due to the very high flyer velocities (~1550 m/s, see Figure 15) and low impact angles in the center of the impact region.

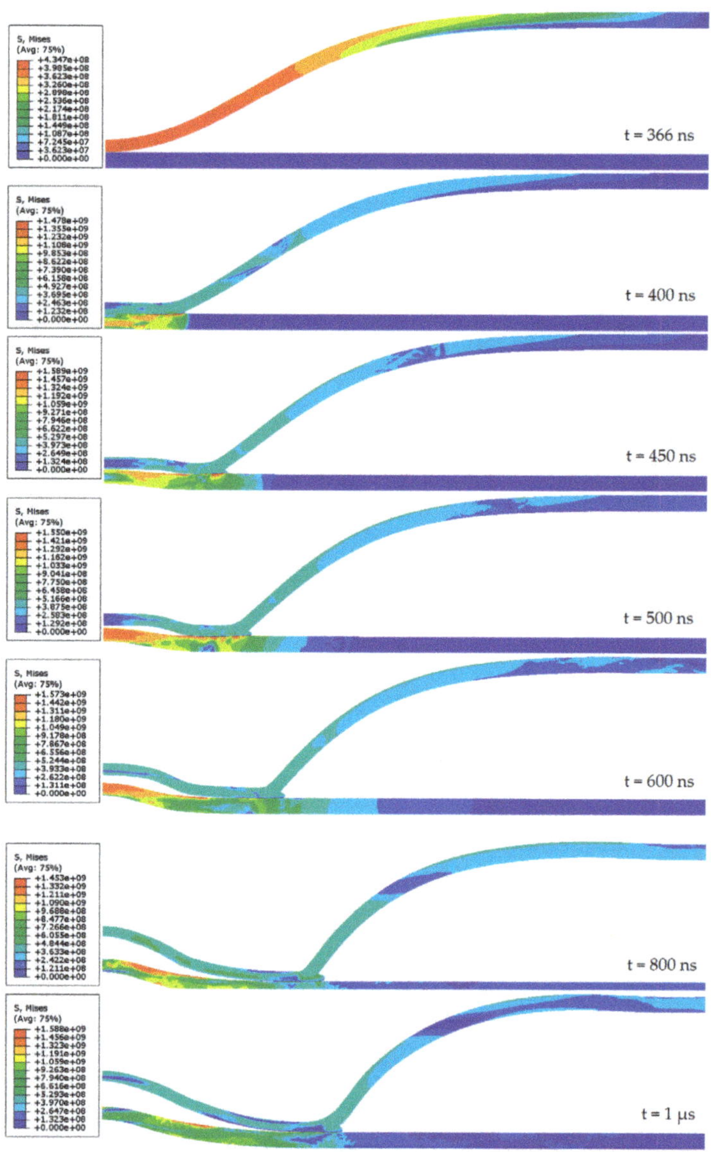

Figure 14. Axisymmetric simulation results: von Mises stress (Pa) contour plot and deformed shape at different times during the collision.

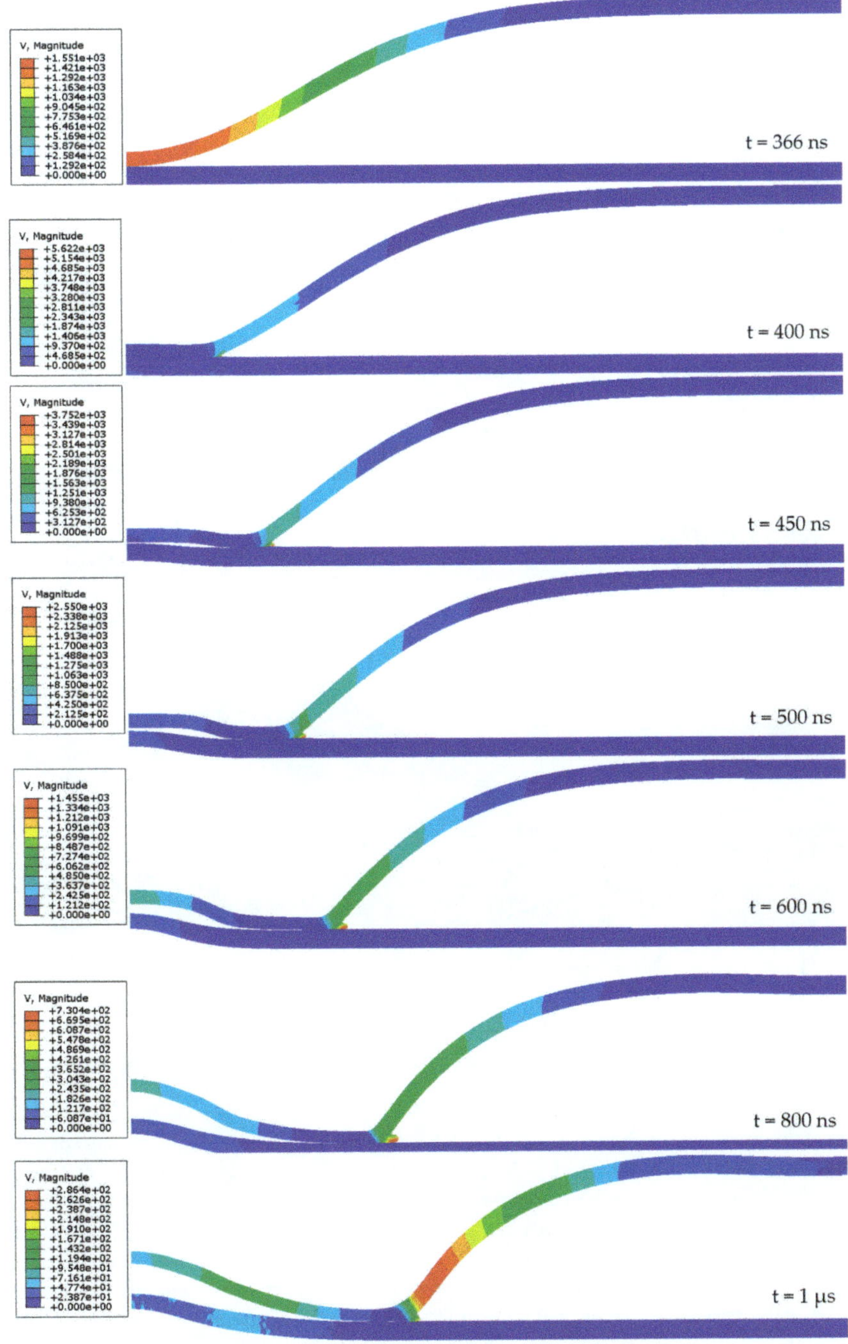

Figure 15. Axisymmetric simulation results: velocity (m/s) contour plot and deformed shape at different times during the collision.

A closer look at the velocities obtained in the collision region (over 5600 m/s, see Figure 16) reveals that jetting velocities were reached in the ALE domain. However, the ALE method is not capable of fully capturing the jetting phenomenon. Note that jetting is one of the known requirements for a successful weld [24,31,32] and happens when metal particles from the foils reach very high velocities upon collision and are ejected from the surface. The resulting 3D images of the deformed foils at 1 µs for this sample case are shown in Figure 17.

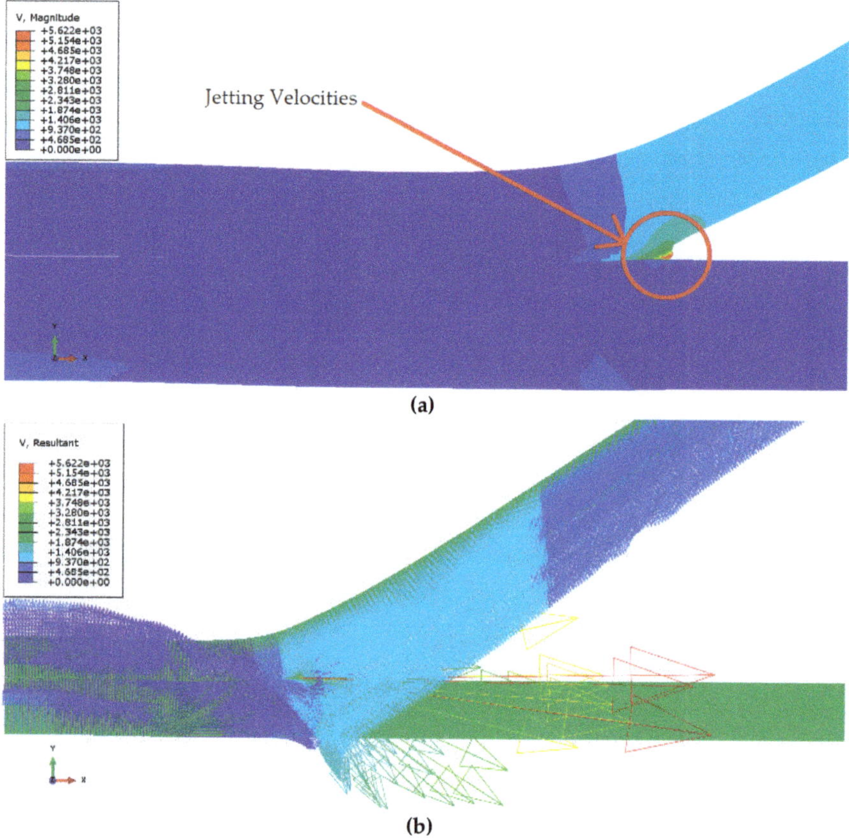

(a)

(b)

Figure 16. Axisymmetric simulation results at 400 ns: (**a**) Jetting velocities (m/s); (**b**) The resultant velocity vectors in the collision region.

One of the most crucial parameters in the successful execution of LIW is the amount of initial gap between the two foils [25]. If this standoff distance is too small, then the flyer will not have enough time to sufficiently accelerate and/or might not reach the required minimum impact angle before making contact with the base. Conversely, if this gap is too large, the flyer might not reach the base plate or might exceed the maximum required impact angle before the collision with the base. Therefore, it is very important to assess the effect of standoff distance on the impact angles and velocities in different regions of the flyer. The axisymmetric ALE simulation results for sample cases with a laser fluence of 31.08 J/cm^2 and standoff distances from 0.1 mm to 0.4 mm are provided in Table 6.

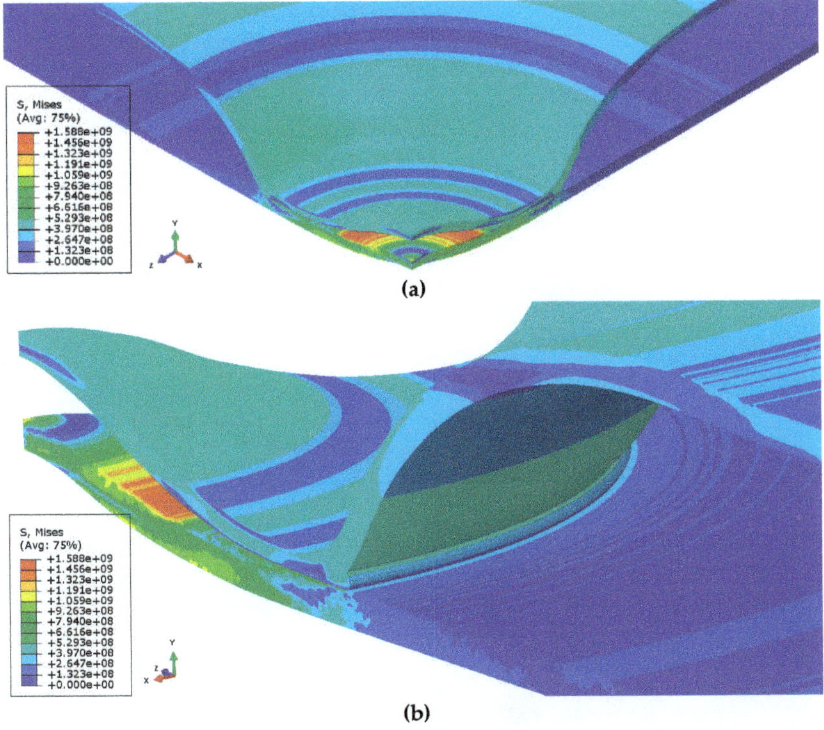

Figure 17. Resultant 3D images of the deformed foils and von Mises stress (Pa) contour plots (the axysimmetric simulation at 1 µs): (**a**) Overview of the geometry; (**b**) Closer view of the springback.

Table 6. The effect of standoff distance based on axisymmetric LIW simulation.

Standoff Distance (mm)	Maximum Impact Velocity (m/s)	Maximum Impact Angle (degrees)	Collision Time (ns)
0.1	1110	5	146
0.2	1370	15	220
0.3	1580	20	293
0.4	1550	30	366

As mentioned earlier, due to the Gaussian profile of the laser beam, the applied pressure load on the flyer is axisymmetric. Therefore, the resulting impact velocities are maximum in the center of the laser spot and decrease in the radial direction away from the center. Impact angles and velocities at the point of collision for the sample case are shown in Figure 15 (at t = 366 ns). The drastic velocity gradients between different regions of the flyer show the importance of incorporating the spatial profile of the laser beam. As seen in Table 5, the maximum velocity is reached at a distance between 0.3 and 0.4 mm away from the initial flyer position. This means that if the standoff distance is large enough, at a certain time the flyer starts to decelerate, showing the importance of considering the temporal profile of the laser beam. If a constant velocity had been assumed, not only would the changes in accelerations and velocities be ignored over time, but so would the gradients between the different regions of the flyer have been neglected. At the point of collision, starting from the center and moving away radially, the impact angle starts at zero degrees, gradually increases to a maximum, and then slowly decreases back to zero close to the edges of the flyer. This shows that assuming an initial shape or a constant impact angle does not give an accurate representation of the LIW process. To support this claim, two different LIW configurations were simulated using an Eulerian technique applying a constant downward initial flyer velocity of 800 m/s. One assumed an initial flyer impact angle and the other assumed a curved initial flyer shape. These additional simulation results are provided for the sake of comparison with the primary work here in which measured spatial and temporal profiles of the laser beam pulse have been incorporated. As seen in Figure 18, the application of a constant initial velocity results in excessive rebounding of the foils upon collision and is therefore not realistic. Moreover, despite using an Eulerian grid, the jetting phenomenon is not observed at any time during the corresponding LIW simulations.

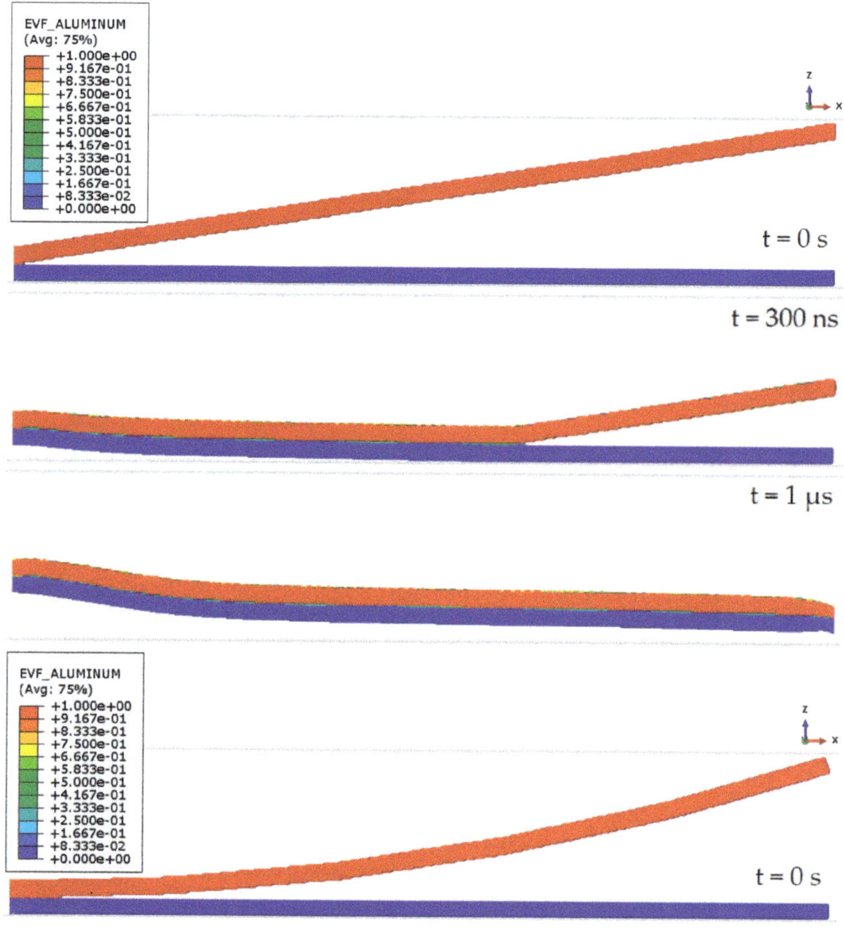

Figure 18. *Cont.*

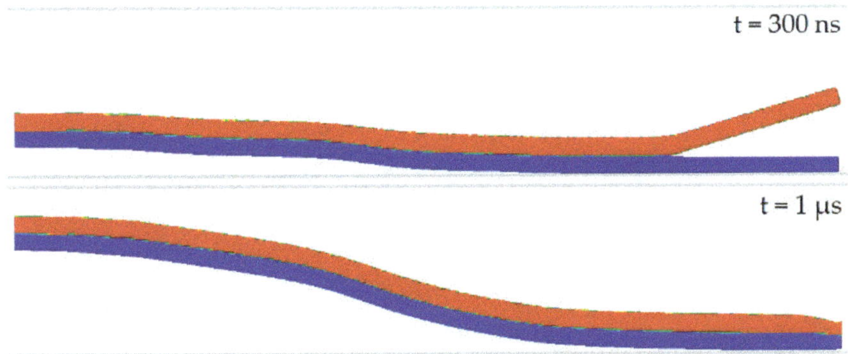

Figure 18. Eulerian volume fraction (EVF) of the aluminum foil in angled and curved LIW orientations at different times in Eulerian simulations assuming an initial flyer shape and constant velocity.

Implementing an Eulerian simulation, the jetting phenomenon and the interlocking of the foils along the weld interface were mimicked. The Eulerian simulation results for a sample case with a 37.30 J/cm^2 laser fluence and a 0.4 mm standoff distance are shown in Figure 19. These results are the volume fraction plots of the aluminum flyer foil. It can be seen that in contrast to some of the reports in the literature [30–32], most of the jet consists of steel target particles. This is despite the fact that aluminum is the softer material compared to steel. Therefore, an experimental investigation into the composition of the jet in the LIW process can be pursued as part of future research.

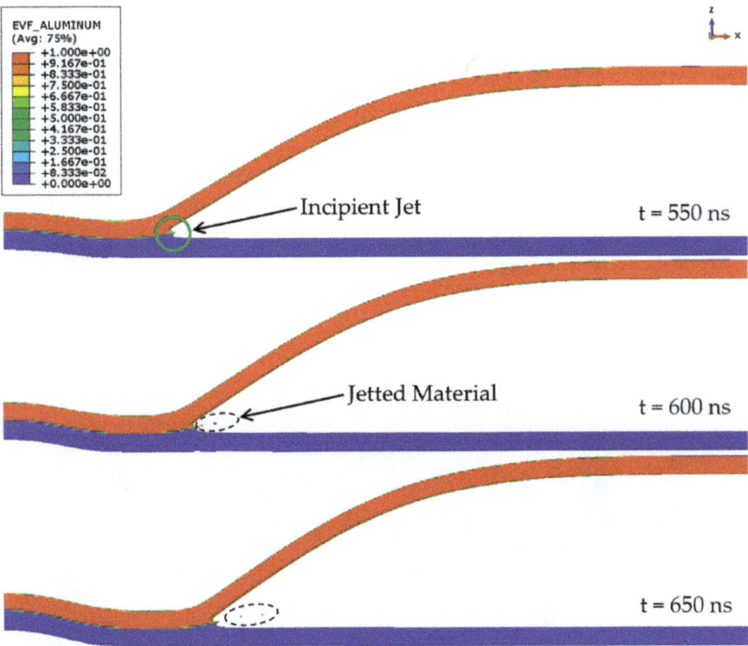

Figure 19. *Cont.*

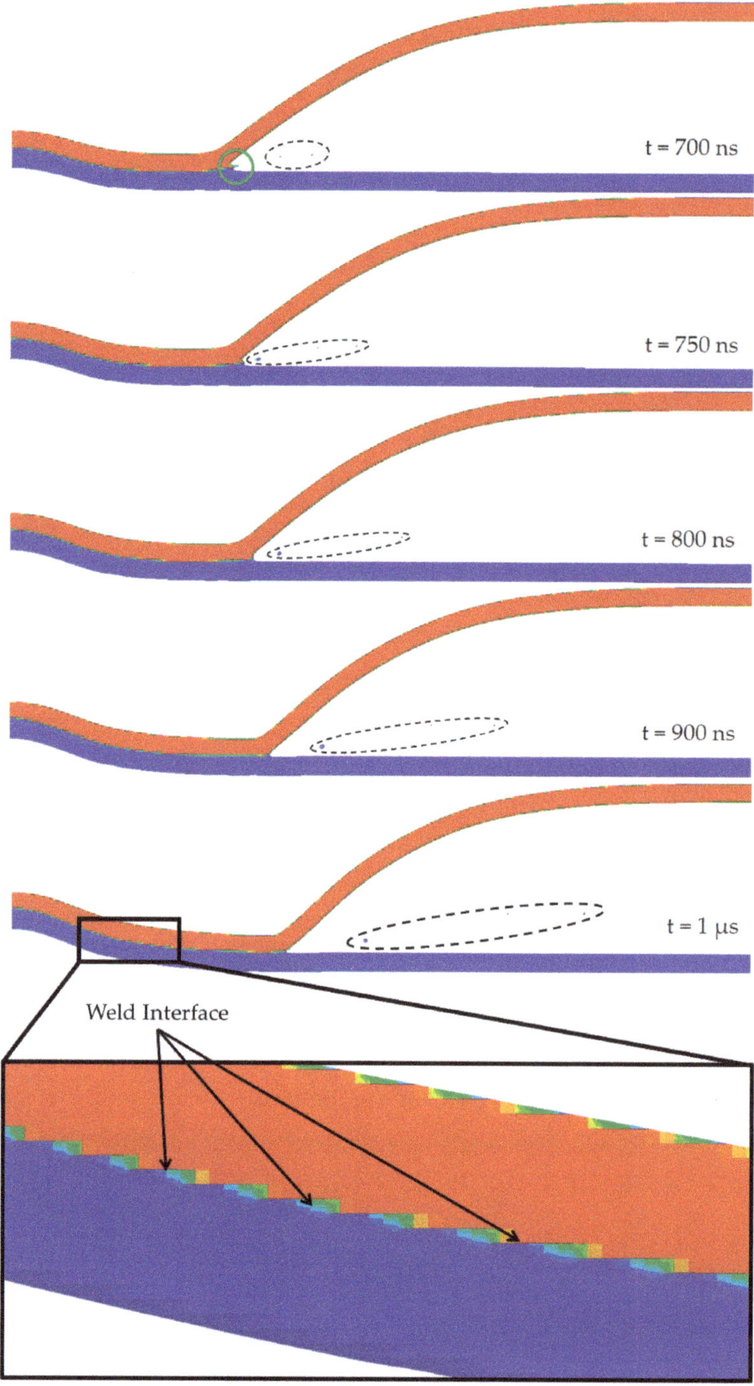

Figure 19. Jetting phenomenon at different times and interlocking of the foils along the weld interface in Eulerian simulation of the LIW process.

As seen in Figure 19, the interlocked foils bounce off in the center of the weld region. This could be attributed to the high fluence of the laser pulse (37.30 J/cm^2) and modeling of the fixed metal specimen as a rigid body, while in reality the metal specimen is made of aluminum. Therefore, the effect of the fixed metal specimen material on rebounding of the foils can also be further studied as part of future research.

5. Conclusions and Future Work

- A more realistic prediction of velocities and deformed shapes in different regions of the foils was achieved through incorporation of the measured temporal and spatial profiles of a Gaussian laser beam pulse pressure, leading to improved simulation of the LIW process in both axisymmetric ALE and Eulerian numerical models.
- LIW experiments were performed using standoff distances of 0.12, 0.26, 0.40, and 0.54 mm, as well as laser fluence values of 31.08 and 37.30 J/cm^2. Irrespective of the laser fluence value, successful welds were obtained only at standoff distances of 0.26 and 0.40 mm.
- Successful welds were obtained without springback in the central region.
- Lap shear test results revealed that the greatest value of maximum force (13.76 N), and thus the strongest weld, was achieved using a standoff distance of 0.26 mm and a laser fluence of 37.30 J/cm^2. In all tests, the failure occurred on the flyer (aluminum) side of the weld.
- Numerical results were compared to experiments, and good agreement was shown between the two.
- The jetting phenomenon and interlocking of the foils along the weld interface were successfully simulated.
- Investigations into the effect of LIW phenomena and factors such as the air medium, jet composition, use of double-sided tape, and the fixed metal specimen material type are deemed as potential topics for future research.

Author Contributions: Conceptualization, S.S.; methodology, S.S., G.H.G., M.I.H., S.F.S. and H.Y.; software, S.S., G.H.G., M.I.H., S.F.S, and H.Y.; validation, S.S.; formal analysis, S.S., G.H.G., S.F.S. and H.Y.; investigation, S.S., G.H.G. and M.I.H.; resources, A.S.M., D.Q.; data curation, S.S.; writing—original draft preparation, S.S.; writing—review and editing, S.S., A.S.M., S.F.S., G.H.G., and H.Y.; visualization, S.S.; supervision, A.S.M.; project administration, A.S.M.; funding acquisition, A.S.M., D.Q.

Funding: This research was funded by The University of Texas at Dallas.

Conflicts of Interest: The authors declare no conflict of interest.

References

1. Crossland, B.; Williams, J.D. Developments in explosive welding. *Aircr. Eng. Aerosp. Technol.* **1968**, *40*, 11–13. [CrossRef]
2. Szecket, A.; Mayseless, M. The triggering and controlling of stable interfacial conditions in explosive welding. *Mater. Sci. Eng.* **1983**, *57*, 149–154. [CrossRef]
3. Mousavi, A.A.; Al-Hassani, S.T.S. Numerical and experimental studies of the mechanism of the wavy interface formations in explosive/impact welding. *J. Mech. Phys. Solids* **2005**, *53*, 2501–2528.
4. Wang, Y.; Beom, H.G. Numerical simulation of explosive welding using the material point method. *Int. J. Impact Eng.* **2011**, *38*, 51–60. [CrossRef]
5. Wang, X.; Zheng, Y. Numerical study of the mechanism of explosive/impact welding using smoothed particle hydrodynamics method. *Mater. Des.* **2012**, *35*, 210–219. [CrossRef]
6. Zhang, Z.L.; Feng, D.L. Investigation of explosive welding through whole process modeling using a density adaptive SPH method. *J. Manuf. Process.* **2018**, *35*, 169–189. [CrossRef]
7. Okagawa, K.; Aizawa, T. Impact seam welding with magnetic pressure for aluminum sheets. *Mater. Sci. Forum* **2004**, *465*, 231–236. [CrossRef]
8. Watanabe, M.; Kumai, S. Interfacial microstructure of magnetic pressure seam welded Al-Fe, Al-Ni and Al-Cu lap joints. *Mater. Sci. Forum* **2006**, *519*, 1145–1150. [CrossRef]

9.   Lee, K.J.; Kumai, S. Interfacial microstructure and strength of steel/aluminum alloy lap joint fabricated by magnetic pressure seam welding. *Mater. Sci. Eng. A* **2007**, *471*, 95–101. [CrossRef]

10.   Ben-Artzy, A.; Stern, A. Wave formation mechanism in magnetic pulse welding. *Int. J. Impact Eng.* **2010**, *37*, 397–404. [CrossRef]

11.   Göbel, G.; Beyer, E.; Kaspar, J. Dissimilar metal joining: Macro-and microscopic effects of MPW. In Proceedings of the 5th International Conference on High Speed Forming, Dortmund, Germany, 24–26 April 2012; pp. 179–188.

12.   Raoelison, R.N.; Sapanathan, T. Interfacial kinematics and governing mechanisms under the influence of high strain rate impact conditions: Numerical computations of experimental observations. *J. Mech. Phys. Solids* **2016**, *96*, 147–161. [CrossRef]

13.   Cui, J.; Li, Y. Joining of tubular carbon fiber-reinforced plastic/aluminum by magnetic pulse welding. *J. Mater. Process. Technol.* **2019**, *264*, 273–282. [CrossRef]

14.   Vivek, A.; Hansen, S.R. Vaporizing foil actuator: A tool for collision welding. *J. Mater. Process. Technol.* **2013**, *213*, 2304–2311. [CrossRef]

15.   Hahn, M.; Weddeling, C. Vaporizing foil actuator welding as a competing technology to magnetic pulse welding. *J. Mater. Process. Technol.* **2016**, *230*, 8–20. [CrossRef]

16.   Vivek, A.; Presley, M. Solid state impact welding of BMG and copper by vaporizing foil actuator welding. *Mater. Sci. Eng. A* **2015**, *634*, 14–19. [CrossRef]

17.   Nassiri, A.; Kinsey, B. Numerical studies on high-velocity impact welding: Smoothed particle hydrodynamics (SPH) and arbitrary Lagrangian–Eulerian (ALE). *J. Manuf. Process.* **2016**, *24*, 376–381. [CrossRef]

18.   Nassiri, A.; Zhang, S. Numerical investigation of CP-Ti & Cu110 impact welding using smoothed particle hydrodynamics and arbitrary Lagrangian-Eulerian methods. *J. Manuf. Process.* **2017**, *28*, 558–564.

19.   Chen, S.; Daehn, G.S. Interfacial microstructures and mechanical property of vaporizing foil actuator welding of aluminum alloy to steel. *Mater. Sci. Eng. A* **2016**, *659*, 12–21. [CrossRef]

20.   Chen, S.; Huo, X. Interfacial characteristics of Ti/Al joint by vaporizing foil actuator welding. *J. Mater. Process. Technol.* **2019**, *263*, 73–81. [CrossRef]

21.   Gupta, V.; Lee, T. A robust process-structure model for predicting the joint interface structure in impact welding. *J. Mater. Process. Technol.* **2019**, *264*, 107–118. [CrossRef]

22.   Groche, P.; Becker, M. Process window acquisition for impact welding processes. *Mater. Des.* **2017**, *118*, 286–293. [CrossRef]

23.   Daehn, G.S.; Lippold, J.C. Low-Temperature Spot Impact Welding Driven without Contact. U.S. Patent No. PCT/US09/36499, 27 December 2011.

24.   Wang, H.; Liu, D.; Taber, G. Laser Impact Welding-Process Introduction and Key Variables. Available online: https://eldorado.tu-dortmund.de/bitstream/2003/29542/6/Wan12.pdf (accessed on 7 October 2019).

25.   Wang, H.; Taber, G. Laser impact welding: Design of apparatus and parametric optimization. *J. Manuf. Process.* **2015**, *19*, 118–124. [CrossRef]

26.   Wang, H.; Vivek, A. Laser impact welding application in joining aluminum to titanium. *J. Laser Appl.* **2016**, *28*, 032002. [CrossRef]

27.   Levieil, B.; Bridier, F. Numerical simulation of low-cycle fatigue behavior of welded joints for naval applications: Influence of residual stresses. *Weld. World* **2017**, *61*, 551–561. [CrossRef]

28.   Wang, X.; Gu, C. Laser shock welding of aluminum/aluminum and aluminum/copper plates. *Mater. Des.* **2014**, *56*, 26–30. [CrossRef]

29.   Peyre, P.; Berthe, L.; Fabbro, R. Laser shock processing of materials: Basics mechanisms and applications. In Proceedings of the 65th Laser Materials Processing Conference, Tokyo, Japan, 2–5 December 2005; pp. 95–109.

30.   Wang, X.; Gu, Y. An experimental and numerical study of laser impact spot welding. *Mater. Des.* **2015**, *65*, 1143–1152. [CrossRef]

31.   Wang, X.; Shao, M. Numerical simulation of laser impact spot welding. *J. Manuf. Process.* **2018**, *35*, 396–406. [CrossRef]

32.   Wang, X.; Li, F. Experimental and numerical study on the laser shock welding of aluminum to stainless steel. *J. Opt. Lasers Eng.* **2019**, *115*, 74–85. [CrossRef]

33.   Wang, X.; Shao, M. Laser impact welding of aluminum to brass. *J. Mater. Process. Technol.* **2019**, *269*, 190–199. [CrossRef]

34.   Wang, X.; Luo, Y. Experimental investigation on laser impact welding of Fe-Based amorphous alloys to crystalline copper. *Materials* **2017**, *10*, 523. [CrossRef]

35. Liu, H.; Gao, S. Investigation on a novel laser impact spot welding. *Metals* **2016**, *6*, 179. [CrossRef]
36. Liu, H.; Jin, H. Investigation on Interface Morphology and Mechanical Properties of Three-Layer Laser Impact Welding of Cu/Al/Cu. *Metallur. Mater. Trans. A* **2019**, *50*, 1273–1282. [CrossRef]
37. Hatamleh, M.I.; Mahadevan, J. Prediction of Residual Stress Random Fields for Selective Laser Melted A357 Aluminum Alloy Subjected to Laser Shock Peening. *ASME J. Manuf. Sci. Eng.* **2019**, *141*, 101011. [CrossRef]
38. Hatamleh, M.I.; Sadeh, S.; Farooq, T. Finite Element Study of Laser Peening on Selective Laser Melted A357 Aluminum Alloy During Tension Test. In Proceedings of the ASME 2018 13th International Manufacturing Science and Engineering Conference, College Station, TX, USA, 18–22 June 2018; p. V004T03A046.
39. Mahadevan, J.S. Probabilistic Material Modeling of Selective Laser Melted A357 Aluminum Alloy Subjected to Laser Shock Peening. Master's Thesis, University of Texas at Dallas, Richardson, TX, USA, 2017.
40. Ding, K.; Ye, L. 1 General introduction. In *Laser Shock Peening: Performance and Process Simulation*; Woodhead Publishing: Cambridge, UK, 2006; p. 2.
41. Fabbro, R.; Fournier, J. Physical study of laser-produced plasma in confined geometry. *J. Appl. Phys.* **1990**, *68*, 775–784. [CrossRef]
42. Hong, X.; Wang, S.B. Confining medium and absorptive overlay: Their effects on a laser-induced shock wave. *J. Opt. Lasers Eng.* **1998**, *29*, 447–455. [CrossRef]
43. Johnson, G.R.; Cook, W.H. Fracture characteristics of three metals subjected to various strains, strain rates, temperatures and pressures. *Eng. Fract. Mech.* **1985**, *21*, 31–48. [CrossRef]
44. Dassault Systèmes. *Abaqus Analysis User's Guide, v6.14*; Dassault Systèmes: Waltham, MA, USA, 2014.
45. Altair Engineering Inc. *Radioss Theory Manual, v14*; Altair Engineering Inc.: Troy, MI, USA, 2015.

Article

Finite Element Analysis of Laser Peening of Thin Aluminum Structures

Kristina Langer [1], Thomas J. Spradlin [1] and Michael E. Fitzpatrick [2,*]

[1] Air Force Research Laboratory, Wright-Patterson Air Force Base, OH 45433, USA;
     kristina.langer@us.af.mil (K.L.); thomas.spradlin.1@us.af.mil (T.J.S.)
[2] Faculty of Engineering, Environment, and Computing, Coventry University, Coventry CV1 5FB, UK
*   Correspondence: Michael.Fitzpatrick@coventry.ac.uk or ab6856@coventry.ac.uk

Received: 20 September 2019; Accepted: 2 December 2019; Published: 6 January 2020

Abstract: Laser shock peening has become a commonly applied industrial surface treatment, particularly for high-strength steel and titanium components. Effective application to aluminum alloys, especially in the thin sections common in aerospace structures, has proved more challenging. Previous work has shown that some peening conditions can introduce at-surface tensile residual stress in thin Al sections. In this study, we employ finite element modeling to identify the conditions that cause this to occur, and show how these adverse effects can be mitigated through selection of peen parameters and patterning.

Keywords: laser peening; residual stress; aluminum alloys

---

1. Introduction

Over the past three decades, laser peening (LP) has emerged as a viable commercial technology for introducing beneficial residual compression into the near-surface regions of metallic components [1,2]. First introduced by Battelle Laboratories, Columbus, Ohio, in the 1970s [3,4], LP uses a high-power-density, short-duration pulsed laser to create a mechanical impact on the surface of a component. The amplitude of the impact, typically at least double the Hugoniot elastic limit (HEL) of the target material, is great enough that the resultant shock wave creates a region of localized plastic deformation in the target material. The elastic springback of the surrounding material around this plastic core creates a state of self-equilibrating residual stress in the component. Good reviews of the LP process can be found in [5–7].

Because the depth of compression resulting from an LP treatment is typically on the order of 1 mm or greater, depending on the component material, geometry, and selected peening parameters, LP has the potential to significantly mitigate fatigue-inducing tensile stresses that can result from applied cyclic loading. This is of particular interest as a potential means for enhancing the fatigue response of military aircraft in which mission changes and extended lifing requirements often tax the fatigue resistance of the airframes beyond their original design specifications [8].

One application of recent interest is LP of thin aluminum components, on the order of 2–3 mm or less such as aircraft skin or web structure, to mitigate surface damage or enhance fatigue response. If LP parameters and process variables are designed appropriately, a favorable compressive state throughout much of the component depth can be induced [9]; however, as is discussed in more detail below, under certain circumstances LP can also induce detrimental tensile stresses into the near-surface regions of thin sections, extending into the component to depths as great as 0.5 mm [10–12].

A number of factors can contribute to the challenge of developing favorable residual stress states in thin structures: First, if the laser power density and subsequent plasma pressure are too high, the resulting distortion in the thin section can reduce the magnitude of the compressive stresses [9,13]. Second, even in the absence of significant distortion, the limited depth of material along the propagation direction of the shock wave constrains the induced plastic field, which in turn can limit the magnitude of the residual stresses. Third, because the plastic wave will typically not completely attenuate through the material depth, reflected waves from the opposite free surface can form, with the effect of altering the residual stresses from compressive to tensile [14]. Fourth, the work hardening resulting from LP affects a greater percentage of the depth of a thin component than a thicker one, and can therefore impact the development of the residual stress fields and also the fatigue response [15]. Finally, the rolling process typically used to produce thin plates induces a crystallographic texture that can influence the development of residual stresses [16].

In early studies, some of these thin section challenges were addressed by splitting the laser beam to peen both sides of the plate simultaneously [9]. While this technique was designed to help minimize overall distortion, the mid-plane collision of the two compressive shock fronts resulted in detrimental, high-magnitude internal tensile stresses [17]. In addition, for many in situ components, such as an aircraft skin, implementation of two-sided peening is not feasible.

More recently, Dorman et al. [10] investigated the use of LP to treat surface scribe marks on thin Al 2024-T351 sheets. Specimens were peened from one side only, with a single layer of square spots patterned in either a line overlaying the scribe mark or a patch covering it, and using an ablative coating on the peen surface and an acoustic damping material on the back surface. The resulting residual stresses were then measured using incremental hole drilling and synchrotron X-ray diffraction. In all cases—regardless of the depth of the scribe mark, the use of an ablative layer, or the intensity of the peening—the near-surface residual stress, measured in the center of the peen spots normal to the peen line, was either tensile or only slightly compressive. The tensile stresses persisted to subsurface depths ranging from 80 μm at the lowest laser power density to 320 μm at the highest, with magnitudes reaching as high as 100 MPa. In addition, in all cases in which the peening pattern was a single line, the resultant residual stress fields were strongly non-biaxial, with higher compression parallel to the peen line axis than in the transverse direction.

Cellard et al. [15] noted similar results in their measurements of LP-induced residual stresses in very thin (1 mm) Ti-17 plates. The specimens were peened from one side only using square laser spots of varying size, intensity, duration, and coverage. Using X-ray diffraction to measure the surface residual stresses, they found that nearly all of the thin specimens had some level of tension at the surface, ranging from 13 to 150 MPa. When the same peening parameters were applied to thick specimens (9 mm), however, nearly all demonstrated compressive residual stress on the surface. They attribute this "thickness effect" to the redistribution of stresses required to maintain equilibrium over a cross-section of the specimen, and suggested that two-sided peening could yield more favorable stress fields in thin specimens.

In contrast, several researchers studying the effects of LP in aluminum sheet have not observed tensile stresses at the surface. Hong and Chengye [18] and Yang et al. [19] peened 2.5-mm-thick Al2024 and observed compressive surface stresses at the centers of round LP spots using conventional XRD. However, as noted in [16], surface XRD measurements can be distorted by rolling-induced texture and thus a secondary stress measurement technique is recommended to reduce uncertainties. Ivetic et al. [20] studied the effects of LP on slightly thicker Al 6082-T6 plates, 3 mm in thickness, and measured near-surface compression using synchrotron X-ray diffraction. However, the depth at which the measurements were recorded was slightly subsurface, 50–250 μm, so that any tensile stresses that might have formed in shallower regions might not have been detected.

In the present work, three-dimensional non-linear finite element modeling is used to investigate the effects of laser peening on residual stresses in thin aluminum plates. We explore the relationship between spot layering and peen patterning, with the objective being to better understand the effects of processing technique on the resulting residual stress fields.

2. LP Process Modeling

Owing to its flexibility and relative ease of use, finite element analysis (FEA) has emerged as a powerful tool for predicting the full residual stress state that results from the laser peening of an arbitrary three-dimensional metallic body. First used by Braisted and Brockman [21] to simulate single LP shots on semi-infinite bodies, FEA techniques have evolved significantly over the past decade as computer processing has become faster and multiprocessing environments have become mainstream. Recent FEA studies of note include the work of Ding and Ye [22], who used three-dimensional FEA to create detailed stress maps for LP steel; Ocaña et al. [23], who developed a multi-shot FEA model capable of realistic LP simulation; Singh et al. [24], who coupled FEA with numerical optimization; Brockman et al. [25], who used a highly refined, fully three-dimensional model to study local variations in surface residual stresses arising from shot patterning; and Hasser et al. [26], who developed a first-ever FEA capability for studying LP-induced surface roughness.

In the present work, a series of finite element analyses were performed using the commercial FEA package Abaqus [27]. The objective of the analyses was to explore computationally the effects of various peening parameters and processing variables on the resulting residual stress fields in thin structures in order to better understand the experimental findings discussed in the previous section: (1) the formation of surface tensile stresses in spot centers that tend to occur in thin plates but not thicker sections, (2) the observed inequality of the residual stresses that occur parallel to and perpendicular to a peen line, and (3) the effects of peen patterning on the resulting stress fields.

The key aspects in using FEA to model an LP shock event are the selection of appropriate input models, namely the material model and the pressure model, and selection of a computational strategy and associated computational parameters that provide for an accurate and efficient solution. In this section, we discuss the material model used to capture high-strain-rate effects, the pressure model used to approximate the laser-induced plasma impact, and the computational model used to develop the stress predictions.

2.1. LP Process

The basic LP process is illustrated in Figure 1. In typical applications, an ablative coating, such as aluminum tape or black paint, along with a transparent overlay, usually water, are applied to the surface of the target material prior to peening. When used together, the confined ablation increases the intensity of the plasma pressure, which results in a higher intensity shock wave in the target material. During the peening process, the component is exposed to a very high intensity (1–9 GW/cm^2) laser pulse, typically from an Nd:YAG or Nd:glass laser system operating at a FWHM of about 6–30 ns per pulse with a beam dimension of less than 10 mm [6,28].

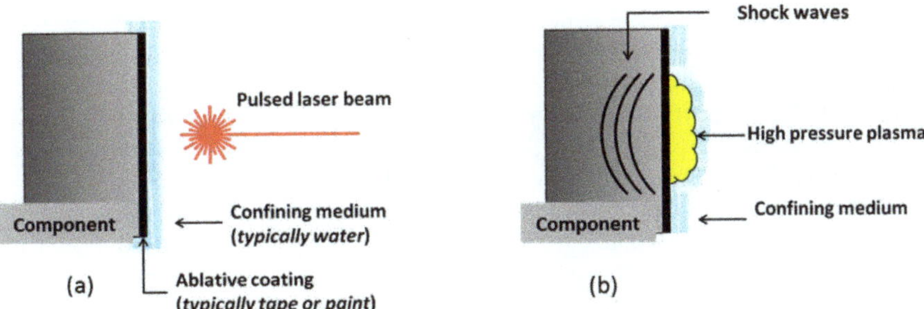

Figure 1. (**a**) A high-intensity, pulsed laser vaporizes an ablative layer on the surface of the component, producing a rapidly expanding plasma. (**b**) The plasma is constrained by a confining medium, which creates a high-amplitude pressure pulse and induces a shock wave in the component.

The resulting pressure pulse generally lasts about 2–3 times the duration of the laser pulse and has a peak magnitude of about 1–10 GPa [28]. The spatial distribution of the plasma pressure can either be uniform or variable, depending upon the laser system employed. In order to generate the near-surface plasticity required to induce a residual stress, the plasma pressure needs to exceed the HEL of the target material for a sufficient amount of time, with an optimal value typically about 2–2.5 times the HEL [29].

2.2. Computational Model

The commercial finite element package ABAQUS [27] was used to solve for the stresses, strains, and displacements that result from the application of laser peening on thin sheets of Al2024-T351. Based on previous work [11,25,30–33], all simulations were run using explicit time integration (ABAQUS/Explicit) to solve the dynamic system. As discussed in [25], using an explicit solver to model both the impact event and the subsequent return to equilibrium is quicker, more scalable, and uses less memory than performing the equilibrium analysis with an implicit solver. Each impact pulse was modeled using two distinct solution steps so that the computational parameters could be adjusted for improved efficiency and convergence. The first solution step, designated as the pulse phase, starts with the initial application of the pressure pulse and ends when no further plastic deformation occurs (2.5 μs was used). The second solution step, termed the equilibrium phase, introduces Rayleigh damping [34] into the model to return the system to a state of near-zero kinetic energy (approximately 10^{-6} to approximate equilibrium: 100 μs was used). One pair of analyses is performed for each laser pulse.

Three-dimensional linear eight-node brick elements with reduced integration (C3D8R in Abaqus terminology) were used throughout, with a mesh size of 50 μm in the laser peened region. This mesh size was selected using the results of convergence studies that showed less than 2% difference in predicted residual stresses with further refinement. External to the peened region the element size was increased gradually from 50 μm to 250 μm using mesh biasing to curtail computational costs.

2.3. Material Model

During the laser peening process, the induced shock waves can produce very high strain rates, on the order of 10^6 s^{-1}, resulting in a material response that is significantly different from that under static or quasi-static loading conditions [35]. To capture these strain rate effects, a Johnson–Cook [36] material model was used. Following earlier work [11], thermal effects resulting from the LP process are considered to be negligible and thus temperature dependence is eliminated from the material model:

$$\bar{\sigma} = \left[A + B(\bar{\varepsilon}^{p})^{n}\right]\left[1 + C\ln\!\left(\dot{\bar{\varepsilon}}^{p}/\dot{\varepsilon}_0\right)\right] \tag{1}$$

Here, $\bar{\sigma}$ is the flow stress; $\bar{\varepsilon}^{p}$ is the effective plastic strain; $\dot{\bar{\varepsilon}}^{p}$ is the effective plastic strain rate; $\dot{\varepsilon}_0$ is a reference plastic strain rate (typically taken to be 1.0 s^{-1}); n is the work hardening exponent; and A, B, and C are empirically-derived constants.

Lesuer [37] used split-Hopkinson-bar testing to measure the deformation response of Al 2024-T3 at moderately high strain rates (10^3–10^4 s^{-1}), and used the data to define the material constants in Equation (1) (Table 1). Although the strain rates typically experienced during LP are an order of magnitude higher, recent flyer-plate testing has confirmed that Lesuer's constants reasonably model the LP material response, with a maximum deviation of less than 10% as compared to the measured values [38]. Comparisons between predicted residual stresses based on a Johnson–Cook material model and measured values have also validated the use this mode for LP simulations [39,40]. Thus, the Johnson–Cook material constants for Al 2024-T3 as shown in Table 1 were used for all simulations in the current work.

Table 1. Johnson–Cook material constants [37].

Material	*A*/MPa	*B*/MPa	*C*	*n*
Al 2024-T3	369	684	0.014	0.93

It should be noted that the constitutive model was not updated in the multi-shot simulations to account for LP-induced cyclic deformation effects [41].

2.4. Pressure Model

In this work, the laser-matter interaction and subsequent plasma formation were not modeled directly; rather, the resulting pressure pulse was applied to the FE model over the area corresponding to the laser pulse as a spatially-uniform, time-varying surface traction. While small transition regions likely exist along the spot perimeters, measurements by the LP vendor indicate the generated pressures are essentially uniform across the spot area.

The confined ablation model developed by Fabbro et al. [42], modified to account for the absorption fraction of the laser energy in the treated plasma, was used to determine the applied pressure $P(x,t)$ = $P(t)$. We can express the energy density at the plasma, $E(t)$, to the nominal energy density of the laser, $I(t)$, using the expression $E(t) = A(t)I(t)$, where $A(t) = 1 - R(t)$ is the absorbed energy and $R(t)$ represents the change in energy density. However, owing to the very short time between the laser pulse and the plasma formation, it is generally assumed that the loss $R(t)$ is negligible and hence $A(t)$ is approximately 1.0 [43–46].

During the heating phase (while the laser is on), this relationship is then described by:

$$E(t) = P(t)\frac{dL}{dt} + \frac{3}{2\alpha}\frac{d}{dt}[P(t)L(t)] \tag{2}$$

while during the cooling phase (after the laser is switched off):

$$P(t) = P(\tau)\left(\frac{L(\tau)}{L(t)}\right)^{\gamma} \tag{3}$$

Here, $L(t)$ is the thickness at time t of the interface between the ablative coating and the transparent overlay, given by:

$$\frac{dL(t)}{dt} = \frac{2}{Z}P(t) \tag{4}$$

and α is the fraction of the internal energy transferred to the workpiece, γ is the adiabatic cooling rate, and Z is the effective acoustic impedance of the interface, defined by:

$$\frac{2}{Z} = \frac{1}{Z_{coating}} + \frac{1}{Z_{overlay}} \tag{5}$$

$Z_{coating}$ and $Z_{overlay}$ are the acoustic impedances of the ablative coating and the transparent overlay, respectively. With water as the overlay ($Z_{overlay} \approx 0.15 \times 10^6$ g cm^{-2} s^{-1}) and aluminum tape as the coating ($Z_{coating} \approx 1.7 \times 10^6$ g cm^{-2} s^{-1}), Z has a value of about 0.3×10^6 g cm^{-2} s^{-1}.

Assuming a Gaussian laser energy density $E(t)$, a typical pressure profile resulting from these equations is shown in Figure 2 for a maximum energy density of 3 GW/cm^2, laser pulse width of 18 ns, $\alpha = 0.09$, and $\gamma = 1.3$. For these parameters, the maximum applied pressure, P_{max} is about 1.95 GPa and the width (FWHM) of the pressure pulse is about 47 ns.

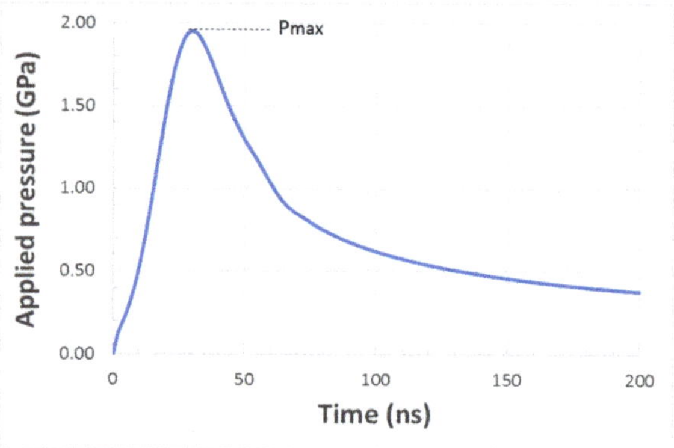

Figure 2. Typical temporal profile of the applied peening pressure.

2.5. Model Confirmation

Residual stress predictions using the model discussed in the previous sections were compared against experimental stress measurements. From previous studies, the value of α was set to 0.09 and γ was 1.3. The measured residual stress profiles were extracted from two laser-peened test samples of 12.7-mm-thick Al2024-T351 aluminum (unclad) using incremental hole drilling (IHD). The IHD diameter was 2 mm and the depth increments ranged from 32 to 128 μm to a total hole depth of 1408 μm. Each sample (IHD-1 and IHD-2) was peened with a single 5 mm × 5 mm square spot located sufficiently far from the boundaries to avoid edge effects and using a power density of 3 GW/cm^2 and a pulse width of 18 ns. The residual stresses were measured at the spot center.

Results from the comparison are shown in Figure 3. The FE results represent the average principal stresses over the equivalent volume of material removed at each IHD step and are reported depthwise at the center of the IHD increment. Numerical integration over the disk-shaped IHD region was used to compute the averages. As is shown, the predicted stresses agree well with the stress measurements, confirming the performance of the model for LP applications.

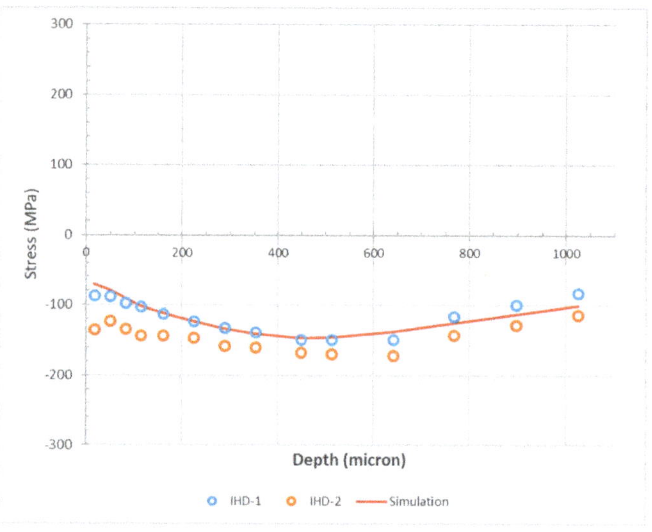

Figure 3. Comparison of predicted residual stresses against measured values.

3. Results and Discussion

The FEA model developed in the previous section was used to investigate the stress distributions resulting from different peening schemes in order to better understand how combinations of peening parameters influence residual stress distributions in thin sections. Only one-sided peening without a backing material was considered, reflecting the potential future application of in situ treatment of thin aircraft components. All simulations used a 5 mm × 5 mm square laser spot.

In general, a favorable state of residual stress for mitigating fatigue has several key characteristics. First, sufficient compression should exist at the surface and in the near-surface regions of critical areas of the component to offset fatigue-inducing tensile stresses. Second, the distribution and magnitude of the compensatory tension that arises from the LP process should not initiate a failure event when fatigue loads are applied. Third, for a line of peen spots that align with and overlay a surface scratch, such as the geometry considered in [10,12], near-surface tensile stresses transverse to the peen line should be absent to prevent premature initiation of a crack from the scratch.

3.1. Single Shot Simulations

Simulations of a single LP shot were used to develop a better understanding of the shock response of laser-peened thin plates under different peening conditions. All thin-section simulations were run on a 2-mm-thick plate of Al2024-T351 (unclad), with the peened area located sufficiently far from the plate boundaries to avoid the effects of reflected waves. Note that due to symmetry, only one half of the plate was modeled.

3.1.1. Effects of Peening Pressure

Figure 4 shows the effects of varying the applied pressure for a single layer of peening with a fairly large (5×5 mm^2) square spot. The contour plots represent a cross-section of the plate taken through the spot center, with the colors representing the range of the in-plane stress component S_{22} normal to the cross-sectional surface. Line plots of stress as a function of depth through the spot center are also shown. Note that in these plots, in contrast to the plots in Figure 3, the stresses are not averaged over a volume; rather, they are extracted directly from the element integration points.

At low pressures (Figure 4a), less than about twice the HEL, surface tensile stresses are absent in the spot center. However, the magnitude of the maximum compressive stress is fairly low and fairly shallow—only about 80 MPa that tapers out after about 0.5 mm depth—and tensile stress persists through the thickness reaching about 60 MPa at a depth just over three-quarters of the way through the thickness.

At pressures about twice the HEL (Figure 4b), small tensile zones appear in the spot centers similar to the experimental findings in [10]. The magnitude of the stress at the surface is about 100 MPa, and the tension persists in a region around the spot center that measures about 250 μm at the surface and extends about 35 μm subsurface. The maximum compressive stresses in this pressure regime are greater than those realized at lower pressures, reaching about −130 MPa, but are not significantly deeper (about 100 μm). It should be noted that the size of the tensile zones predicted by these simulations are generally smaller than what was reported in [10]. This likely results from the pressure pulse approximation assumption of a spatially uniform distribution along with averaging effects in the experimental values.

Increasing the applied pressure to three times the HEL (Figure 4c) increases the size of the surface tensile zone slightly but also drives the compressive stresses deeper into the component thickness. Near to the spot centerline the compression persists throughout the plate thickness; however, the stress state is not uniform across the spot width, with tensile pockets forming subsurface around the compressed region. At very high applied pressures, about four times the HEL (Figure 4d), through-thickness compression is no longer achievable, with tensile stresses forming at the back surface.

It should be noted that the specific results for this series of single-shot simulations depend on not only the maximum applied pressure, but also on the shape and length of the pressure pulse and the thickness of the plate. For thicker or thinner sheets, and for different pulse shapes, the precise points at which the nature of the residual stresses change will vary; however, the basic trends are similar.

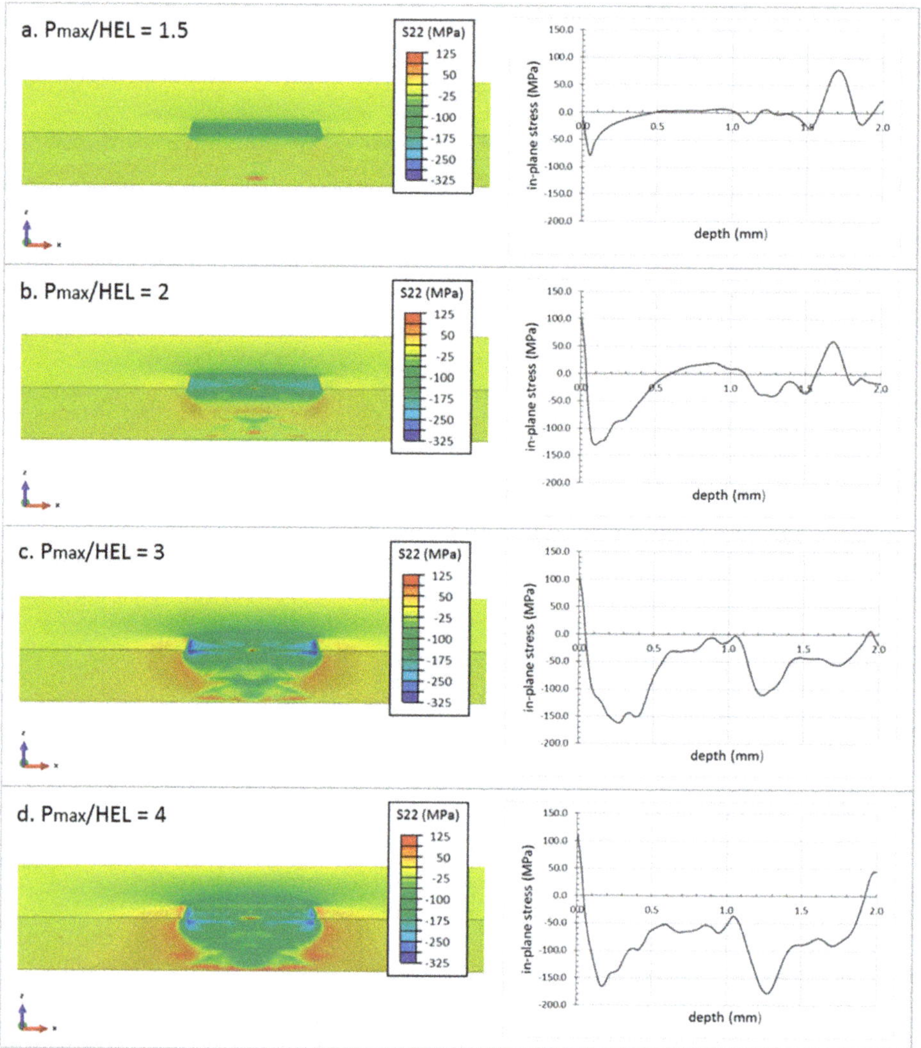

Figure 4. Finite element analysis (FEA)-computed in-plane residual stress from a single square laser peening (LP) shot (5 × 5 mm^2) on a 2-mm-thick Al2024-T351 plate at various pressure levels.

A comparison between the LP response of a thicker plate (10 mm) and the response of a thin section (2 mm) is shown in Figure 5. Both simulations used a maximum applied pressure of 2.5 times the HEL, with the plate thickness as the only difference. As can be seen from the line plot in Figure 5, the near-surface tensile stresses around the spot center in the thin section are completely absent from the thick section. In addition, because the thick section does not experience stress reversals resulting from wave reflections off the back surface, the maximum residual compressive stresses are significantly greater in magnitude than that of a thin section, by a factor of about two.

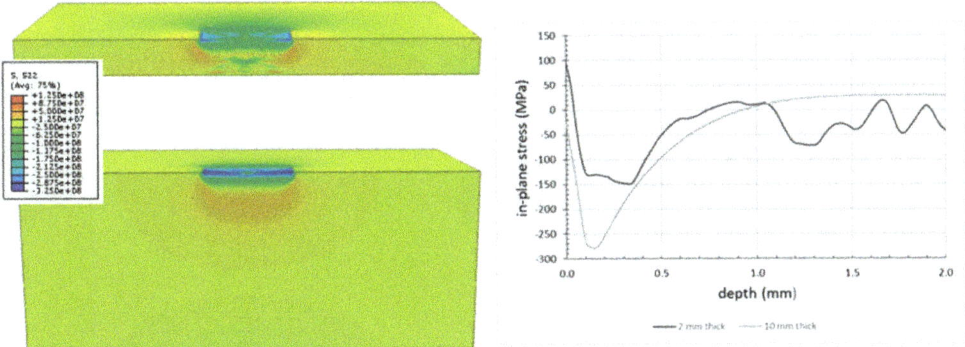

Figure 5. Effect of plate thickness on predicted in-plane residual stress (single square LP shot, 5×5 mm^2, on a 2-mm-thick Al2024-T351 plate).

3.1.2. Effects of Peen Layers

When additional layers of peening are considered on a thin section, as shown in Figure 6, the near-surface tensile stresses are mitigated. For the case shown, with an applied pressure of 2.5 times the HEL, a second layer of peening reduces the size and magnitude of the surface tensile field by more than 50%. Adding a third layer of peening completely suppresses the near-surface tension and a fourth layer drives the surface into compression. The maximum subsurface compressive stress is only minimally affected by additional layers, increasing by less than 50 MPa from one layer to four, but the depth of compression more than doubles. At the same time, however, the additional peen layers result in larger tensile stresses on the back surface.

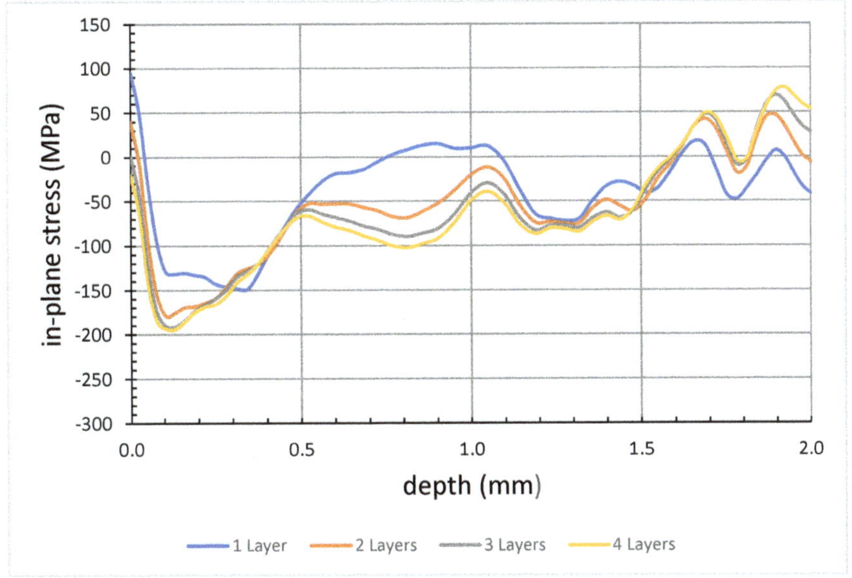

Figure 6. Effect of peen layering on predicted in-plane residual stress in thin sections (single square LP shot, 5×5 mm^2, on a 2-mm-thick Al2024-T351 plate).

3.1.3. Observations

It should be noted that in all single-shot simulations, the predicted in-plane residual stresses are equi-biaxial, as would be expected due to symmetry. The S_{22} stresses in the plane normal to the y-axis are equivalent to the S_{11} stresses in the plane normal to the x-axis.

The results of these single-shot simulations suggest that a favorable state of residual stress in thin sections can be realized by selecting a peening pressure high enough to achieve the desired compression without the detrimental subsurface tensile stresses, along with sufficient peen layers to suppress the near-surface tension. As discussed in the following sections, however, a third parameter, namely, the peen patterning, can also strongly influence the resulting stresses in thin sections, and, as will be discussed, observations for single shots do not necessarily hold in multi-shot scenarios.

3.2. Simulation of a Peened Line

Although the single shot simulations provide insight into the effects of applied pressure and layering, they cannot capture the interactions between adjacent and overlapping spots, or the effects of shot order and patterning. To investigate these effects, the FEA model used for the single shot simulations—a 2 mm thick plate of Al2024-T351—was extended lengthwise to accommodate the simulation of peened line under various peening conditions.

3.2.1. Effects of Adjacent Spots

Figure 7 shows the simulation results for a scenario similar to that considered in [10], namely, a single layer of peen spots applied edge-to-edge along a line with minimal overlap. The applied pressure used in the simulations was 2.5 times the HEL. As is shown, the center of each peen spot in the line exhibits a surface tensile stress normal to the peen line with a magnitude of about 110 MPa and extending about 50 μm subsurface. The corresponding stress for a single peen spot is shown for comparison. Although the magnitude of the surface tensile stress is similar for the single spot and the peen line, the magnitude of the maximum compression resulting from the line configuration is about 20% lower than that of a single spot. The overall depth of compression is smaller in the line configuration as well, by about 0.25 mm.

Although the near-surface tensile stresses are contained only in the central regions near the spot center, significant tension (with a peak value of about 100 MPa) builds up subsurface in the regions between adjacent spots. For many fatigue applications these large subsurface tensile stresses can be undesirable, as they can lead to subsurface crack initiation that is undetectable by visual inspection of the component surface.

Figure 7 also compares the through-thickness profiles for the in-plane stresses taken parallel to (S_{11}) and transverse to (S_{22}) the peen line; as is shown, the stress states in the two directions are noticeably different, with as much as a 50 MPa variation in some locations. In general, the stresses transverse to the peened line are less compressive (or more tensile) than those in the longitudinal direction. This could pose a concern for certain applications in which the peen line was designed to overlay a surface scratch. Because the stresses are of lower magnitude (or even more tensile) in the critical direction (i.e., the direction that would tend to encourage crack formation and opening), the true potential of the LP treatment would not be realized, and could even exacerbate the situation. It should also be noted that through the spot centers, the stresses transverse to the peen line are generally more tensile than those predicted by single shot simulations.

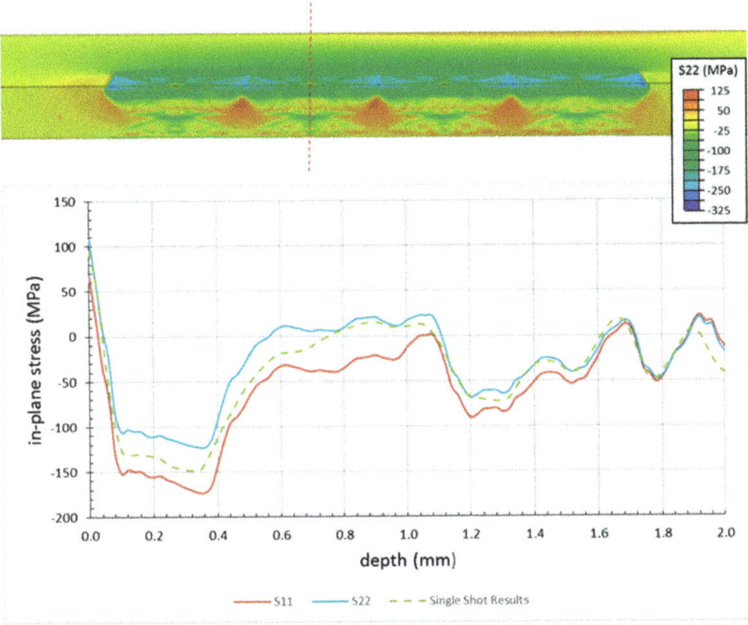

Figure 7. Comparison of stress profiles between a single peen spot and a peen line.

3.2.2. Effects of Peen Layers

As was the case for single shots, augmenting the peen line with additional layers can help alleviate the surface tensile stresses at the spot centers. In Figure 8, second and third layers of peen spots are applied along the peen line at a 50% offset between one layer and the next, as shown in the schematic. The line plots illustrate the in-plane stress through the center of a spot on the topmost layer (denoted Location L1). Similar to the case of a single peen spot, the surface tensile zones in the spot centers are reduced as additional layers are applied; however, the effects of adjacent spots are very apparent. A second layer flips the surface stress to compression (from 108 MPa to −243 MPa) while a third layer flips it again (from −243 MPa to 46 MPa). The end result is that although the stress at this location is reduced with additional peen layers, shallow tensile regions—about 20 μm deep and 20 μm in width—persist in the center of each spot that comprises the topmost layer of the plate.

It should be noted that while Figure 8 suggests that limiting the peening to two layers would alleviate the surface tensile stress at Location L1, detrimental stresses form elsewhere. Figure 9 illustrates this by extracting stress plots at location L2 after one and two layers of peening. As is shown, the surface stress at this location decreases more than 50% with the addition of the second layer (from 108 MPa to 48 MPa), but the tension persists.

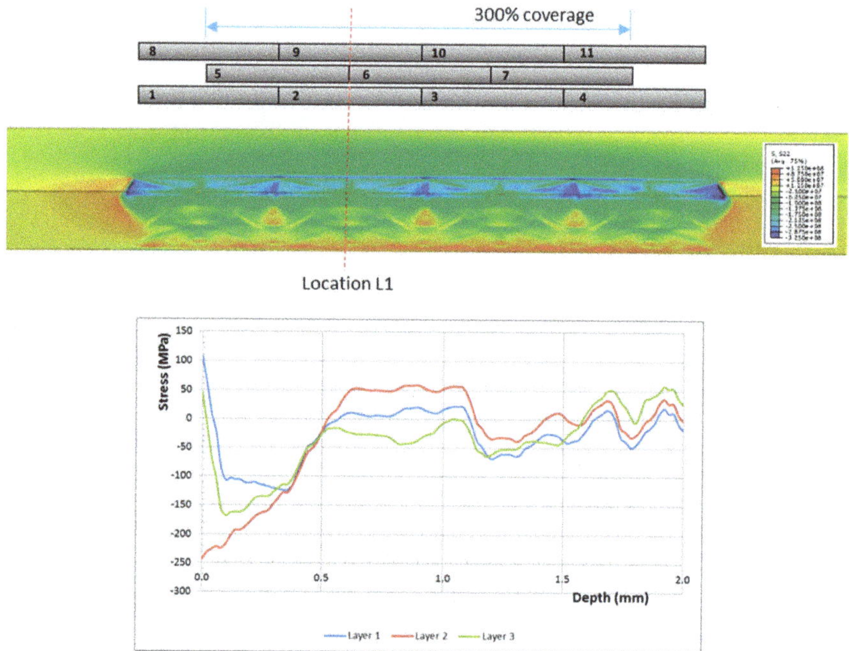

Figure 8. Effect of peen layering on predicted in-plane residual stress in thin sections. Stresses were extracted from the FEA model at Location L1 after one, two, and three layers of peening.

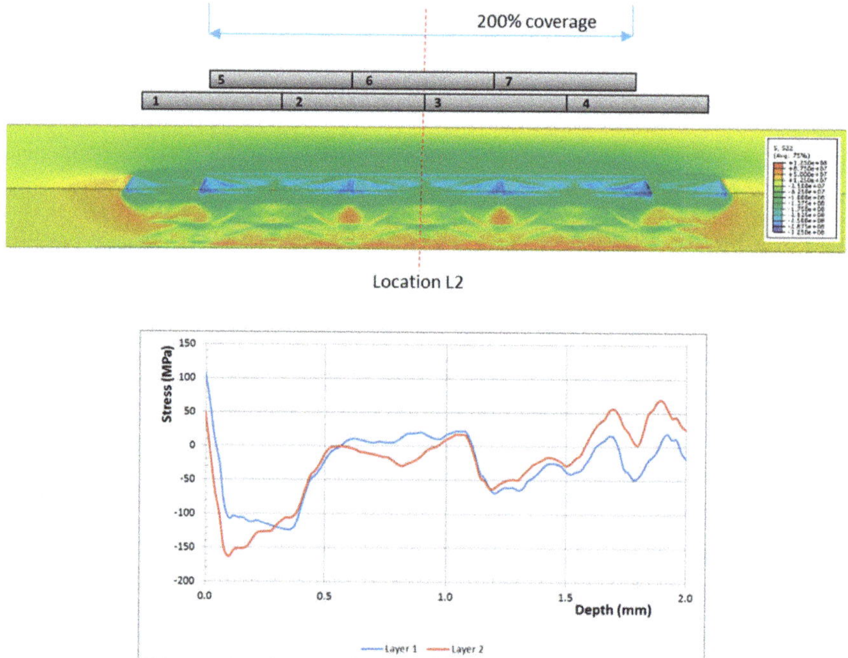

Figure 9. Effect of peen layering on predicted in-plane residual stress in thin sections. Stresses were extracted from the FEA model at Location L2 after one and two layers of peening.

3.2.3. Effects of Peen Patterning

In Figure 10, we consider the effects of altering the peen patterning while still maintaining a coverage of 300% along the central portion of the peen line. Two additional scenarios were evaluated: Pattern 2, which is similar in application to the scenario considered in Section 3.2.2 (denoted in this section as Pattern 1) but reduces the offset between layers from one-half the spot width to one-third (Figure 10a); and Pattern 3, which uses a running overlap (Figure 10b) pattern. As is shown, both of these patterns achieve exactly three layers of peening along the central portion of the peen line.

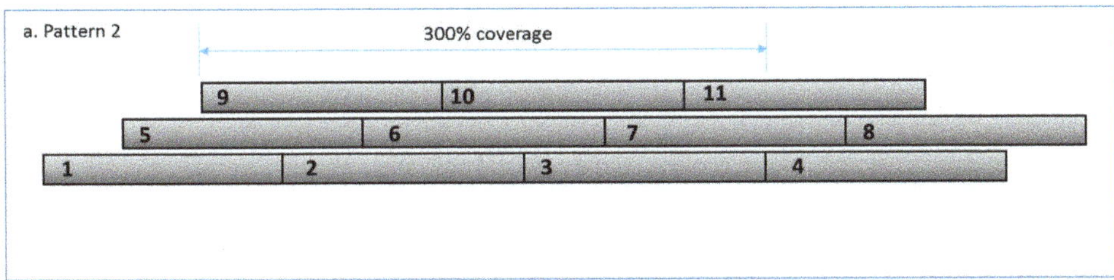

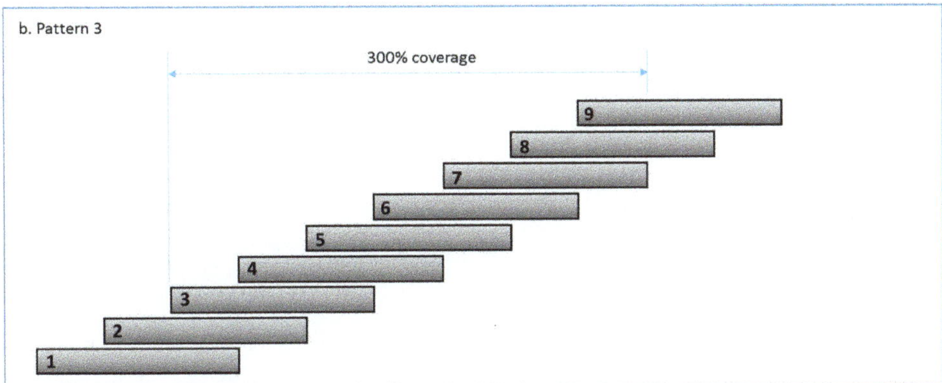

Figure 10. Schematics for achieving 300% peen coverage: (**a**) Pattern 2; (**b**) Pattern 3. Figure 8 shows Pattern 1.

The corresponding contour and line plots for peen Patterns 2 and 3 are shown in Figure 11. As is illustrated, decreasing the offset from one-half to one-third the spot width does not alleviate the tensile stresses at the surface. As was observed with Pattern 1, small regions of tension persist in the spot centers of the topmost layer for Pattern 2. However, the build-up of subsurface tensile stresses is significantly reduced by reducing the offset.

The running peen pattern (Pattern 3), on the other hand, does result in a complete suppression of surface and near-surface tensile stress (Figure 11b), to a depth of almost 1 mm. For the peening parameters used in the simulations, the surface stress in the spot centers decreases from about 60 MPa to about −135 MPa, while the subsurface tension at mid-section decreases from a maximum of 100 MPa for Pattern 1 to about 0 MPa for Pattern 3. Note that the more compressive surface stresses observed from the running overlap pattern result from the beneficial interactions of subsequent peen spots applied over previously induced fields; for other combinations of peening parameters and specimen geometry other surface conditions may arise.

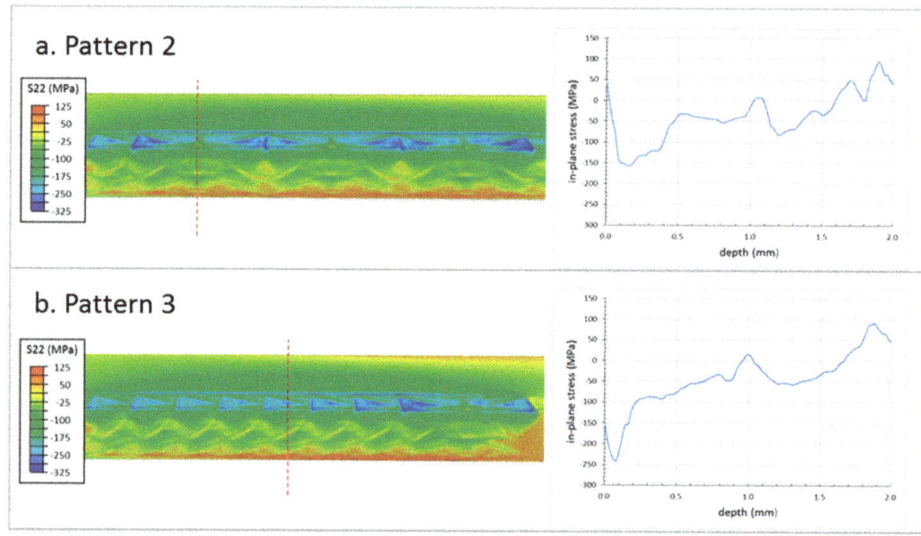

Figure 11. Effects of peen layering on predicted in-plane residual stress: (**a**) Pattern 2; (**b**) Pattern 3. The patterns are shown in Figure 10.

4. Conclusions

In this work, finite element analysis was used to study the effects of peen patterning on the residual stress fields induced by laser peening of thin aluminum plates. The accuracy of the model was confirmed using experimental measurements of residual stress obtained by incremental hole drilling. The following conclusions can be drawn:

1. In thin plates of aluminum, above a threshold value of peak applied pressure, laser peening can lead to undesirable local "hotspots" of tensile residual stress as a consequence of the interaction of the shock wave with the geometry. In thin sections, reflected stress waves occur that, in combination with the smaller amount of constraining material, can lead to tensile stress on the peened surface through localized reverse yielding.
2. Mitigation of the tensile regions can be affected by layering of the peen spots, where additional layers can reduce the tensile zones and generate compression. However, for a given specimen or component geometry, the resultant stress fields are dependent upon the peening parameters and the precise geometry of the patterning that is applied.
3. In thin structures, the residual stresses resulting from applying multiple peen spots to a single location do not necessarily match the results obtained from applying multiple peen spots using offset peen patterning.
4. Peen patterning can affect the biaxiality of the residual stress field, most notably if a line of peen spots is generated to overlay a surface scratch, for example, where the lowest magnitude of residual stress is found to be perpendicular to the peen line.

Author Contributions: All authors contributed to the conceptualization, methodology, and review and editing of the manuscript. T.J.S. developed the fundamental models and parametric studies. K.L. undertook the advanced model development and simulation. M.E.F. was responsible for providing calibration and validation data, along with funding acquisition. K.L. and M.E.F. were responsible for preparation of the initial draft of the manuscript. All authors have read and agreed to the published version of the manuscript.

Funding: This research study was sponsored by the Air Force Office of Scientific Research, Air Force Material Command, USAF, under grant number FA8655-12-1-2084, and the Air Force Research Laboratory's Aerospace Systems Directorate. The U.S. Government is authorized to reproduce and distribute reprints for Government purpose notwithstanding any copyright notation thereon. The views and conclusions contained herein are those of the authors and should not be interpreted as necessarily representing the official policies or endorsements, either expressed or implied, of the Air Force Office of Scientific Research or the U.S. Government. MEF is grateful for funding from the Lloyd's Register Foundation, a charitable foundation helping to protect life and property by supporting engineering-related education, public engagement and the application of research.

Conflicts of Interest: The authors declare no conflict of interest.

References

1. Tenaglia, R.D.; Lahrman, D.F. Preventing Fatigue Failures with Laser Peening. In *AMPTIAC Quarterly*; Defense Technical Information Center: Fort Belvoir, VA, USA, 2003; pp. 3–7.

2. Bair, R. Application of Surface Residual Stresses for Durability and Damage Tolerance Improvement in F-22 Wing Attachment Lugs. In Proceedings of the Aircraft Structural Integrity Program (ASIP) Conference, Jacksonville, FL, USA, 1–3 December 2009.

3. Fairand, B.P.; Wilcox, B.A.; Gallagher, W.J.; Williams, D.N. Laser shock-induced microstructural and mechanical property changes in 7075 aluminum. *J. Appl. Phys.* **1972**, *43*, 3893–3895. [CrossRef]

4. Fairand, B.P.; Clauer, A.H.; Jung, R.G.; Wilcox, B.A. Quantitative assessment of laser-induced stress waves generated at confined surfaces. *Appl. Phys. Lett.* **1974**, *25*, 431–433. [CrossRef]

5. Montross, C.S.; Wei, T.; Ye, L.; Clark, G.; Mai, Y.W. Laser shock processing and its effects on microstructure and properties of metal alloys: A review. *Int. J. Fatigue* **2002**, *24*, 1021–1036. [CrossRef]

6. Ding, K.; Ye, L. *Laser Shock Processing, Process Performance and Simulation*; CRC Press: Boca Raton, FL, USA, 2006.

7. Wu, J.; Zhao, J.; Qiao, H.; Liu, X.; Zhang, Y.; Hu, T. Acoustic wave detection of laser shock peening. *Opto Electron. Adv.* **2018**, *1*, 180016. [CrossRef]

8. US Air Force Scientific Advisory Board. *Sustaining Air Force Aging Aircraft into the 21st Century*, Report SAB-TR-11-01. 01 August 2011.

9. Clauer, A.H.; Lahrman, D.F. *Laser Shock Processing as a Surface Enhancement Process*; Durable Surfaces: Malvern, PA, USA, 2001; pp. 121–142.

10. Dorman, M.; Toparli, M.B.; Smyth, N.; Cini, A.; Fitzpatrick, M.E.; Irving, P.E. Effect of laser shock peening on residual stress and fatigue life of clad 2024 aluminium sheet containing scribe defects. *Mater. Sci. Eng. A* **2012**, *548*, 142–151. [CrossRef]

11. Bhamare, S.; Ramakrishnan, G.; Mannava, S.R.; Langer, K.; Vasudevan, V.K.; Qian, D. Simulation-based optimization of laser shock peening process for improved bending fatigue life of Ti-6Al-2Sn-4Zr-2Mo alloy. *Surf. Coat. Technol.* **2013**, *232*, 464–474. [CrossRef]

12. Smyth, N.A.; Toparli, M.B.; Fitzpatrick, M.E.; Irving, P.E. Recovery of fatigue life using laser peening on 2024-T351 aluminium sheet containing scratch damage: The role of residual stress. *Fatigue Fract. Eng. Mater. Struct.* **2019**, *42*, 1161–1174. [CrossRef]

13. Toparli, M.B.; Fitzpatrick, M.E. Residual Stresses Induced by Laser Peening of Thin Aluminium Plates. In *Materials Science Forum*; Trans Tech Publications: Zurich, Switzerland, 2011; pp. 504–509.

14. Achintha, M.N.; Nowell, D.; Shapiro, K.; Withers, P.J. Eigenstrain modelling of residual stress generated by arrays of laser shock peening shots and determination of the complete stress field using limited strain measurements. *Surf. Coat. Technol.* **2013**, *216*, 68–77. [CrossRef]

15. Cellard, C.; Retraint, D.; François, M.; Rouhaud, E.; Le Saunier, D. Laser shock peening of Ti-17 titanium alloy: Influence of process parameters. *Mater. Sci. Eng. A* **2012**, *532*, 362–372. [CrossRef]

16. Toparli, M.B.; Fitzpatrick, M.E. Through Thickness Residual Stress Measurements by Neutron Diffraction and Hole Drilling in a Single Laser-Peened Spot on a Thin Aluminium Plate. In *Materials Science Forum*; Trans Tech Publications: Zurich, Switzerland, 2014; pp. 167–172.

17. Ding, K.; Ye, L. FEM simulation of two sided laser shock peening of thin sections of Ti-6Al-4V alloy. *Surf. Eng.* **2003**, *19*, 127–133. [CrossRef]

18. Hong, Z.; Chengye, Y. Laser shock processing of 2024-T62 aluminum alloy. *Mater. Sci. Eng. A* **1998**, *257*, 322–327. [CrossRef]

19. Yang, J.M.; Her, Y.C.; Han, N.; Clauer, A. Laser shock peening on fatigue behavior of 2024-T3 Al alloy with fastener holes and stopholes. *Mater. Sci. Eng. A* **2001**, *298*, 296–299. [CrossRef]

20. Ivetic, G.; Meneghin, I.; Troiani, E.; Molinari, G.; Ocana, J.; Morales, M. Fatigue in laser shock peened open-hole thin aluminium speciemens. *Mater. Sci. Eng. A* **2012**, *534*, 573–579. [CrossRef]

21. Braisted, W.; Brockman, R. Finite element simulation of laser shock peening. *Int. J. Fatigue* **1999**, *21*, 719–724. [CrossRef]

22. Ding, K.; Ye, L. Three-dimensional dynamic finite element analysis of multiple laser shock peening processes. *Surf. Eng.* **2003**, *19*, 351–358. [CrossRef]

23. Ocaña, J.L.; Morales, M.; Molpeceres, C.; Torres, J. Numerical simulation of surface deformation and residual stresses fields in laser shock processing experiments. *Appl. Surf. Sci.* **2004**, *238*, 242–248. [CrossRef]

24. Singh, G.; Grandhi, R.; Stargel, D.; Langer, K. Modeling and optimization of a laser shock peening process. In Proceedings of the 12th AIAA/ISSMO Multidisciplinary Analysis and Optimization Conference, Victoria, BC, Canada, 10–12 September 2008; pp. 5838–5850.

25. Brockman, R.A.; Braisted, W.R.; Olson, S.E.; Tenaglia, R.D.; Clauer, A.H.; Langer, K.; Shepard, M.J. Prediction and characterization of residual stresses from laser shock peening. *Int. J. Fatigue* **2012**, *36*, 96–108. [CrossRef]

26. Hasser, P.J.; Malik, A.S.; Langer, K.; Spradlin, T.J. Simulation of Surface Roughness Effects on Residual Stress in Laser Shock Peening. In Proceedings of the ASME 2013 International Manufacturing Science and Engineering Conference Collocated with the 41st North American Manufacturing Research Conference: American Society of Mechanical Engineers, Madison, WI, USA, 10–14 June 2013.

27. *ABAQUS Version 6.14 User's Manual*; Dassault Systèmes: Providence, RI, USA, 2015.

28. Peyre, P.; Fabbro, R. Laser Shock Processing: A Review of the Physics and Applications. *Opt. Quantum Electron.* **1995**, *27*, 1213–1229.

29. Peyre, P.; Fabbro, R.; Merrien, P.; Lieurade, H.P. Laser shock processing of aluminium alloys. *Application to high cycle fatigue behaviour. Mater. Sci. Eng. A* **1996**, *210*, 102–113.

30. Spradlin, T.J.; Grandhi, R.V.; Langer, K. Experimental validation of simulated fatigue life estimates in laser-peened aluminum. *Int. J. Struct. Integr.* **2011**, *2*, 74–86. [CrossRef]

31. Zhou, Z.; Gill, A.S.; Qian, D.; Mannava, S.R.; Langer, K.; Wen, Y.; Vasudevan, V.K. A finite element study of thermal relaxation of residual stress in laser shock peened IN718 superalloy. *Int. J. Impact Eng.* **2011**, *38*, 590–596. [CrossRef]

32. Zhou, Z.; Bhamare, S.; Ramakrishnan, G.; Mannava, S.R.; Langer, K. Thermal relaxation of residual stress in laser shock peened Ti-6Al-4V alloy. *Surf. Coat. Technol.* **2012**, *206*, 4619–4627. [CrossRef]

33. Spradlin, T.; Grandhi, R.; Langer, K. Modified Symmetry Cell Approach for Simulation of Surface Enhancement Over Large Scale Structures. In Proceedings of the 2011: ICSP-11, South Bend, IN, USA, 12–15 September 2011; pp. 111–116.

34. Hatamleh, M.I.; Mahadevan, J.S.; Malik, A.S.; Qian, D. Variable Damping Profiles for Laser Shock Peening Simulation Using Modal ANalysis and the SEATD Method. In Proceedings of the ASME 12th International Manufacturing Science and Engineering Conference MSEC2017, Los Angeles, CA, USA, 4–8 June 2017.

35. Meyers, M.A. *Dynamic Behavior of Materials*; John Wiley & Sons: New York, NY, USA, 1994.

36. Johnson, G.R.; Cook, W.H. A constitutive model and data for metals subjected to large strains, high strain rate and high temperatures. In Proceedings of the 7th International Symposium on Ballistics, The Hague, The Netherlands, 19–20 April 1931; pp. 541–547.

37. Lesuer, D.R. *Experimental Investigations of Material Models for T-6Al-4V Titanium and 2024-T3 Aluminum*; Lawrence Livermore National Laboratory: Berkeley, CA, USA, 2000.

38. Langer, K.; Olson, S.; Brockman, R.; Braisted, W.; Spradlin, T.; Fitzpatrick, M.E. High strain-rate material model validation for laser peening simulation. *J. Eng.* **2015**, *13*, 150–157. [CrossRef]

39. Amarchinta, H.K.; Grandhi, R.V.; Langer, K.; Stargel, D.S. Material model validation for laser shock peening process simulation. *Model. Simul. Mater. Sci. Eng.* **2009**, *17*, 015010. [CrossRef]

40. Amarchinta, H.K.; Grandhi, R.V.; Clauer, A.H.; Langer, K.; Stargel, D.S. Simulation of residual stress induced by a laser peening process through inverse optimization of material models. *J. Mater. Process. Technol.* **2010**, *210*, 1997–2006. [CrossRef]

41. Angulo, I.; Cordovilla, F.; García-Beltrán, A.; Smyth, N.S.; Langer, K.; Fitzpatrick, M.E.; Ocaña, J.L. The effect of material cyclic deformation properties in residual stress generation by laser shock processing. *Int. J. Mech. Sci.* **2019**, *156*, 370–381. [CrossRef]

42. Fabbro, R.; Fournier, J.; Ballard, P.; Devaux, D.; Virmont, J. Physical study of laser-produced plasma in confined geometry. *J. Appl. Phys.* **1990**, *68*, 775–784. [CrossRef]

43. Ocana, J.L.; Morales, M.; Molpeceres, C.; Torres, J.; Porro, J.A.; Gomez, G.; Rubio, C. Predictive assessment and experimental characterization of the influence of irradiation parameters on surface deformation and residual stresses in laser shock processed metallic alloys. *High Power Laser Ablation V* **2004**, *5548*, 642–653.
44. Morales, M.; Ocana, J.L.; Molpeceres, C.; Porro, J.A.; García-Beltrán, A. Model based optimization criteria for the generation of deep compressive residual stress fields in high elastic limit metallic alloys by ns-laser shock processing. *Surf. Coat. Technol.* **2008**, *202*, 2257–2262. [CrossRef]
45. Ocaña, J.L.; Morales, M.; Molpeceres, C.; Porro, J.A. Laser Shock Processing of Metallic Materials: Coupling of Laser-Plasma Interaction and Material Behaviour Models for the Assessment of Key Process Issues. *High Power Laser Ablation* **2010**, *1278*, 902–913.
46. Morales, M.; Correa, C.; Porro, J.; Molpeceres, C.; Luis Ocaña, J. Thermomechanical modelling of stress fields in metallic targets subject to laser shock processing. *Int. J. Struct. Integr.* **2011**, *2*, 51–61. [CrossRef]

Article

Integrated Numerical-Experimental Assessment of the Effect of the AZ31B Anisotropic Behaviour in Extended-Surface Treatments by Laser Shock Processing

Ignacio Angulo *, Francisco Cordovilla, Ángel García-Beltrán, Juan A. Porro, Marcos Díaz and José Luis Ocaña

UPM Laser Centre. ETS Ingenieros Industriales. Universidad Politécnica de Madrid. C/José Gutiérrez Abascal, 2, 28006 Madrid, Spain; francisco.cordovilla.baro@upm.es (F.C.); agarcia@etsii.upm.es (Á.G.-B.); japorro@etsii.upm.es (J.A.P.); marcos.diaz@upm.es (M.D.); jlocana@etsii.upm.es (J.L.O.)

* Correspondence: ignacio.angulo@upm.es; Tel.: +34-910676700

Received: 5 December 2019; Accepted: 22 January 2020; Published: 29 January 2020

Abstract: In recent years, an increasing interest in designing magnesium biomedical implants has been presented due to its biocompatibility, and great effort has been employed in characterizing it experimentally. However, its complex anisotropic behaviour, which is observed in rolled alloys, leads to a lack of reliable numerical simulation results concerning residual stress predictions. In this paper, a new model is proposed to focus on anisotropic material hardening behaviour in Mg base (in particular AZ31B as a representative alloy) materials, in which the particular stress cycle involved in Laser Shock Processing (LSP) treatments is considered. Numerical predictions in high extended coverage areas obtained by means of the implemented model are presented, showing that the realistic material's complex anisotropic behaviour can be appropriately computed and—much more importantly—it shows a particular non-conventional behaviour regarding extended areas processing strategies.

Keywords: laser shock processing; anisotropy; residual stress; FEM analysis; Mg AZ31B alloy

---

1. Introduction

Since 1930, an increasing interest in magnesium alloys has been documented in the aerospace industry [1,2] as a light-weight material. It is attractive for the biomedical industry in implant design, since it is degradable, with outstanding mechanical properties in comparison with polymers [3–5]. This has led to a great number of studies regarding mechanical properties enhancement [6–8], corrosion resistance [5,7–11] and microstructure modifications [12–14], among others. Single Mg crystal responses to shock loading in various conditions and orientations is also documented, including the study of the spall fracture [15], confirming the main mechanisms of hexagonal-closed-packed (hcp) crystals subject to uniaxial shock compression. In addition, the shock loading response of a single Mg crystal is compared with the one observed in a Mg alloy. The spall fracture results, from reference [15], were complemented in a recent publication with additional experiments at very high strain rates (up to $10^{-7} \cdot \text{s}^{-1}$) [16].

Magnesium alloys are presented in two different forms: The first ones are called casting alloys, which have defects such as inclusions or pores. The second ones are called wrought alloys, whose mechanical properties are enhanced with respect to casting alloys. This group includes rolled and extruded alloys. This paper is centered on the study of a rolled Mg AZ31B alloy, in which the grain is deformed along the rolling direction leading to a relevant asymmetry in its mechanical behaviour [17]. This anisotropic behaviour is motivated by the fact that different deformation modes are active depending on the loading path [18].

In general terms, a great effort has been employed in the experimental characterization of physical properties of magnesium alloys, but there is still a lack of simulation procedures regarding the accurate prediction of the residual stresses field. This may be motivated by the complex anisotropy and the atypical hardening behaviour of rolled Mg AZ31B alloy, in which great differences are observed in stress–strain curves in the normal direction (ND), rolling direction (RD) and transverse direction (TD). This anisotropic hardening has been studied by Tucker in Mg AZ31B alloy for different strain rates up to 4300 s^{-1}, in which the effect of the specimen temperature in results is also documented [19,20]. Jäger et al. [21] developed quasi static tensile behaviour experiments of hot rolled Mg AZ31 alloys up to 673 K. Biaxial tests were performed at room temperature to study the mechanical formability and elongation to failure [22]. Most of the researchers oriented these studies to understand the high-complexity physics involved in the different slip modes presented, twinning and detwinning in alternative paths, dislocations reorientation and the influence of grain size and texture in the mechanical properties in different loading orientations.

Regarding the developed mechanical constitutive models, several approaches to characterize stress–strain curves have been proposed in different situations [23,24]. Koh used Voce–Swift type isotropic hardening law to model the tensile stress–strain curves in RD and TD directions with temperatures ranging from 373 K to 573 K, with the purpose of evaluating the increased formability in hot rolled specimens [23]. Several authors used a visco-plastic self-consistent model to characterize Mg AZ31 alloys in a large variety of conditions [24]. However, a great number of constants need to be calibrated to fix these models. Thus, the identification of the particular stress cycles, temperatures and strain rates involved in LSP is necessary in order to develop a material model valid for residual stress determination.

In Laser Shock Processing (LSP), a high intensity pulsed laser beam irradiates the material's surface, forcing a sudden vaporization of the irradiated metallic target. The generated plasma expansion develops in a high intensity shockwave that propagates through the material leading to high strain rate deformations. The stresses in a material subject to LSP are characterized by alternative stress states in different orientations in the deviatoric plane. A hydrostatic pressure is simultaneously applied, which increases as the deviatoric stress does [25,26]. The hydrostatic pressure in metallic materials is usually neglected. However, several authors included an equation of state (Mie–Grüneisen) in their models [27,28], which correlates the applied pressure with the material's density. Nevertheless, the material's compressibility in the range of pressures involved in the considered alloy (from 0 to 5 GPa approximately) is properly defined by the Hooke's Law inherent to the elasto-plastic behaviour. Therefore, the hypothesis of neglecting the hydrostatic pressure is assumed and LSP will be considered as a process in which alternative stresses in different directions are applied.

Experimental evidence shows that anisotropic hardening behaviour occurs when alternative cycles are involved: In general, once a material is deformed by a compressive state, further lower amplitude tensile states result in plastic straining (Bauschinger effect). This is usually modelled by a kinematic hardening rule. Although the explicit consideration of the yield surface displacement within a saturated isotropic hardening rule is normally used to model aluminum alloys subject to LSP treatments in order to prevent the prediction of an undefined expansion of the yield surface [26], for the case of the considered alloy, reasonably good agreement is still possible for low-density treatments using a cyclicity-independent (i.e., conventional Johnson-Cook) hardening model.

In this paper, Hill's yield surface is used to properly model material hardening in both direct and reverse deformation paths. The reasonably good agreement between numerical residual stress predictions and experimental results for the highly anisotropic material considered (AZ31B alloy) confirms the appropriateness of the explicit consideration of anisotropic deformation behaviour in generic highly deformable materials.

2. Materials and Methods

2.1. Material Description

The material used in the present study is a 5 mm thickness plate of Mg AZ31B alloy (3 wt.% Al, 1 wt.% Zn, balance Mg) provided by Magnesium Elektron (Magnesium Elecktron Ltd., Manchester, UK). The Young modulus, E, is equal to 45 GPa. The Poisson ratio is 0.35. The anisotropic stress–strain curves along ND, RD and TD at quasi-static and dynamic conditions were adapted from Tucker's results [19], in which the stress–strain curves are represented for different strain rates. Considering that each of the stress–strain points in the original publication are not provided, a graph digitalization software was used to obtain each of the experimental results (Figure 1), which is necessary in order to develop and calibrate a material model.

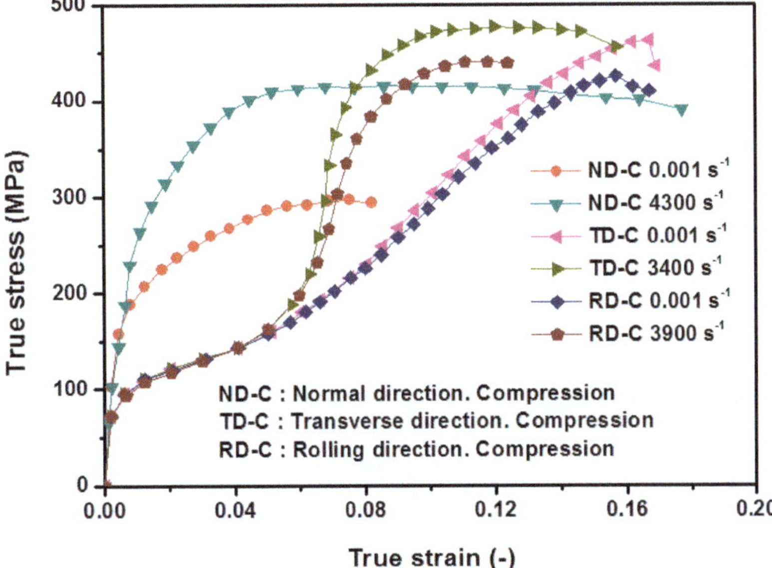

Figure 1. Stress–strain curves in rolled Mg AZ31B alloy at different strain rates adapted from [19].

2.2. Methodology

2.2.1. Anisotropic Hardening Formulation based on Hill's Yield Surface to Model Alternative Loading Paths Presented in LSP

Classic J_2 plasticity theory assumes that the yield condition in stress space is governed by the second deviatoric stress invariant, J_2. This implies that the elastoplastic surface in stress space is considered as independent of the hydrostatic pressure and the third deviatoric stress invariant, J_3. The second invariant, J_2, is defined as a function of a generic stress state as follows:

$$J_2 = \frac{1}{6}\left[(\sigma_{22} - \sigma_{33})^2 + (\sigma_{11} - \sigma_{33})^2 + (\sigma_{11} - \sigma_{22})^2\right] + (\sigma_{23})^2 + (\sigma_{31})^2 + (\sigma_{12})^2 \tag{1}$$

The most widely implemented yield criterion based on J_2 theory used to model metallic materials is the von Mises criterion. It assumes that the material yields when the distortion energy per unit of volume reaches a limit. This limit is defined as the distortion energy achieved in a uniaxial tensile test, in which the yield condition is presented when $\sigma_1 = \sigma_y$ and $\sigma_2 = \sigma_3 = 0$, where σ_y is the yield stress of the material and σ_1, σ_2 and σ_3 are the three principal stresses. Thus, the yield surface results in a coaxial cylinder with the hydrostatic axis ($\sigma_1 = \sigma_2 = \sigma_3$) in the principal stresses space:

$$\left(\sigma_y\right)^2 = \frac{1}{2}\left[(\sigma_{22} - \sigma_{33})^2 + (\sigma_{11} - \sigma_{33})^2 + (\sigma_{11} - \sigma_{22})^2\right] + 3\left((\sigma_{23})^2 + (\sigma_{31})^2 + (\sigma_{12})^2\right) \tag{2}$$

$$\left(\sigma_y\right)^2 = \frac{1}{2}\left[(\sigma_2 - \sigma_3)^2 + (\sigma_1 - \sigma_3)^2 + (\sigma_1 - \sigma_2)^2\right] \tag{3}$$

In 1950, Hill extended the von Mises yield criterion to define anisotropic yield surfaces (Equation (4)), in which the parameters involved F, G, H, L, M and N can be defined as functions of the plastic strain. Thus, the material's elasto-plastic anisotropic behaviour can then be fully modelled. However, it is easier for researchers to define the yield stresses in three different directions, σ_{yi} ($i = 1$ to (3), expressed as a function of a generic yield stress, σ_y, where R_{ii} are the first three potentials (Equation (5)). The other three potentials, R_{ij} ($i, j = 1$ to 3, $i \neq j$, $i < j$), affect the material's behavior when shear stresses are presented. These potentials can be correlated with Hill's original parameters [29].

$$\left(\sigma_y\right)^2 = \frac{1}{2}\left[F(\sigma_{22} - \sigma_{33})^2 + G(\sigma_{11} - \sigma_{33})^2 + H(\sigma_{11} - \sigma_{22})^2\right] + 3\left(L(\sigma_{23})^2 + M(\sigma_{31})^2 + N(\sigma_{12})^2\right) \tag{4}$$

$$\sigma_{yi} = R_{ii}\sigma_y \tag{5}$$

In LSP, the plasma expansion develops in a high intensity shockwave that propagates through the material, which experiences a phenomenon called uniaxial strain [26], in which the material is subject to alternative loading paths (plastic loading and unloading). In the case of materials of highly anisotropic behaviour, as the considered case of AZ31B alloy, the proper modelling of these alternative loading paths requires an extended yield surface definition, in which the material's compressive behaviour along ND, RD and TD is properly modelled. Since the yield stress of the present alloy is relatively small, low-density treatments are expected to be the suitable ones to enhance its mechanical properties. Thus, the yield surface displacement, usually modelled by means of a kinematic hardening rule, will not be considered.

In the present paper, the plastic loading is modelled by means of the compressive stress–strain curve along the normal direction (ND), which is coincident with the direction of the applied pressure. Therefore, a biaxial compressive state in rolling and transverse directions (RD and TD) is considered during plastic unloading. Intuition suggests the use of the tensile stress–strain curve in the normal direction (ND) during plastic unloading. However, the effect of anisotropy in RD and TD would be neglected as a consequence, which is not realistic since different behaviour is expected along these directions.

2.2.2. Model Calibration Based on the Anisotropic Stress-Strain Curves of Mg AZ31B Alloy

The anisotropic stress–strain behaviour of Mg AZ31B alloy has been documented by many authors [30,31]. Several of them have focused their studies on warm temperatures (373 K to 573 K) since magnesium formability is relatively low at room temperature. Others have characterized the stress–strain curves in the ND subjected to uniaxial compressive states [20]. Concerning the cyclic behaviour, several authors studied the hysteresis loops in a wide variety of conditions: temperature, strain amplitude, directions and strain rates [17,30].

A great percentage of the stress–strain characterization works are centered on the RD and TD. The ND direction is harder to study due to its short length. In consequence, a lack of results is presented especially in the tensile direction. In spite of these difficulties, quasi static ND tensile stress–strain curves have been characterized in cyclic behaviour studies, in which the specimens were obtained from a 50 mm thickness rolled plate [18,30]. However, the dynamic behaviour at high strain rates is not characterized. In this section, an anisotropic model is calibrated based on the experimental dynamic characterization along ND, RD and TD.

The particularities of LSP processes need to be considered in order to select the most suitable experimental results to define the material model. Results need to include high strain rate dependence, while the effect of temperature and high compressive stress triaxialities could be ignored. In LSP, the explicit consideration of asymmetry in stress–strain curves may be considered for high-density treatments [26]. Due to the relatively small yield stress of AZ31B alloy, low-density treatments are expected to suitable and, consequently, the explicit consideration of asymmetry is neglected in the present study.

Regarding thermal effects at the material's surface, although the generated plasma usually reaches high temperatures, it is maintained during a very short period of time. Thus, the material is not expected to experience large temperature variations and room temperature is used for model calibration. For instance, the thermally affected depth for a steel alloy subject to LSP is estimated at 30 μm [32], while the plastically affected depth typically reaches more than 1 mm. As a result, thermal effects are usually neglected in LSP processes and the hypothesis of considering it as a purely mechanical process is generally adopted [32].

The experimental results that best fit with the above LSP characteristics definition are the ones presented by Tucker et al. [19]. These results include the quasi static and high strain rate behaviour in ND (only compression), RD (both tensile and compressive) and TD (both tensile and compressive) directions.

The plastic loading will be modelled by means of the compressive stress–strain curve in the ND. Therefore, compressive stress–strain curves in RD and TD characterize the biaxial stress state during plastic unloading. The yield stress of the present alloy is relatively small compared with most of the metallic materials subject to LPS treatments. As a consequence, low-density treatments are expected to be the suitable ones. This implies that the effect of the cyclic plasticity, modelled by means of a kinematic hardening rule, could be ignored. Thus, the yield surface in the present model will experience an anisotropic expansion in the stress space when the material is plastically deformed.

Regarding the stress–strain curve calibration along ND, a Voce type hardening law [33] has been used, including strain rate dependence (Equation (6)), where σ_{ND} is the compressive yield stress in the ND, σ_{ND0} is the first compressive yield stress, Q is the yield stress saturated increment, $\dot{\varepsilon}_0$ is a reference strain rate, b and b_0 are constants to be calibrated, $F(\dot{\varepsilon}_r)$ is a strain rate function that models the saturation equivalent plastic strain at different strain rates, and $\Delta\varepsilon_p$ is the equivalent plastic strain.

Considering that experimental results are obtained up to 4300 s^{-1}, the function $F(\dot{\varepsilon}_r)$ is set as constant for further strain rates, since there are no experimental data beyond 4300 s^{-1}. It is documented by some authors that extrapolating experimental results for higher strain rates usually leads to overestimations in the compressive residual stress predictions [34]. Table 1 presents the calibrated constants obtained for the best fit option. Figure 2 shows a comparison between the experimental results and the numerical predictions.

$$\sigma_{ND}(\Delta\varepsilon_p) = \sigma_{ND0} + Q\,F(\dot{\varepsilon}_r)\left(1 - e^{-(F(\dot{\varepsilon}_r)b + b_0)(\Delta\varepsilon_p)}\right) \tag{6}$$

$$F(\dot{\varepsilon}_r) = \begin{cases} 1 & \dot{\varepsilon}_r < \dot{\varepsilon}_0 \\[2mm] \dfrac{\dot{\varepsilon}_r}{\dot{\varepsilon}_1} + 1 & \dot{\varepsilon}_0 \leq \dot{\varepsilon}_r \leq \dot{\varepsilon}_1 \\[2mm] 2 & \dot{\varepsilon}_r > \dot{\varepsilon}_1 \end{cases} \tag{7}$$

Table 1. Calibrated constants for the analytic predictions of compressive stress–strain curves (ND).

Parameter	Value
σ_{ND0} (MPa)	178
Q (MPa)	125
b_0 (–)	19
b (–)	18
$\dot{\varepsilon}_0$ (s^{-1})	0.001
$\dot{\varepsilon}_1$ (s^{-1})	4300

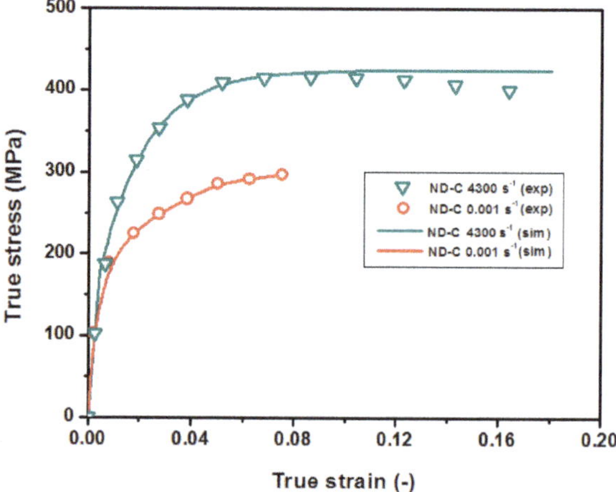

Figure 2. Calibrated model vs. experimental stress–strain curves along ND.

The material experiences softening at a certain plastic strain (approximately when the strain is equal to 0.07 at high strain rates), which is not possible to model with a purely Voce expression (monotonically increasing function). Koh used a Swift–Voce equation to predict softening [23]. However, in LSP, the equivalent plastic strain, $\Delta \varepsilon_p$, usually exceeds the material's ultimate tensile stress with no failure. This is due to the high compressive stress triaxialities, η, involved in LSP treatments, ensuring the complete protection of the material. In 2004, Bao and Wierbzicky [35] postulated a cut-off value for the compressive stress triaxiality beyond which failure will never occur ($\eta < -1/3$). This is the particular situation presented in LSP treatments. Therefore, a saturated hardening curve is considered to model the stress-strain scenarios beyond the ultimate tensile stress. Thus, Voce's equation itself predicts correctly this particular situation. It is important to point out that the isotropic models, which are widely accepted in LSP simulations (i.e., Johnson–Cook model) predict an unbounded expansion of the yield stress when high-density treatments are applied. This leads to overestimated compressive residual stress predictions, which are not observed experimentally. Thus, the use of a Voce equation (with saturated yield stress) may represent more accurate predictions than classic isotropic hardening models widely implemented in LSP simulations [26].

Regarding the constitutive equations to model the plastic unload, both RD and TD compressive stress–strain curves are considered. These curves present an atypical behaviour at an approximately equivalent plastic strain of 0.059, beyond which a new deformation mechanism is activated and yield stress increases abruptly. In order to model this behaviour, the superposition of two Voce equations will be used: The first one is characterized by a low amplitude. The second one, with higher amplitude, is activated once the equivalent plastic strain reaches the critical value. Equations (8)–(11) represent the stress–strain behaviour in RD and TD, both for quasi static and dynamic conditions. The dynamic behaviour from quasi static conditions up to the highest calibrated strain rates is obtained by linear interpolation. The dynamic behaviour beyond the experimental data is set as constant. The calibrated parameters for the best fit option are shown in Table 2. Figures 3 and 4 show a comparison between the experimental results and the corresponding analytical predictions along RD and TD respectively.

$$\sigma_{RDe}\left(\Delta\varepsilon_p\right) = min\left(\sigma_{D0} + Q_D\left(1 - e^{-b_D(\Delta\varepsilon_p)}\right) + \partial\left(\Delta\varepsilon_p\right)\cdot Q_{RD2}\left(1 - e^{-b_{De}(\Delta\varepsilon_p)}\right), \sigma_{RDmax}\right) \tag{8}$$

$$\sigma_{TDe}\left(\Delta\varepsilon_p\right) = min\left(\sigma_{D0} + Q_D\left(1 - e^{-b_D(\Delta\varepsilon_p)}\right) + \partial\left(\Delta\varepsilon_p\right)\cdot Q_{TD2}\left(1 - e^{-b_{De}(\Delta\varepsilon_p)}\right), \sigma_{TDmax}\right) \tag{9}$$

$$\sigma_{RDd}\left(\Delta\varepsilon_p\right) = \sigma_{D0} + Q_D\left(1 - e^{-b_D(\Delta\varepsilon_p)}\right) + \partial\left(\Delta\varepsilon_p\right)\cdot Q_{RDd}\left(1 - e^{-b_{RDd}(\Delta\varepsilon_p)}\right) \tag{10}$$

$$\sigma_{TDd}\left(\Delta\varepsilon_p\right) = \sigma_{D0} + Q_D\left(1 - e^{-b_D(\Delta\varepsilon_p)}\right) + \partial\left(\Delta\varepsilon_p\right)\cdot Q_{TDd}\left(1 - e^{-b_{TDd}(\Delta\varepsilon_p)}\right) \tag{11}$$

$$\partial\left(\Delta\varepsilon_p\right) = \begin{cases} 0, & \Delta\varepsilon_p < \left(\Delta\varepsilon_p\right)_c \\ 1, & \Delta\varepsilon_p \geq \left(\Delta\varepsilon_p\right)_c \end{cases} \tag{12}$$

where σ_{RDe} and σ_{TDe} are the quasi static yield stresses along RD and TD respectively. σ_{RDd} and σ_{TDd} are the dynamic yield stresses. $\left(\Delta\varepsilon_p\right)_c$ is the critical equivalent plastic strain beyond which the high amplitude Voce function is activated.

Table 2. Calibrated constants for the analytic predictions of stress–strain curves.

Parameter	Value
σ_{D0} (MPa)	60
σ_{RDmax} (MPa)	426
σ_{TDmax} (MPa)	463
Q_D (MPa)	98
Q_{RD2} (MPa)	720
Q_{TD2} (MPa)	810
Q_{RDd} (MPa)	280
Q_{TDd} (MPa)	320
b_D (–)	50
b_{De} (–)	5
b_{RDd} (–)	80
b_{TDd} (–)	95
$\left(\Delta\varepsilon_p\right)_c$ (–)	0.059

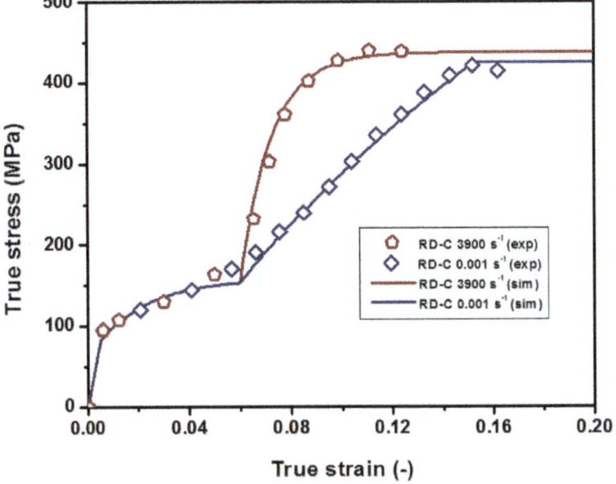

Figure 3. Calibrated model vs. experimental stress–strain curves along RD.

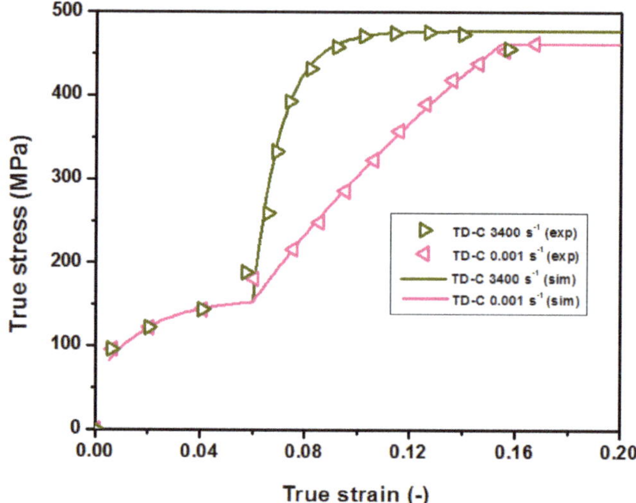

Figure 4. Calibrated model vs. experimental stress–strain curves along TD.

The calibrated stress–strain curves following the presented procedure are plotted together in Figure 5. The reference compressive yield stress, σ_y, is set equal to σ_{D0}. Thus, potentials, R_{ij}, are obtained dividing the corresponding yield stress in ND, TD and RD by the reference yield stress, σ_{D0}. The determination of Hill's coefficients is remarkably conditioned by the process, according to reference [36].

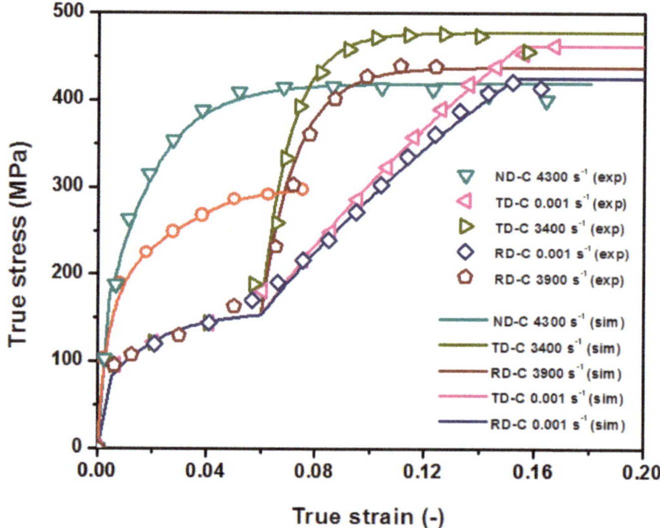

Figure 5. Calibrated model vs. adapted experimental stress–strain curves for ND, RD and TD.

2.2.3. Experimental Determination of the Spatial Pressure Pulse Profiles

In LSP, the high intensity laser beam applied to the surface forces a sudden vaporization generating a plasma, whose expansion develops in pressures about 5 GPa on the material's irradiated area. This pressure generates a shockwave that propagates through the material, leading to plastic deformations and a final residual stress state. Thus, both the characterization of the plasma pressure and the simulation of the shockwave propagation through the material are key issues to model properly LSP treatments.

Regarding the plasma characterization, several authors adopt a simplified expression to estimate the maximum pressure as a function of laser intensity, in which the hypothesis of adiabatic expansion is considered [37]. However, in the present study, the plasma characterization is provided by a detailed calculation of plasma pressure and thermal flux (HELIOS code) as described in previous publications by the authors [38–40].

The laser device used in the present study is a Q-switched Nd:YAG (Spectra-Physics, Santa Clara, USA) operating at 10 Hz and providing 9 ns FWHM, 2 J per pulse with near Gaussian spatial profile. The spot diameter was set to 1.5 mm. This leads to a peak laser intensity, $I = 28.5$ GW/cm^2. The material's reflectivity was calculated with the aid of Wu and Shin model [41,42], which is set as an input parameter to the HELIOS code. The in-time pressure profile is represented in Figure 6.

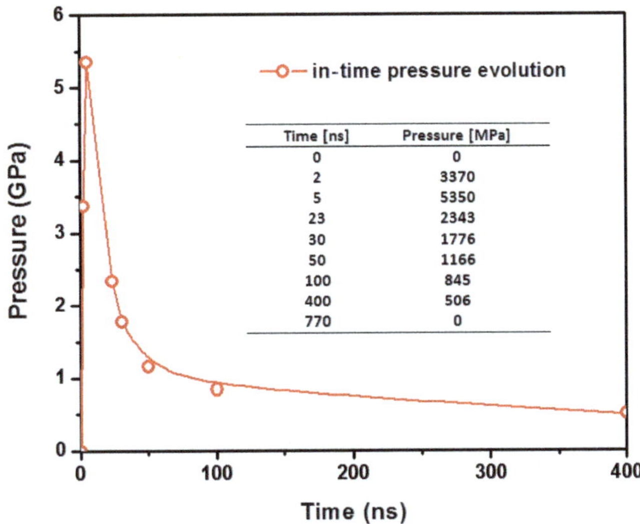

Time [ns]	Pressure [MPa]
0	0
2	3370
5	5350
23	2343
30	1776
50	1166
100	845
400	506
770	0

Figure 6. In-time pressure evolution calculated by means of HELIOS code.

The deformation profiles after the application of 1 to 5 concentric pulses in rolled Mg AZ31B alloy were obtained in the Laser Center UPM by means of a confocal microscope LEICA DCM 3D (Leica Microsystems GmbH, Wetzlar, Germany). In Figure 7, a characteristic map obtained for the application of a single pulse is presented.

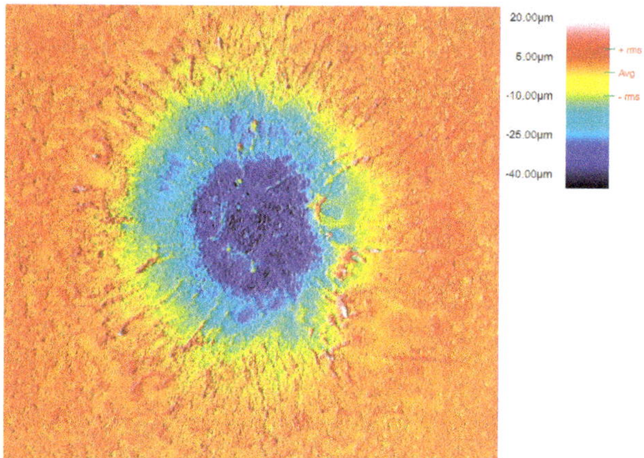

Figure 7. Confocal image of the deformed surface after the application of the first pulse.

The general form of the spatial distribution is defined by Equation (13), where $P(t)$ is the temporal pressure profile, a_ϕ, R_c and H are constants to be calibrated, and r represents the distance from a generic surface point to the center of the laser spot. In Figure 8, a comparison between the experimental softened profiles and its corresponding numerical predictions for the best fit option is plotted. Table 3 presents these calibrated constants.

$$P(r,t) = P(t)e^{-a_\phi\left(\frac{r}{R_c}\right)^{2H}}$$

(13)

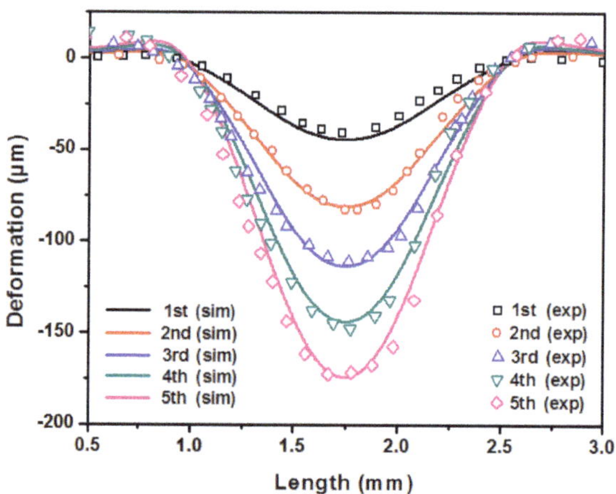

Figure 8. Experimental deformation profiles vs. numerical predictions.

Table 3. Calibrated constants for spatial pressure definition.

Parameter	Value
a_ϕ (–)	1.9
R_c (mm)	0.8
H (–)	1.3

3. Results

3.1. Experimental Characterization of the Residual Stresses for Different Input Parameters

The residual stresses have been obtained experimentally by means of the hole drilling method (ASTM E837-13 standard) [43], in which the material is drilled and the resulting deformation as the material is removed is measured by a precision strain gage up to 1 mm thickness (HBM K-RY61-1.5/120R-3 prewired). The specimens subject to LSP treatment were obtained from a 5 mm thickness rolled Mg AZ31B plate. Four different treatment strategies were implemented experimentally, covering a treated squared area of 900 mm^2 (Figure 9 shows a schematic representation):

(1) Setting a laser spot diameter of 1.5 mm, with a distance between successive pulses of 0.66 mm, developing in an Equivalent Overlapping Density, EOD [44], of 225 pp/cm^2, and an Equivalent Local Overlapping Factor, ELOF [44], of 4. The peening direction (PD) is set coincident with the transverse direction, TD.

(2) Setting a laser spot diameter of 1.5 mm, with a distance between successive pulses of 0.66 mm, developing in an Equivalent Overlapping Density, EOD, of 225 pp/cm^2, and an Equivalent Local Overlapping Factor, ELOF, of 4. The peening direction (PD) is set coincident with the rolling direction, RD.

(3) Setting a laser spot diameter of 1.5 mm, with a distance between successive pulses of 0.50 mm, developing in an Equivalent Overlapping Density, EOD, of 400 pp/cm^2, and an Equivalent Local Overlapping Factor, ELOF, of 7. The peening direction (PD) is set coincident with the transverse direction, TD.

(4) Setting a laser spot diameter of 1.5 mm, with a distance between successive pulses of 0.50 mm, developing in an Equivalent Overlapping Density, EOD of 400 pp/cm^2, and an Equivalent Local Overlapping Factor, ELOF, of 7. The peening direction (PD) is set coincident with the rolling direction, RD.

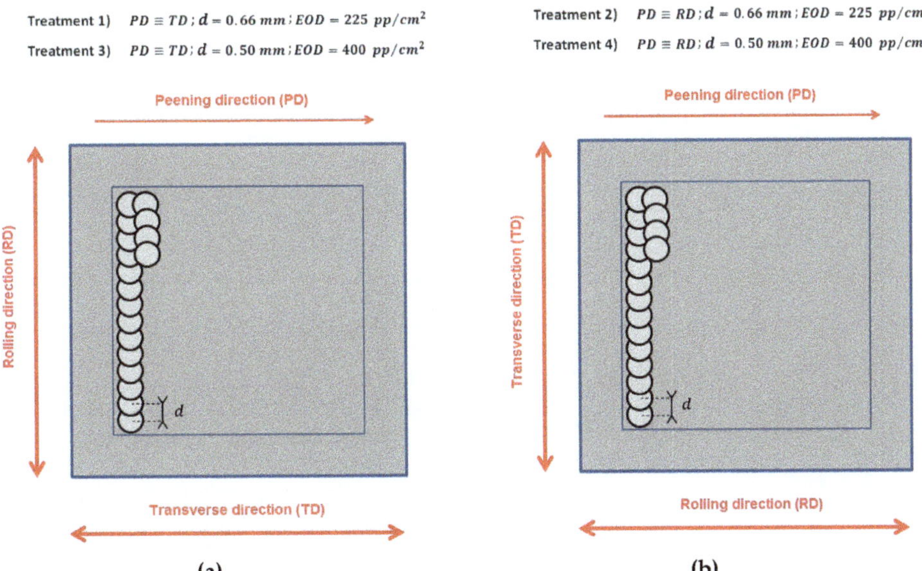

Figure 9. (**a**) Schematic representation of treatments (1) and (3) (PD = TD); (**b**) Schematic representation of treatments (2) and (4) (PD = RD).

The EOD corresponds to the number of applied pulses per unit of area, while the ELOF represents the average number of pulses at a given location. All the studied configurations can be considered as low-density treatments according to their respective EOD's and ELOF's. This is consistent with the proposed material model, in which the cyclic plasticity modelling is neglected. Higher density treatment-modelling would require further research in the proper calibration of stress–strain asymmetry.

The experimental results from the material's surface up to 1 mm for configurations (1), (2), (3) and (4) are represented in Figures 10 and 11. In these figures, the in-depth minimum, S_{min}, and maximum in plane, S_{max} are represented with their corresponding uncertainties, calculated with the aid of Hdrill software. Several conclusions can be extracted from the experimental results:

(1) For all the experimental measurements developed, the minimum stress, S_{min}, is almost coincident with the stress along the PD, while the maximum in-plane stress, S_{max}, is similar to the stress along a perpendicular direction to the ND and PD This result is independent of the relative orientation between the PD and the RD.

(2) While significant differences are observed between treatment (1) and the others, great similarities are presented between configurations (2), (3) and (4). This suggests that the residual stresses tend to reach a saturation value for relatively low-density treatments, which is as expected considering the relatively low yield stress of the present alloy in comparison with most of metallic materials.

(3) Setting the peening direction (PD) coincident with the rolling direction (RD) leads to greater peak compressive residual stresses for low-density treatments (Figure 10).

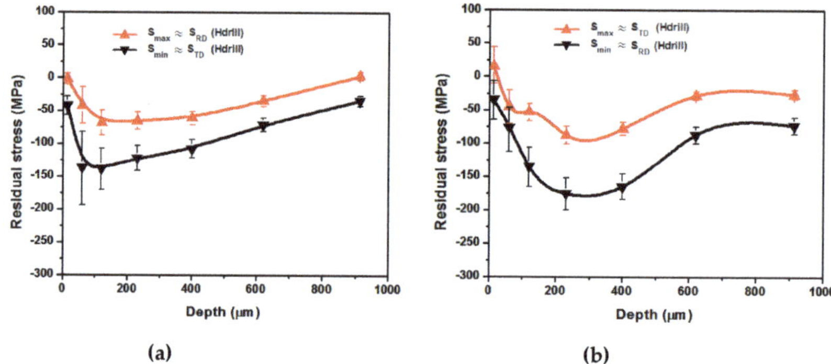

(a) (b)

Figure 10. (a) Experimental residual stresses for EOD = 225 pp/cm^2, PD = TD. (b) Experimental residual stresses for EOD = 225 pp/cm^2, PD = RD.

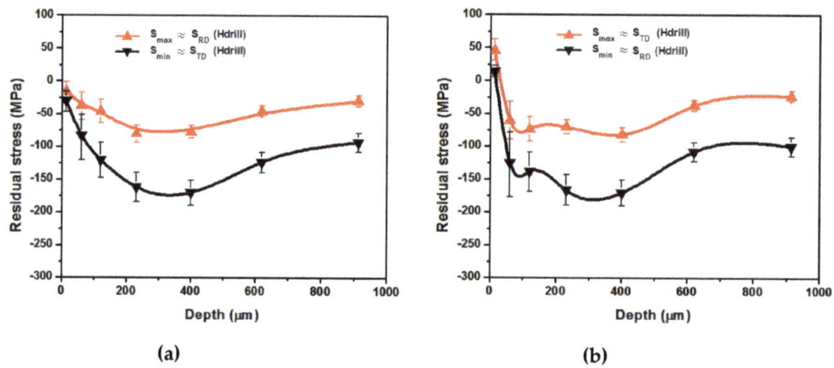

(a) (b)

Figure 11. (a) Experimental residual stresses for EOD = 400 pp/cm^2, PD = TD; (b) Experimental residual stresses for EOD = 400 pp/cm^2, PD = RD.

3.2. Realistic Modelling Results for Extended Surface High-Coverage LSP Treatments

The advances in computational resources in the last decade have motivated an increase in the number of publications regarding numerical predictions of extended LSP treatments in steel, aluminum and titanium alloys, with different loading strategies. However, there is still a lack of results concerning magnesium alloys which may be caused by the significant anisotropy presented. In this section, the corresponding numerical predictions of four particular treatments are obtained by means of two different models: (1) Considering the anisotropic model presented in Section 2.2, in which the stress–strain behaviour along ND, RD and TD directions is fully characterized. (2) Considering a purely isotropic model, in which both the plastic loading and unloading are modelled by the compressive stress–strain curve in ND direction. The purpose of comparing these two models is to show the impact in numerical predictions of the explicit consideration of anisotropy, since conventional LSP models consider a unique stress–strain curve. Results confirm that representative differences are predicted between the anisotropic model and the isotropic one. Better agreement with experimental results is obtained by means of the anisotropic model.

In realistic high-coverage extended treatments, such as the ones presented in this study, the peening direction motivates anisotropy in the residual stress predictions even in isotropic materials. Concretely, higher compressive residual stresses are usually predicted numerically and obtained experimentally along the PD [45]. This effect is also presented in the considered anisotropic alloy, in which results show that the minimum principal stress direction is coincident with the PD. The relative orientation between the PD and RD motivates variations in S_{max} and S_{min} which cannot be predicted by purely isotropic hardening models.

The numerical predictions have been extracted and averaged from a representative squared area of $2\ mm^2$ up to 1 mm depth, which represents a similar volume than the one removed by means of the hole drilling method. Figures 12–15 show a comparison between analytical predictions and experimental results for strategies (1), (2), (3) and (4) respectively. Table 4 shows a comparison between experimental results and the analytical predictions obtained by means of the anisotropic model. $(d_{max})_i$ is the depth at which the maximum peak compressive residual stress is presented. $min(S_{min})_i$ represents the peak compressive residual stress. $i = isot$ and $i = anisot$ corresponds to the analytical predictions of isotropic and anisotropic models respectively, and $i = exp$ to the experimental results.

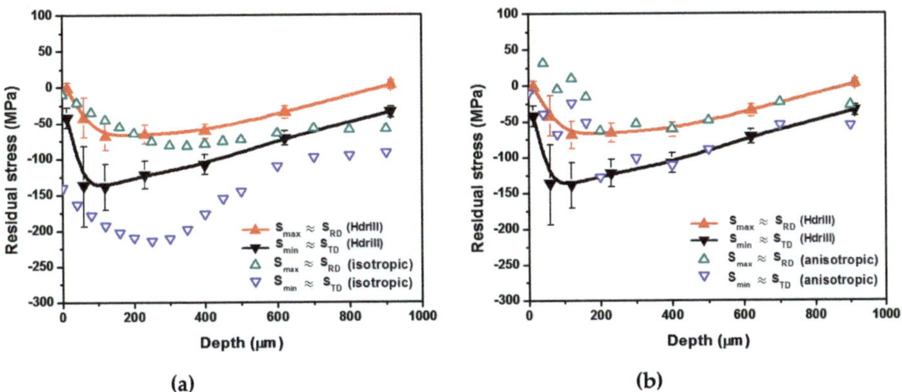

Figure 12. Experimental results vs. numerical predictions for strategy (1). (**a**) Analytical isotropic model predictions. (**b**) Anisotropic model predictions.

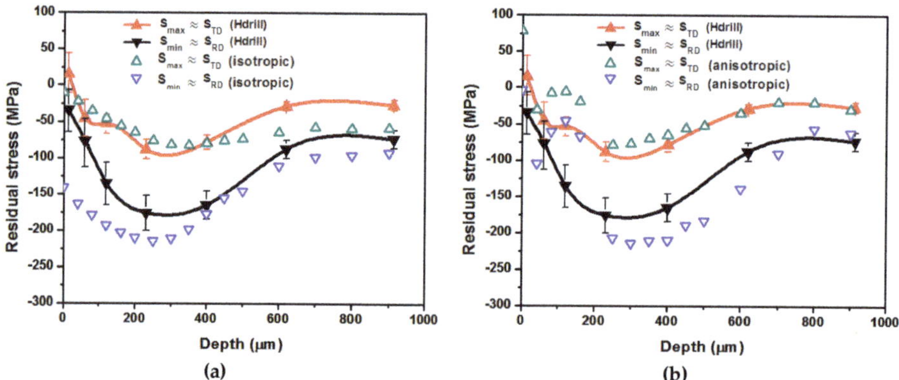

Figure 13. Experimental results vs. numerical predictions for strategy (2). (**a**) Analytical isotropic model predictions. (**b**) Anisotropic model predictions.

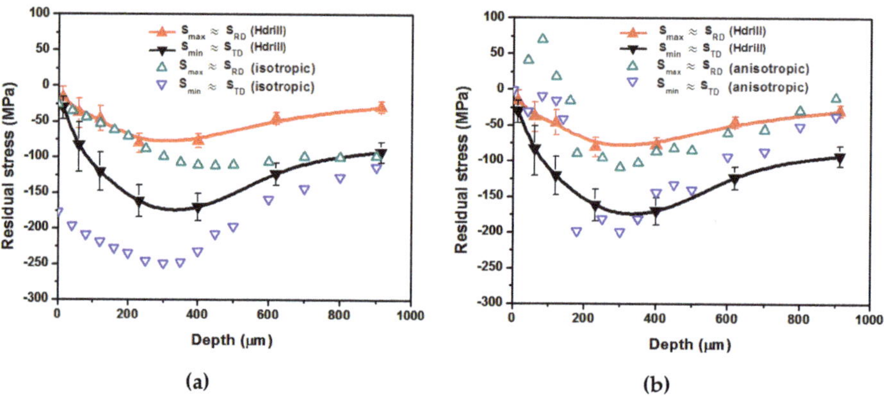

Figure 14. Experimental results vs. numerical predictions for strategy (3). (**a**) Analytical isotropic model predictions. (**b**) Anisotropic model predictions.

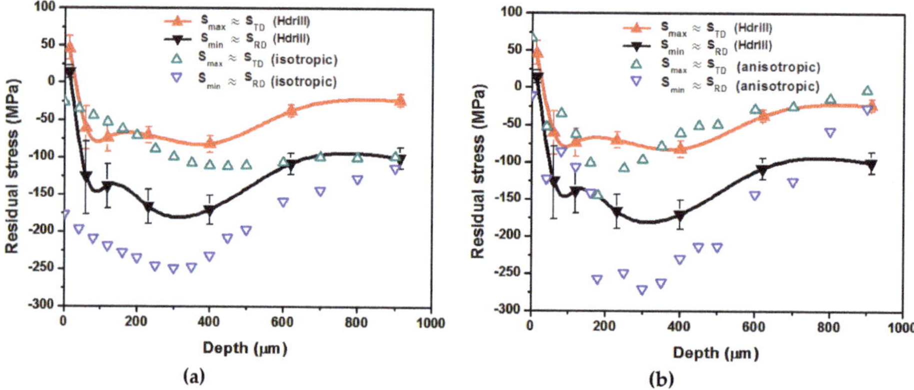

Figure 15. Experimental results vs. numerical predictions for strategy (4). (**a**) Analytical isotropic model predictions; (**b**) Anisotropic model predictions.

Table 4. Experimental results vs. analytical predictions for treatments (1), (2), (3) and (4).

Feature	(1)	(2)	(3)	(4)
EOD (pp/cm^2)	225	225	400	400
Peening direction (–)	TD	RD	TD	RD
$(d_{max})_{isot}$ (μm)	250	250	299	299
$(d_{max})_{anisot}$ (μm)	197	297	297	300
$(d_{max})_{exp}$ (μm)	100	300	334	315
$min(S_{min})_{isot}$ (MPa)	−215	−215	−249	−249
$min(S_{min})_{anisot}$ (MPa)	−130	−213	−199	−268
$min(S_{min})_{exp}$ (MPa)	−136	−178	−174	−181

Regarding the principal stress directions, the PD sets the minimum stress curve, S_{min}. The maximum in plane stress, S_{max}, is perpendicular to the ND and the PD. This is predicted correctly by both models. However, the effect of the relative orientation between the PD and the RD in the stress curves, which is significant between the low-density treatments (strategies (1) and (2)), can only be simulated by the anisotropic model.

Concerning the comparison between experimental results and the analytical predictions, the anisotropic model provides a reasonably accurate calculation the effect of the relative orientation between the PD and RD for strategies (1) and (2). A comparison between Figures 12 and 13 shows that setting the PD coincident to the RD leads to greater peak compressive residual stresses. In addition, the depth at which this peak value is achieved also increases.

The anisotropic model provides, in general, more accurate predictions than the isotropic one. In particular, the isotropic model predicts high compressive residual stresses at material's surface, which is not consistent with experimental results. In treatments where the EOD is set equal to 400 pp/cm^2 (strategies (3) and (4)), experimental evidence shows that the residual stresses tend to saturate. Reasonably good agreement is obtained by means of the anisotropic model for strategy (3) while an overestimation of the compressive residual stresses is predicted for configuration (4). This is not surprising since the model is conceived for low-density treatments, which are the most suitable ones for the present alloy.

4. Discussion

This paper provides a complete methodology to obtain residual stress predictions in anisotropic materials subject to low-density LSP treatments. In summary: Firstly, the material's stress–strain curves in ND, RD and TD directions were properly modelled both at low and high strain rates. Then, the spatial-temporal pressure pulse distribution was calibrated with the aid of experimentally determined deformation profiles. Finally, the average in-depth residual stresses were obtained for four different treatment strategies, showing the effect of the treatment density and the relative orientation between the PD and RD in results. A reasonably good agreement is presented between the experimental results and the analytical predictions obtained by means of the proposed anisotropic model.

Experimental results suggest that the PD sets the minimum principal stress direction, which implies that the anisotropy in residual stress distribution is mainly motivated by the treatment strategy itself. In fact, the minimum stress, S_{min}, is obtained along the PD for the four treatment strategies experimented. This is properly simulated by both the isotropic and the anisotropic models. Nevertheless, the relative orientation between the PD and the RD motivates differences in the minimum stress curve, S_{min}, and the maximum stress, S_{max}, in low-density treatments. This effect cannot be predicted by the isotropic model, since a unique stress–strain curve is defined.

Regarding the comparison between the numerical predictions and the experimental results, better agreement is obtained by means of the implemented anisotropic model. This is observed specially at material's surface, where the isotropic model predicts high compressive residual stresses that are not presented in experimental results. The residual stress distribution from 200 μm to 700 μm is predicted with great accuracy for configurations (1), (2) and (3). The greatest differences between the analytical predictions of the anisotropic model and the experimental results are presented near to the material's surface, where numerical results tend to predict higher tensile residual stresses than the ones observed experimentally. This may be caused by the evaporation of the material's surface during the application of successive pulses, leading to removal of already deformed material. This effect is not considered in LSP simulations [25–27,32,34,45].

The presented procedure is particularized for Mg AZ31B alloy. However, it is applicable to any material of interest subject to relatively low-density LSP treatments. In addition, the effect of the explicit consideration of mechanical anisotropy may differ depending on the selected material, since it is motivated by variations in the stress–strain curves along different loading paths.

5. Conclusions

This paper provides a complete guideline to model anisotropic behaviour of metallic materials subject to LSP treatments. Experimental evidence, confirmed by the corresponding numerical predictions, leads to the identification of the material's stress–strain anisotropy as a key factor to appropriately compute residual stress predictions in low-density treatments. The main conclusions are as follows:

(1) The relevance of including anisotropic behaviour in material modelling in residual stress predictions has been demonstrated in the particular case of Mg AZ31B alloy. Conventional isotropic models are not able to predict accurately the in-depth residual stress distribution. In addition, incorrect principal stress directions are obtained by means of purely isotropic models.

(2) The experimental results show that the anisotropy in residual stresses after treatment are motivated mainly by the treatment strategy. The minimum stress, S_{min}, is similar to the stress along the PD, while the maximum in plane residual stress, S_{max}, is similar than the one presented along a perpendicular direction to the ND and the PD. Both the implemented isotropic and anisotropic models predict properly this effect. Nevertheless, better agreement between numerical predictions and experimental results is achieved by the proposed anisotropic model.

(3) S_{min} and S_{max} curves are influenced by the relative orientation between the PD and the RD. Setting the PD coincident to the RD leads to greater compressive residual stresses for low-density treatments (EOD = 225 pp/cm^2). This effect cannot be computed properly by isotropic models. Therefore, the anisotropic hardening model is required for this purpose.

Regarding further developments, several issues may be considered:

(1) Modelling the material loss during target irradiation. The application of successive pulses leads to the evaporation of already deformed surfaces, which may modify residual stresses especially at material's surface.

(2) Analyzing the effect of the anisotropic modelling in residual stress predictions with a large variety of different input parameters, including additional relative orientations between the PD and material's anisotropic directions.

Author Contributions: Conceptualization, Á.G.-B. and J.L.O.; Data curation, F.C.; Investigation, J.A.P. and M.D.; Methodology, I.A.; Project administration, J.L.O.; Software, I.A. and F.C.; Supervision, J.L.O.; Validation, J.A.P., Á.G.-B. and M.D.; Writing—original draft, I.A. and F.C.; Writing—review and editing, I.A. and F.C. All authors have read and agreed to the published version of the manuscript.

Funding: Work partly supported by MINECO (Spain; Projects MAT2012-37782 and MAT2015-63974-C4-2-R).

Conflicts of Interest: The authors declare no conflict of interest.

References

1. Alderliesten, R.; Rans, C.; Benedictus, R. The applicability of magnesium based fibre metal laminates in aerospace structures. *Compos. Sci. Technol.* **2008**, *68*, 2983–2993. [CrossRef]
2. Czerwinski, F. Controlling the ignition and flammability of magnesium for aerospace applications. *Corros. Sci.* **2014**, *86*, 1–16. [CrossRef]

3. Staiger, M.P.; Pietak, A.M.; Huadmai, J.; Dias, G. Magnesium and its alloys as orthopedic biomaterials: A review. *Biomaterials* **2006**, *27*, 1728–1734. [CrossRef] [PubMed]

4. Zeng, R.; Dietzel, W.; Witte, F.; Hort, N.; Blawert, C. Progress and challenge for magnesium alloys as biomaterials. *Adv. Eng. Mater.* **2008**, *10*, 3–14. [CrossRef]

5. Witte, F.; Hort, N.; Vogt, C.; Cohen, S.; Kainer, K.U.; Willumeit, R.; Feyerabend, F. Degradable biomaterials based on magnesium corrosion. *Curr. Opin. Solid State Mater. Sci.* **2008**, *12*, 63–72. [CrossRef]

6. Ma, R.; Zhao, Y.; Wang, Y. Grain refinement and mechanical properties improvement of AZ31 Mg alloy sheet obtained by two-stage rolling. *Mater. Sci. Eng.* **2017**, *691*, 81–87. [CrossRef]

7. Pan, F.; Wang, Q.; Jiang, B.; He, J.; Chai, Y.; Xu, J. An effective approach called the composite extrusion to improve the mechanical properties of AZ31 magnesium alloy sheets. *Mater. Sci. Eng.* **2016**, *655*, 339–345. [CrossRef]

8. Kim, S.H.; Bae, S.W.; Lee, S.W.; Moon, B.G.; Kim, H.S.; Kim, Y.M.; Park, S.H. Microstructural evolution and improvement in mechanical properties of extruded AZ31 alloy by combined addition of Ca and Y. *Mater. Sci. Eng.* **2018**, *725*, 309–318. [CrossRef]

9. Cui, L.Y.; Gao, S.D.; Li, P.P.; Zeng, R.C.; Zhang, F.; Li, S.Q.; Han, E.H. Corrosion resistance of a self-healing micro-arc oxidation/polymethyltrimethoxysilane composite coating on magnesium alloy AZ31. *Corros. Sci.* **2017**, *118*, 84–95. [CrossRef]

10. Bagherifard, S.; Hickey, D.J.; Fintová, S.; Pastorek, F.; Fernandez-Pariente, I.; Bandini, M.; Guagliano, M. Effects of nanofeatures induced by severe shot peening (SSP) on mechanical, corrosion and cytocompatibility properties of magnesium alloy AZ31. *Acta Biomater.* **2018**, *66*, 93–108. [CrossRef]

11. Jian, S.Y.; Chu, Y.R.; Lin, C.S. Permanganate conversion coating on AZ31 magnesium alloys with enhanced corrosion resistance. *Corros. Sci.* **2015**, *93*, 301–309. [CrossRef]

12. Mao, B.; Liao, Y.; Li, B. Gradient twinning microstructure generated by laser shock peening in an AZ31B magnesium alloy. *Appl. Surf. Sci.* **2018**, *457*, 342–351. [CrossRef]

13. Ge, M.Z.; Xiang, J.Y. Effect of laser shock peening on microstructure and fatigue crack growth rate of AZ31B magnesium alloy. *J. Alloys Compd.* **2016**, *680*, 544–552. [CrossRef]

14. Zhang, X.; Mao, B.; Siddaiah, A.; Menezes, P.L.; Liao, Y. Direct laser shock surface patterning of an AZ31B magnesium alloy: Microstructure evolution and friction performance. *J. Mater. Process. Technol.* **2020**, *275*, 116333. [CrossRef]

15. Kanel, G.I.; Garkushin, G.V.; Savinykh, A.S.; Razorenov, S.V.; De Resseguier, T.; Proud, W.G.; Tyutin, M.R. Shock response of magnesium single crystals at normal and elevated temperatures. *J. Appl. Phys.* **2014**, *116*, 143504. [CrossRef]

16. De Rességuier, T.; Hemery, S.; Lescoute, E.; Villechaise, P.; Kanel, G.I.; Razorenov, S.V. Spall fracture and twinning in laser shock-loaded single-crystal magnesium. *J. Appl. Phys.* **2017**, *121*, 165104. [CrossRef]

17. Matsuzuki, M.; Horibe, S. Analysis of fatigue damage process in magnesium alloy AZ31. *Mater. Sci. Eng.* **2009**, *504*, 169–174. [CrossRef]

18. Park, S.H.; Hong, S.G.; Yoon, J.; Lee, C.S. Influence of loading direction on the anisotropic fatigue properties of rolled magnesium alloy. *Int. J. Fatigue* **2016**, *87*, 210–215. [CrossRef]

19. Tucker, M.T.; Horstemeyer, M.F.; Gullett, P.M.; El Kadiri, H.; Whittington, W.R. Anisotropic effects on the strain rate dependence of a wrought magnesium alloy. *Scr. Mater.* **2009**, *60*, 182–185. [CrossRef]

20. Ulacia, I.; Dudamell, N.V.; Gálvez, F.; Yi, S.; Pérez-Prado, M.T.; Hurtado, I. Mechanical behavior and microstructural evolution of a Mg AZ31 sheet at dynamic strain rates. *Acta Mater.* **2010**, *58*, 2988–2998. [CrossRef]

21. Jäger, A.; Lukáč, P.; Gärtnerová, V.; Bohlen, J.; Kainer, K.U. Tensile properties of hot rolled AZ31 Mg alloy sheets at elevated temperatures. *J. Alloys Compd.* **2004**, *378*, 184–187. [CrossRef]

22. Chino, Y.; Kimura, K.; Mabuchi, M. Deformation characteristics at room temperature under biaxial tensile stress in textured AZ31 Mg alloy sheets. *Acta Mater.* **2009**, *57*, 1476–1485. [CrossRef]

23. Koh, Y.; Kim, D.; Seok, D.Y.; Bak, J.; Kim, S.W.; Lee, Y.S.; Chung, K. Characterization of mechanical property of magnesium AZ31 alloy sheets for warm temperature forming. *Int. J. Mech. Sci.* **2015**, *93*, 204–217. [CrossRef]

24. Kabirian, F.; Khan, A.S.; Gnäupel-Herlod, T. Visco-plastic modeling of mechanical responses and texture evolution in extruded AZ31 magnesium alloy for various loading conditions. *Int. J. Plast.* **2015**, *68*, 1–20. [CrossRef]

25. Ballard, P.; Fournier, J.; Fabbro, R.; Frelat, J. Residual stresses induced by laser-shocks. *J. Phys. IV* **1991**, *1*, C3-487–C3-494. [CrossRef]

26. Angulo, I.; Cordovilla, F.; García-Beltrán, A.; Smyth, N.S.; Langer, K.; Fitzpatrick, M.E.; Ocaña, J.L. The effect of material cyclic deformation properties on residual stress generation by laser shock processing. *Int. J. Mech. Sci.* **2019**, *156*, 370–381. [CrossRef]

27. Engebretsen, C.C.; Palazotto, A.; Langer, K. Strain Rate Dependent FEM of Laser Shock Induced Residual Stress. In *Challenges in Mechanics of Time-Dependent Materials, Volume 2, Proceedings of the 2018 Annual Conference on Experimental and Applied Mechanics, Greenville, SC, USA, 4–7 June 2018*; Springer International Publishing: New York, NY, USA, 2019; pp. 109–114.

28. Langer, K.; Olson, S.; Brockman, R.; Braisted, W.; Spradlin, T.; Fitzpatrick, M.E. High strain-rate material model validation for laser peening simulation. *J. Eng.* **2015**, *13*, 150–157. [CrossRef]

29. Hill, R. A theory of the yielding and plastic flow of anisotropic metals. *Proc. R. Soc. Lond. Ser. A Math. Phys. Sci.* **1948**, *193*, 281–297. [CrossRef]

30. Park, S.H.; Hong, S.G.; Bang, W.; Lee, C.S. Effect of anisotropy on the low-cycle fatigue behavior of rolled AZ31 magnesium alloy. *Mater. Sci. Eng.* **2010**, *527*, 417–423. [CrossRef]

31. Yang, F.; Duan, Q.Q.; Yang, Y.S.; Wu, S.D.; Li, S.X.; Zhang, Z.F. Fatigue properties of rolled magnesium alloy (AZ31) sheet: Influence of specimen orientation. *Int. J. Fatigue* **2011**, *33*, 672–682. [CrossRef]

32. Peyre, P.; Chaieb, I.; Braham, C. FEM calculation of residual stresses induced by laser shock processing in stainless steels. *Mater. Sci. Eng.* **2007**, *15*, 205. [CrossRef]

33. Voce, E. The relationship between stress and strain for homogeneous deformation. *J. Inst. Met.* **1948**, *74*, 537–562.

34. Amarchinta, H.; Granhi, R.; Clauer, A.; Langer, K.; Stargel, D. Simulation of residual stress induced by a laser peening process through inverse optimization of material models. *J. Mater. Process. Technol.* **2010**, *210*, 1997–2006. [CrossRef]

35. Bao, Y.; Wierzbicki, T. On fracture locus in the equivalent strain and stress triaxiality space. *Int. J. Mech. Sci.* **2004**, *46*, 81–98. [CrossRef]

36. Basu, S.; Dogan, E.; Kondori, B.; Karaman, I.; Benzerga, A.A. Towards designing anisotropy for ductility enhancement: A theory-driven investigation in Mg-alloys. *Acta Mater.* **2017**, *131*, 349–362. [CrossRef]

37. Fabbro, R.; Fournier, J.; Ballard, P.; Devaux, D.; Virmont, J. Physical study of laser-produced plasma in confined geometry. *J. Appl. Phys.* **1990**, *68*, 775–784. [CrossRef]

38. MacFarlane, J.J.; Golovkin, I.E.; Woodruff, P.R. Helios-cr–a 1-d radiation-magnetohydrodynamics code with inline atomic kinetics modeling. *J. Quant. Spectrosc. Radiat. Transf.* **2006**, *99*, 381–397. [CrossRef]

39. Morales, M.; Porro, J.A.; Blasco, M.; Molpeceres, C.; Ocana, J.L. Numerical simulation of plasma dynamics in laser shock processing experiments. *Appl. Surf. Sci.* **2009**, *255*, 5181–5185. [CrossRef]

40. Morales, M.; Porro, J.A.; García-Ballesteros, J.J.; Molpeceres, C.; Ocaña, J.L. Effect of plasma confinement on laser shock microforming of thin metal sheets. *Appl. Surf. Sci.* **2011**, *257*, 5408–5412. [CrossRef]

41. Wu, B.; Shin, Y.C. A self-closed thermal model for laser shock peening under the water confinement regime configuration and comparisons to experiments. *J. Appl. Phys.* **2005**, *97*, 113517. [CrossRef]

42. Wu, B.; Shin, Y.C. From incident laser pulse to residual stress: A complete and self-closed model for laser shock peening. *J. Manuf. Sci. Eng.* **2007**, *29*, 117–125. [CrossRef]

43. American Society for Testing and Materials. *Standard Test Method for Determining Residual Stresses by the Hole-Drilling Strain-Gage Method*; ASTM E837-13a; ASTM International: West Conshohocken, PA, USA, 2013. [CrossRef]

44. Ocaña, J.L.; Morales, M.; Porro, J.A.; Blasco, M.; Molpeceres, C.; Iordachescu, D.; Rubio-González, C. Induction of engineered residual stresses fields and associate surface properties modification by short pulse laser shock processing. In *Materials Science Forum, Proceedings of the International Conference on Processing & Manufacturing of Advanced Materials, Berlin, Germany, 25–29 August 2009*; Trans Tech Publications: Zurich, Switzerland, 2010; Volume 638, pp. 2446–2451.

45. Ocaña, J.L.; Correa, C.; García-Beltrán, A.; Porro, J.A.; Díaz, M.; Ruiz-de-Lara, L.; Peral, D. Laser Shock Processing of thin Al2024-T351 plates for induction of through-thickness compressive residual stresses fields. *J. Mater. Process. Technol.* **2015**, *223*, 8–15. [CrossRef]

Article

Effect of Laser Peening Process Parameters and Sequences on Residual Stress Profiles

Zina Kallien [1,*], Sören Keller [1], Volker Ventzke [1], Nikolai Kashaev [1] and Benjamin Klusemann [1,2]

[1] Helmholtz-Zentrum Geesthacht, Institute of Materials Research, Materials Mechanics, Max-Planck-Straße 1, 21502 Geesthacht, Germany; soeren.keller@hzg.de (S.K.); volker.ventzke@hzg.de (V.V.); nikolai.kashaev@hzg.de (N.K.); benjamin.klusemann@hzg.de (B.K.)

[2] Leuphana University of Lüneburg, Institute of Product and Process Innovation, Universitätsallee 1, 21335 Lüneburg, Germany

* Correspondence: zina.kallien@hzg.de

Received: 9 April 2019; Accepted: 30 May 2019; Published: 4 June 2019

Abstract: Laser Peening (LP) is a surface modification technology that can induce high residual stresses in a metallic material. The relation between LP process parameters, in particular laser sequences, as well as pulse parameters and the resulting residual stress state was investigated in this study. The residual stress measurements, performed with the hole drilling technique, showed a non-equibiaxial stress profile in laser peened AA2024-T3 samples with a clad layer for certain parameter combinations. Shot overlap and applied energy density were found to be crucial parameters for the characteristic of the observed non-equibiaxial residual stress profile. Furthermore, the investigation showed the importance of the advancing direction, as the advancing direction influences the direction of the higher compressive residual stress component. The direction of higher residual stresses was parallel or orthogonal to the rolling direction of the material. The effect was correlated to the microstructural observation obtained via electron backscattered diffraction. Additionally, for peening with two sequences of different advancing directions, the study showed that the order of applied advancing directions was important for the non-equibiaxiality of the resulting residual stress profile.

Keywords: laser peening; laser shock peening; residual stresses; shot pattern; energy density; overlap; hole drilling; AA 2024; cladded aluminum

---

1. Introduction

The improvement of fatigue life of lightweight structures plays an important role in the construction of aircrafts. In order to fulfill the requirement of long-living lightweight structures, advanced materials with innovative processing technologies have to be utilized. Aluminum alloys like AA2024-T3 are frequently used, for example, in commercial aircrafts [1]. AA2024 is characterized by a high strength and a comparatively low density. Despite the advanced development of modern alloys, corrosion and fatigue remain factors limiting aircrafts' working life [2]. Modification technologies like laser peening (LP), also named laser shock peening (LSP), treat the surface of the material locally, leading to residual stresses in the material which can improve corrosion and fatigue resistance [3,4]. The LP process is contact-free, highly controllable and achieves deeper residual stress profiles in comparison to shot peening [5].

To generate tailored residual stress distributions, it is necessary to gain knowledge of the residual stress influencing LP process parameters. The result of the LP process is influenced by laser parameters like the energy input and the size of the laser focus [6,7] as well as the geometry of the laser focus [8]. The achievable maximum residual stress values are limited by the material's yield strength. Multiple investigations [9–11] have observed increased residual stresses for higher energy densities.

Multiple shots on one area lead to higher and deeper residual stresses [5,9,12]. Furthermore, as shown by Toparli and Fitzpatrick [13], a shot overlap also leads to higher stresses. The overlap rate may influence the anisotropy of the residual stress profile [13–15]. To sum up, the energy density, directly defined by the focus size and chosen energy input, the number of shots, and the applied overlap rate determine how much energy an area experiences. This is crucial for the residual stress distribution.

The shot pattern describes the strategy for the positioning of the laser pulses on the peening area. Saliminarizi et al. [16] compared a row-wise peening pattern to a spiral peening pattern with regard to the surface properties of the material. As a result, surface roughness was found to be dependent on the peening pattern and overlap rate. Xu et al. [15] have shown that the choice of scanning path has an influence on residual stresses in a 316L stainless steel blade. Correa et al. [14,17] showed that the choice of the advancing direction influences the residual stress profile and the improvement of fatigue attributes. Furthermore, the same research group [18] proposed that randomly applied shots provide a more equibiaxial stress profile than the row-wise application of shots.

The present study relates the applied shot pattern to the residual stresses developing in the material. The systematic investigation showed the influence of the advancing direction as well as the applied overlap on the residual stress profile. Moreover, different shot patterns are combined by peening two sequences with different advancing directions, and possible mechanisms for the resulting residual stress profiles are discussed. The experimental setup for this work as well as material used and measurement technology are presented in the following section. The influence of the clad layer and the use of aluminum foil during peening on the resulting residual stresses is also shown in this study. Moreover, the importance of the choice of the advancing direction for the residual stress state as well as the effect of shot overlap is presented for one peening sequence. Additionally, the residual stresses for peening with two sequences with different shot patterns for each sequence are pointed out. For unpeened materials as well as for materials treated by particular shot patterns, local orientation changes are investigated using electron backscattered diffraction (EBSD). Furthermore, the effect of different energy densities on the material's surface is investigated.

2. Materials and Methods

2.1. Laser Peening

During the LP process, the pulsed laser is focused on the material's surface. The energy input of a laser pulse leads to vaporization of the material at the surface. Thermal expansion of the developing plasma induces a shock wave propagating into the material. Local plastic deformation occurs and causes residual stresses. A transparent overlay increases the pressure and the duration of the plasma and, thus, the efficiency of the process [3]. In the present work, water is used as a transparent overlay. An Nd:YAG laser (wavelength 1064 nm) with the full width at half maximum (FWHM) of 20 ns and a frequency of 10 Hz was employed. A squared laser focus of 1 mm × 1 mm with 1.5 J of energy is used to study the effect of the laser sequence. Additionally, the energy input is changed to 1 J and 3 J for investigating the influence of the energy density. For each specimen, two peening areas which had no interacting effects on the residual stress profile are used. Possible interacting effects were tested for quadratic specimens (40 mm × 40 mm) with only one peening area, which did not show variation in the residual stress profile in comparison to a specimen (40 mm × 80 mm) peened with the same process parameters in two different areas.

For the experimental investigation of the process, AA2024-T3clad specimens (40 mm × 80 mm) with thicknesses of 4.8 mm are investigated (see Figure 1a,b). The specimens feature a clad layer of approximately 0.15 mm thickness on both sides. For some specimens, the aluminum clad layer on one side is removed by milling 0.2 mm, leading to a specimen thickness of 4.6 mm. This allows the investigation of the effect of the clad layer on residual stresses. The milled specimens are named AA2024-T3 in the following. Influences on residual stresses due to the reduced thickness of the milled specimens are assumed negligible. The dimension of the peening areas on the specimens are

16 mm × 17 mm for the investigation of the laser sequence. For particular experiments, an adhesive Al foil of approximately 0.05 mm thickness is applied to the untreated specimen surface with special attention to avoid any air inclusions between the foil and specimen. After peening, the adhesive foil is removed to allow residual stress measurements of the specimen only.

An overview of the different applied peening strategies in this study is given in Figure 1c–l. The two basic peening strategies are characterized by the choice of the advancing direction in relation to the rolling direction (see Figure 1c,h). For the investigation of the influence of shot overlap, the overlap is chosen to be 50 percent only in one direction. Consequently, an overlap can be applied parallel or orthogonal to the chosen advancing direction. The generation of the overlap is schematically shown in Figure 1d,e,i,j. The two basic patterns are also peened with a row-wise change in direction (see Figure 1f,k). The applied strategies for peening with two sequences are presented in Figure 1g,l.

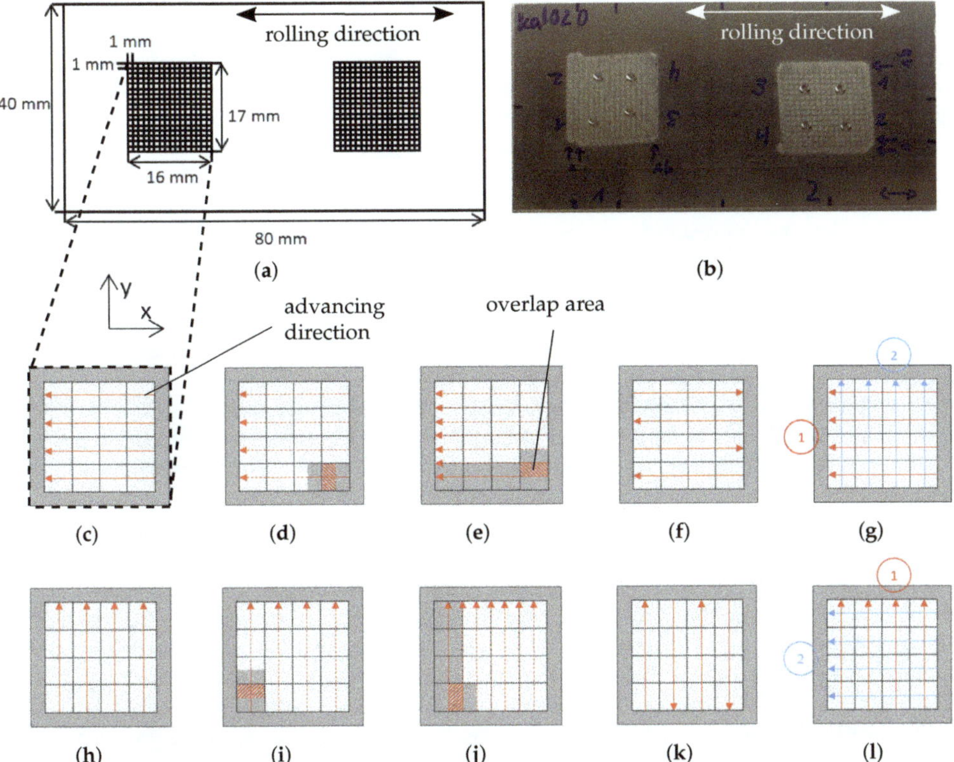

Figure 1. (a) An illustration of the specimen dimensions with a schematic peening area; (b) a peened specimen after residual stress measurements; (c–l) the schematic shot patterns used in this study: (c–f) one peening sequence with an advancing direction parallel to the rolling direction: (c); (d) overlap in the advancing direction; (e) overlap orthogonal to the advancing direction; (f) a row-wise change in direction; (g) the application of two orthogonal shot patterns, first parallel and second orthogonal to the rolling direction; (h–k) one peening sequence with an advancing direction orthogonal to the rolling direction: (h); (i) overlap in the advancing direction; (j) overlap orthogonal to the advancing direction; (k) a row-wise change in direction; (l) the application of two orthogonal shot pattern, first orthogonal and second parallel to the rolling direction.

2.2. Residual Stress Measurement via Incremental Hole Drilling Technique

The PRISM system by Stresstech (Rennerod, Germany) was used for the residual stress measurements. The basis of the system is the incremental hole drilling technique and the electronic speckle pattern interferometry (ESPI). The procedure can be divided into three steps [19]:

1. Drilling the hole increment;
2. Recording surface deformation using ESPI; and
3. Calculation of the residual stresses based on the integral method.

All three steps were performed for each increment of the hole, leading to a residual stress profile over the material depth. Coherent light illuminates the surface around the hole, and the light reflection results in a shift of each pixel depending on the roughness of the surface. The superposition of reflected light and reference beam lead to the speckle pattern. The phase shift after the reflection defines the pixel intensity. The comparison of intensity before and after each drilling increment is an indicator for the displacement [20]. Accordingly, thousands of pixels are considered for determining the displacement using the ESPI technique. Subsequently, the integral method is applied. The hole drilling technique assumes constant residual stresses parallel to the material's surface as well as purely elastic deformations during the drilling. Restrictions and necessary assumptions for the incremental hole drilling method using ESPI are explained in detail [21]. In this regard, Chupakhin et al. [22] presented a correction method for equibiaxial residual stress profiles based on an artificial neural network to account for possible plasticity effects.

For all performed measurements, holes of 2 mm diameter and 1 mm depth were drilled incrementally. One hole was drilled in 19 increments. Increments close to the surface were smaller than below, according to the expected residual stress gradient. The position of the holes in the peened area is exemplarily shown in Figure 1b. For most investigated combinations of shot patterns, eight measurements were performed to have sufficient statistics. Four measurements are performed for the peening experiments with Al foil, with overlap, with a 1 J pulse energy as well as the peening experiment where the advancing direction is orthogonal to the rolling direction with a row-wise change in direction. Therefore, the average value as well as the minimum and maximum measured values are shown for these results.

2.3. Determining Local Orientation via EBSD

The crystal orientations of rolled AA2024-T3clad within unpeened materials as well as LP treated regions close to the sheet surface are investigated using a scanning electron microscope (SEM) (JSM-6490LV with EBSD by EDAX, Jeol Ltd., Tokyo, Japan) combined with EBSD in order to clarify the question of how the crystal orientations of AA2024-T3clad have been changed by the LP treatment. The EBSD analysis is performed for an unpeened material as well as for two areas treated with one LP sequence at 1.5 J, where one was peened with the advancing direction parallel and the other one orthogonal to the rolling direction (RD). The specimens are prepared by means of multi-stage grinding and subsequent final polishing whereby the prepared plane was defined by the rolling direction and the direction of material thickness. The LP-treated samples are cut in the center of the peened areas. The specimens are analyzed at 30 kV, a beam current of 0.25 nA, an emission current of 78 µA, a magnification of 300×, a working distance of 14 mm, a step size of 0.70 µm, and a sample tilt of 70°. The area directly below the clad layer is analyzed in order to generate the inverse pole figures. Consequently, the clad layer is not included in the analysis with the inverse pole figures. The calculation of [001] and [010] inverse pole figures to determine the crystal directions was conducted on the basis of the generalized spherical harmonic expansion (GSHE) method and an assumed triclinic sample symmetry. [001] corresponds to the transverse direction (TD) and [010] to the thickness direction, the normal direction of the rolled AA2024-T3clad sheet.

3. Results and Discussion

3.1. Residual Stresses in Base Material

The residual stress profiles for the unpeened AA2024-T3clad and AA2024-T3 (milled clad layer) specimens are given in Figure 2. Near the surface, both specimens show a relatively high standard deviation, whereas the milled specimen shows an even larger variation which might be related to the milling process. With increasing depth, the deviation decreases strongly. For the AA2024-T3clad specimen, compressive residual stresses are measured within the clad layer. Deeper residual stresses are tensile in the order of 20 MPa with less than a 10 MPa difference between both in-plane stress components. The specimen AA2024-T3 without a clad layer shows slightly higher compressive residual stresses directly underneath the material surface, whereas stresses within the material are slightly lower than for AA2024-T3clad. The minimal differences in residual stresses in both unpeened specimens can be related to the milling process.

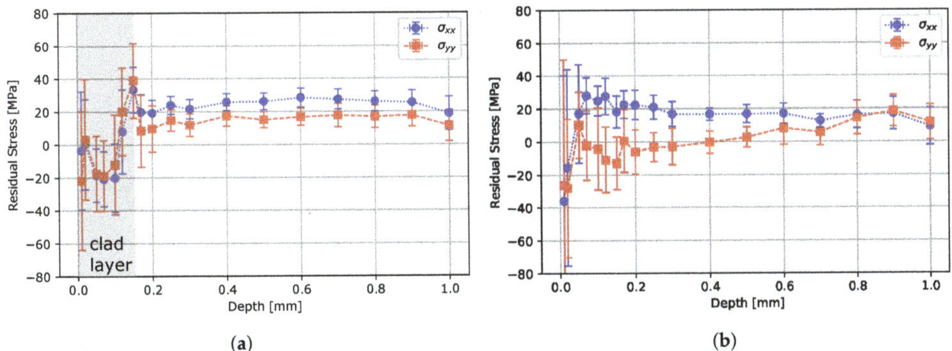

Figure 2. The residual stress profile of unpeened materials: (**a**) AA2024-T3clad, the gray area indicates the clad layer; (**b**) AA2024-T3 (milled).

3.2. Influence of Clad Layer and Aluminum Foil

To investigate the influence of the clad layer, the specimens were peened parallel to the rolling direction. The resulting residual stresses show a non-equibiaxial profile (see Figure 3). The higher compressive residual stresses, σ_{yy}, are orthogonal to the rolling direction. In comparison to AA2024-T3, AA2024-T3clad shows a different residual stress state near the surface. The clad layer led to an offset in the maximum residual stresses to a deeper depth, which corresponds to the clad layer thickness. The residual stresses within the clad layer are relatively low. Overall, the compressive residual stresses are slightly increased for AA2024-T3clad.

In industrial practice, an Al foil is often used. This leads to a better surface quality because the foil acts as thermal protection. Consequently, the energy input does not melt the surface of the material [23]. Slightly higher compressive residual stresses were measured for peening with Al foil of AA2024-T3 as well as AA2024-T3clad (see Figure 3). Especially for the measurement points in a depth larger than 0.4 mm, the differences are slightly more pronounced. Xu et al. [24] observed that a coating can increase the induced residual stresses, e.g., because of different absorption attributes of the laser energy. However, due to the fact that both the clad layer and foil consist of pure aluminum, no differences in absorption are expected. In this study, the increase detected is marginal compared to the absolute value of the residual stresses so that no significant influence of the Al foil on the residual stresses could be highlighted. Therefore, the effect is not discussed further.

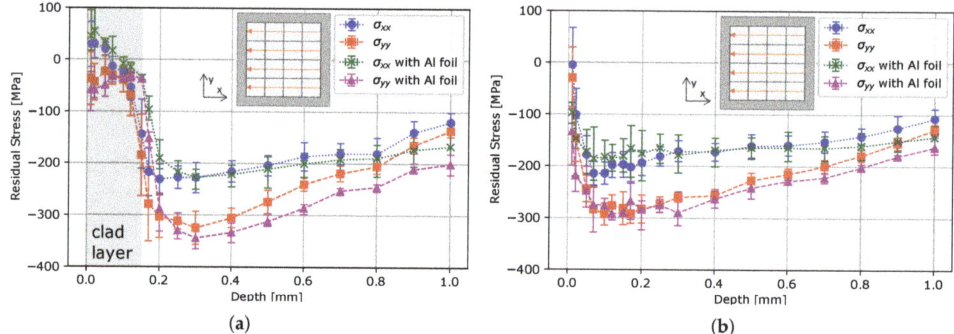

Figure 3. Laser Peening (LP) with and without Al foil: The advancing direction is chosen parallel to the rolling direction of the specimens: (**a**) AA2024-T3clad, the gray area indicates the clad layer; (**b**) AA2024-T3 (milled).

3.3. Influence of Laser Pulse Pattern

The measurement results for different patterns and overlap, as presented schematically in Figure 1c–l, are compared to investigate the influence of the chosen laser pulse pattern on the residual stresses. Firstly, the effect of the chosen advancing direction, including row-wise changes, is discussed. This is followed by a discussion on the effect of overlap and, finally, peening with two sequences of orthogonal advancing directions.

3.3.1. Influence of Advancing Direction

As already observed from the previous results, the residual stresses are non-equibiaxial for a single peening sequence where the stresses transversing to the advancing direction are larger than the parallel ones. The difference between both in-plane stress components is more pronounced for peening parallel to the rolling direction as compared to peening orthogonal to the rolling direction (see Figure 4a,b). The advancing direction determined which in-plane stress component dominates the residual stress profile. However, the absolute values of the maximum residual stresses as well as the values in a material depth of 1 mm are not significantly different for the two applied advancing directions. Especially for nonsymmetrical peening areas and structures, the choice of advancing direction could have an even more significant effect in this regard, which was seen in previous studies [14,15]. As illustrated in Figure 5, stress profiles for bidirectional peening (see Figure 1f,k) are not significantly different as compared to unidirectional peening (see Figure 1c,h). Therefore, a row-wise change in the advancing direction does not influence the resulting residual stress profile.

In order to correlate the observed residual stress profiles for different advancing directions to possible local orientation effects, EBSD measurements were performed. Unpeened AA2024-T3clad (see Figure 6) exhibits an orientation band between <1 0 1>//[001] and <1 1 2>//[001] whereby the crystal direction <1 0 1>//[001] is characterized by the highest axial intensity (see Figure 7a). In the normal direction, parallel to the material thickness, the crystal direction <0 0 1>//[010] is pronounced, followed by <1 0 1>//[001]. The incident laser beam mainly interacted with the (1 0 0) crystal planes at the beginning of the LP process.

The LP sequence parallel to the rolling direction leads to the formation of crystal directions such as <1 1 1>//[001], <1 0 2>//[001], and <0 0 1>//[001] (see Figure 7b). Furthermore, the crystal direction <0 0 1>//[010] parallel to the sheet normal direction shows a weakening whereas the axial intensity of <1 0 1>//[010] increased after the LP treatment. For instance, the weakening of <0 0 1>//[010] and the gain of <1 0 1>//[010] compared with orientations of the base material means a rotation around the [010] axis within 45°. The appearance of <1 1 2>//[010] having low axial intensity is the result of a split-up.

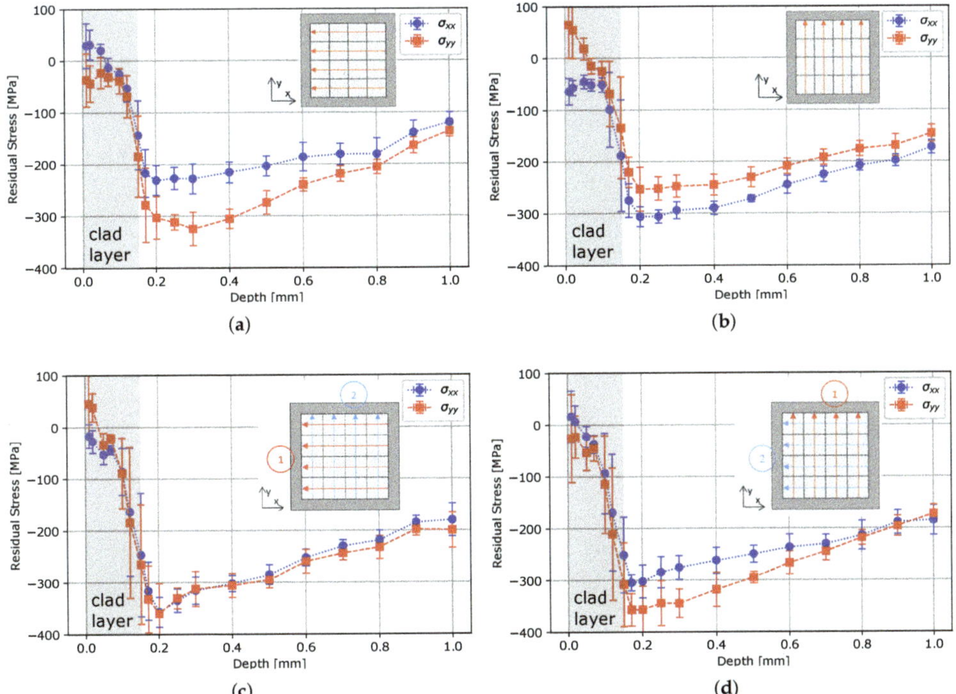

Figure 4. AA2024-T3clad: (**a**) one peening sequence with the LP advancing direction parallel to the rolling direction of the specimens; (**b**) one peening sequence with the LP advancing direction orthogonal to the rolling direction of the specimens; (**c**) two peening sequences with the first peening sequence advancing direction chosen parallel to the rolling direction and second peening sequence advancing direction chosen orthogonal to the rolling direction; (**d**) two peening sequences with the first peening sequence advancing direction chosen orthogonal to the rolling direction and the second peening sequence advancing direction chosen parallel to the rolling direction. The gray area indicates the clad layer.

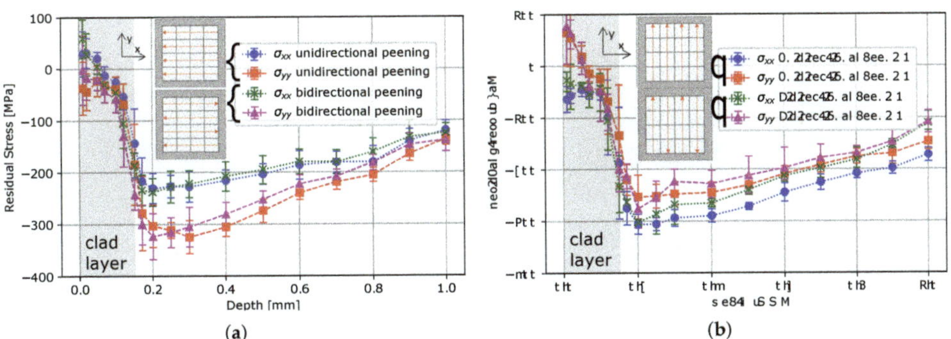

Figure 5. LP with unidirectional and bidirectional advancing directions for AA2024-T3clad: (**a**) advancing direction parallel to the rolling direction; (**b**) advancing direction orthogonal to the rolling direction. The gray area indicates the clad layer.

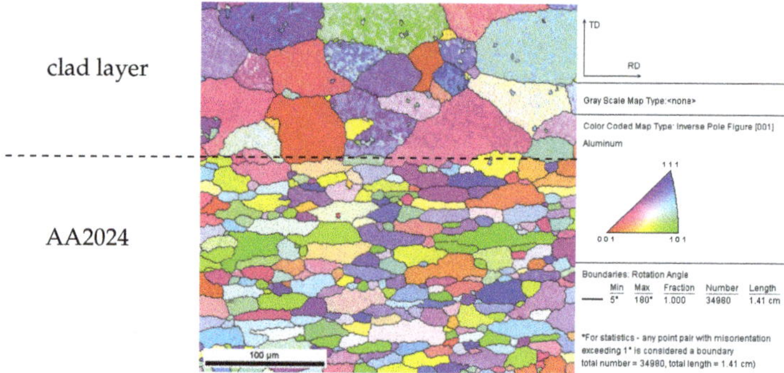

Figure 6. Electron backscattered diffraction (EBSD) micrographs of unpeened AA2024-T3clad: the TD-direction is defined as the direction of the material's thickness ([010]), and RD-direction is defined as the rolling direction ([100]).

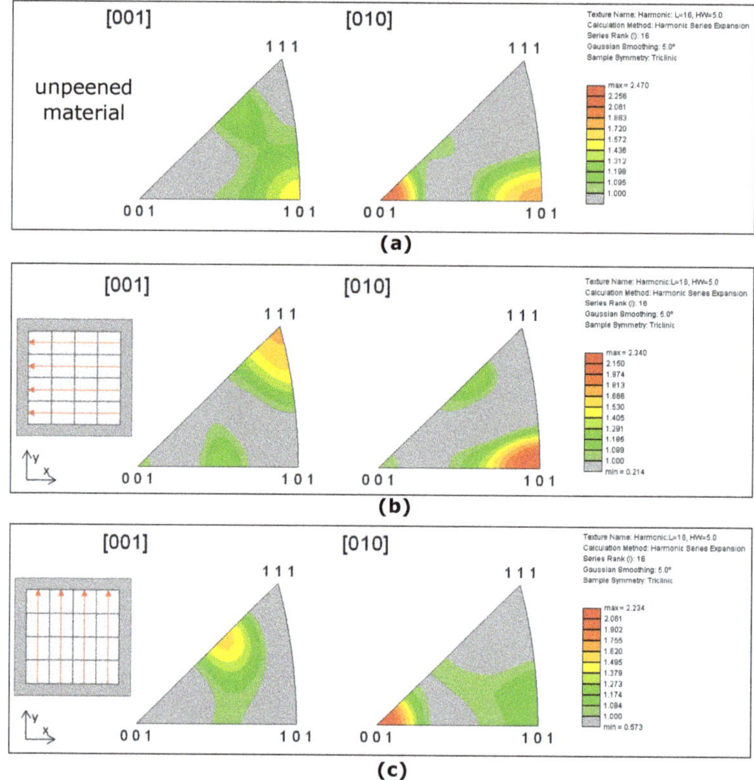

Figure 7. Inverse pole figures of unpeened and peened AA2024-T3clad specimens: The [001] direction is orthogonal to the rolling direction; the [010] direction corresponds to the specimens' thickness direction: (**a**) unpeened material; one peening sequence was applied where the advancing direction was parallel to the rolling direction (**b**) or orthogonal to the rolling direction (**c**).

The LP sequence orthogonal to the rolling direction of the AA2024-T3clad sheet results in the formation of an orientation band between <1 1 2>//[001] and <1 0 2>//[001] (Figure 7c). This means a shift of the <1 0 1>//[001] crystal direction to <1 0 2>//[001] as well as an increase of axial intensity of the <1 1 2>//[001] crystal direction within the LP treated region. The <0 0 1>//[010] crystal direction has been retained, whereas the <1 0 1>//[010] showed a weakening compared to the base material and additionally a connection to <1 1 3>//[010].

The comparison to the base material shows that a significant change has taken place due to local plastic deformations induced by the propagating shock waves induced by LP. Kashaev et al. [25] reached a similar conclusion. However, different changes in local orientation were observed between the two applied advancing directions. The material seems to react differently depending on the choice of advancing direction which might be an explanation for the differences in the residual stress profiles. Especially, the weakening of initial axial intensities or pole densities seem to play a significant role. The results indicate that LP induced texture changes can be correlated to the development of misorientations within the microstructure as seen by the angle spreading of axial intensity distributions.

3.3.2. Influence of Overlap

The residual stress measurement results of the experiments with overlap are presented in Figure 8. The overlap rate was chosen to be 50 percent in one direction, as schematically shown in Figure 1d,e,i,j. An overlap of 50 percent, no matter if the overlap was parallel or orthogonal to the chosen advancing direction, led to a stronger non-equibiaxial residual stress state. The compressive stresses measured parallel to the advancing direction as well as the stress values at 1 mm depth were not influenced. The residual stresses orthogonal to the advancing direction were significantly higher for a peening with overlap than without. However, the reason for this observation is that the overall energy input the area receives is higher for peening with overlap, leading to increased residual stresses. The observed phenomenon agrees with the findings by Toparli and Fitzpatrick [13]. In this study, the effect of the overlap is stronger for a peening parallel to the rolling direction of the specimens (Figure 8b) than for peening orthogonal to the rolling direction (Figure 8c). The phenomenon that the non-equibiaxiality is more pronounced for peening parallel to the rolling direction was observed for all performed peening experiments. An overlap seems to intensify this effect. However, it should be noted that the measured residual stress values exceed 80 percent of the material's yield strength. Consequently, the results can only be interpreted qualitatively; otherwise correction methods, such as those developed by Chupakhin et al. [22] for equibiaxial residual stress profiles, need to be adapted for the special application case.

3.3.3. Influence of Order of Different Shot Patterns

In this study, peening with two sequences means that the peening area is peened twice by advancing directions orthogonal to each other as schematically illustrated in Figure 1g,l. Figure 4c shows the residual stress profile for a specimen which was peened via an advancing direction parallel to the rolling direction and afterwards via an advancing direction orthogonal to the rolling direction. An equibiaxial stress profile was observed. In contrast, for a peening strategy where, first, the advancing direction is orthogonal and then the second one is parallel to the rolling direction, the stress state was non-equibiaxial (see Figure 4d). The stresses orthogonal to the rolling direction were higher compared to the ones parallel to the rolling direction. This is in contrast to the residual stress distribution only after the first sequence (see Figure 4a,b).

In accordance with the expectations, specimens peened with two sequences showed higher maximum residual stresses in comparison to one sequence peening. Comparing the stress profiles after two peening sequences (see Figure 4c,d) with the ones after one peening sequence (see Figure 4a,b), it was found that the compressive residual stresses orthogonal to the advancing direction of the second sequence were more significantly affected by the application of the second sequence than the parallel stresses. Therefore, the application of a second peening sequence influenced the stresses orthogonal to the chosen advancing direction, whereas the parallel stresses were hardly changed.

Overall, the measured residual stress profiles show that the order in which the advancing directions are applied on the surface is very important. The choice of peening pattern can control the stress value, the direction in which the highest stresses are induced, and by that, the difference between both in-plane stress components.

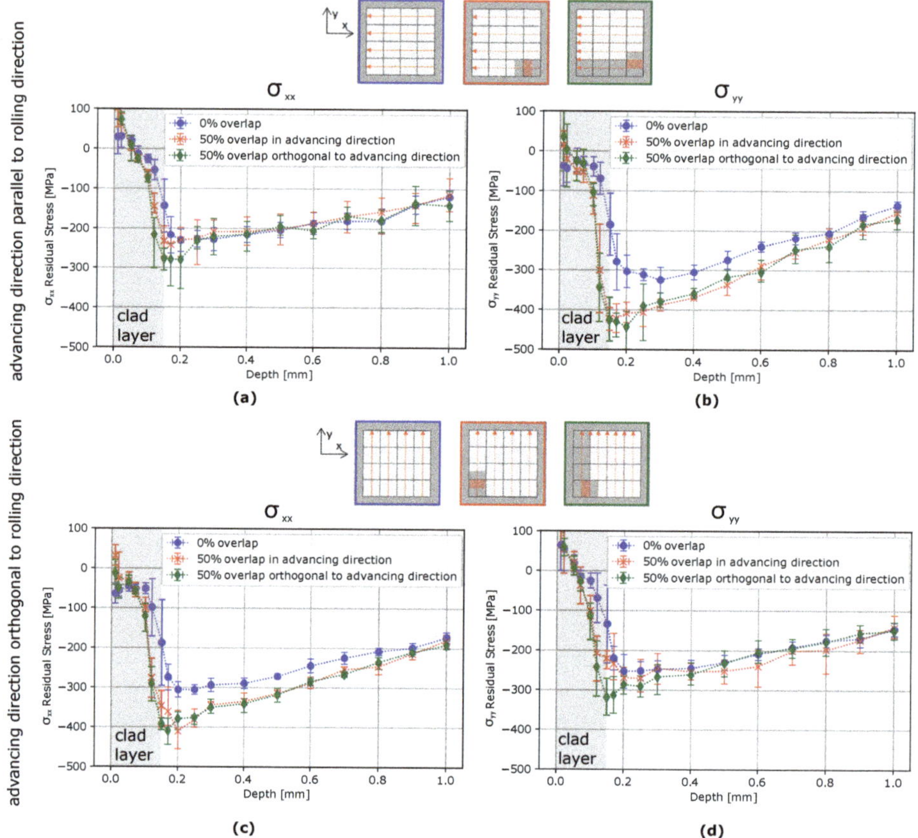

Figure 8. Residual stress profiles of AA2024-T3clad after LP for different advancing directions by varying the overlap, parallel or orthogonal to the advancing direction of the peening: No overlap, as well as 50 percent overlap are considered. Advancing direction either parallel ((**a**) σ_{xx}; (**b**) σ_{yy}) or orthogonal ((**c**) σ_{xx}; (**d**) σ_{yy}) to rolling direction. The gray area indicates the clad layer.

3.4. Influence of Laser Pulse Energy

In order to study the influence of the laser energy, first, single shots were investigated for AA2024-T3 and AA2024-T3clad specimens for laser pulse energies of 1 J, 1.5 J, and 3 J via light optical microscope (Leica DMI 5000M, Leica Microsystems GmbH, Wetzlar, Germany). The apparent influenced surface area by one single shot is significantly larger than the applied focus size of 1 mm × 1 mm, as illustrated by micrographs from the light microscope in Figure 9. For one shot with 1 J, the apparent influenced area is approximately 3.07 mm^2 for AA2024-T3 and approximately 3.70 mm^2 for AA2024-T3clad, indicating that the effect is slightly more significant for the cladded specimen. The micrograph obtained from scanning electron microscope (SEM) of the peened area for AA2024-T3clad (Figure 10) indicates that the affected surface area most probably represents molten surface material. The higher the applied laser energy, the larger is the apparent influenced area on the specimens' surface by one single shot (see Figure 9). The shape of the influenced area is no longer a square. With increasing laser pulse energy, the influenced area of the single shot becomes more circular.

Assuming that the influenced area correlates with the pressure affected area, this leads to overlapping effects even if the laser spots are positioned without overlap. Consequently, for higher laser energies, the overlap effect becomes more significant. This might be one reason for the increasing difference between both in-plane stress components (see Figure 11). Overall, higher laser pulse energies lead to higher compressive residual stresses at the depth of one millimeter, whereas the maximum value of compressive stresses does not seem to be affected significantly. In previous studies [8,10,11], it was observed that an increased energy density leads to deeper residual stresses which agrees with the results obtained by varying the energy input from 1 J to 3 J. However, the increase to 3 J did not lead to increased maximum residual stresses for the performed experiments. One aspect to consider in this regard is that the residual stresses induced by an energy input of 1 J are already close to the material's yield strength, which is approximately 345 MPa for AA2024-T3 [26].

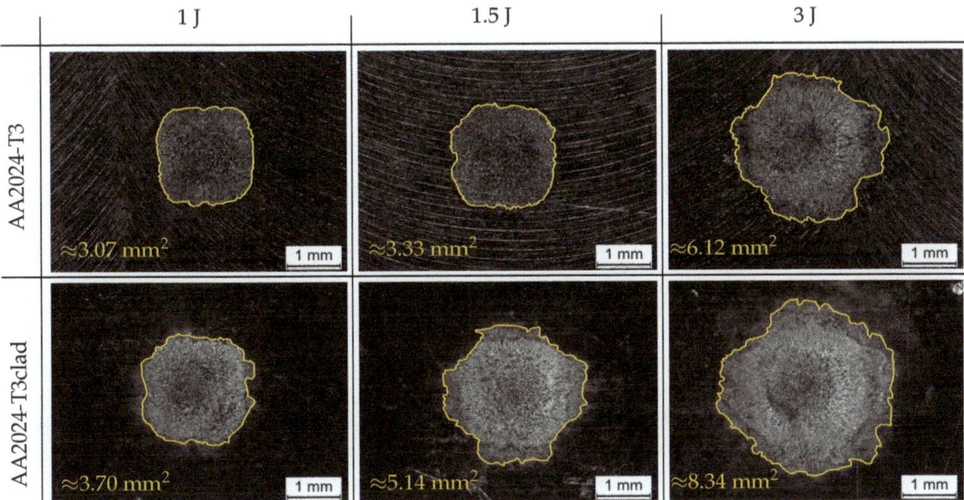

Figure 9. Light optical micrographs for AA2024-T3clad and AA2024-T3 for single LP shots with a 1 mm × 1 mm squared laser focus with three laser energies: 1 J, 1.5 J, and 3 J. The size of the affected area on the specimens' surface is indicated.

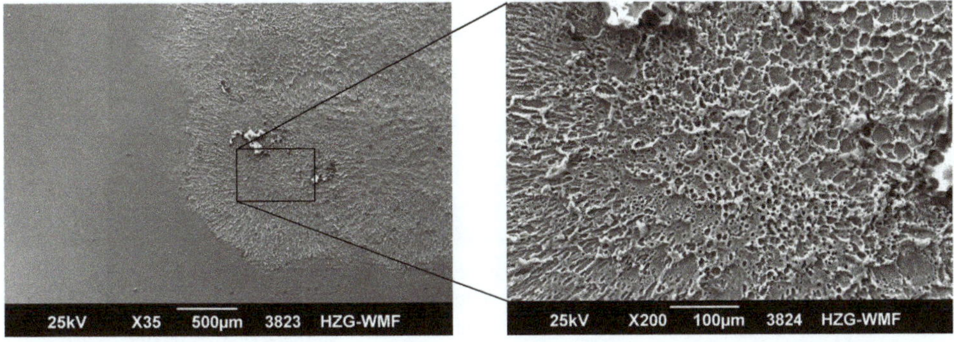

Figure 10. SEM micrographs of an AA2024-T3clad peened area applying two sequences by using a laser focus size of 1 mm × 1 mm and a laser pulse energy of 1.5 J. For the first sequence, the advancing direction was chosen to be orthogonal and, in the second sequence, parallel to the rolling direction (see Figure 11). The surface area indicates molten surface material for the peened area.

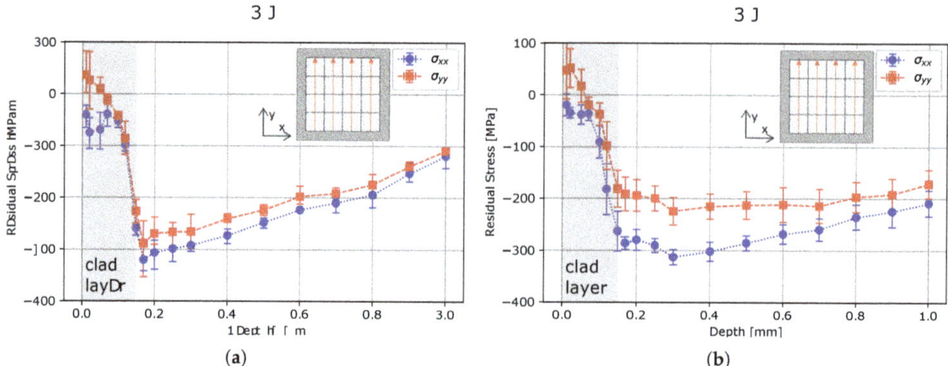

Figure 11. Residual stress profiles for AA2024-T3clad specimens with the advancing direction chosen orthogonal to the rolling direction: (**a**) 1 J and (**b**) 3 J. The gray area indicates the clad layer.

Experiments with increased energy input and experiments with overlap both showed the similar phenomenon of an increased difference between both in-plane stress components. The experiments with adjusted shot overlap of 50 percent qualitatively showed that the overlap rate is related to the non-equibiaxial stress profile, which agrees with the findings by Toparli and Fitzpatrick [13]. Therefore, a potential increase of the surface pressure-affected area related to the laser power density may influence the difference between both in-plane stress components. Increased laser pulse energy leads to a larger affected area, which might correlate with a different spatial pressure distribution. However, this effect is accompanied by an increased maximum pressure due to the increased pulse energy. These two effects superimpose for LP with increased pulse energies.

Besides the difference between both in-plane stress components, measurements for a higher laser power density show an increase in stress value at a 1 mm depth. This cannot be observed for the experiments with overlap in this study. We assume that the shape of the apparent influenced area is more circular at high energy densities and that the pressure distribution might not be uniform all over the affected area. Therefore, an increase in energy density could only be interpreted as a weak form of peening with overlap. In addition, as the experiments show a possible dependency of the difference between the stress components depending on the rolling direction, a possible anisotropic material behavior might be a mechanism to consider in explaining the difference between both in-plane stress components. Further investigation is needed for a substantiated interpretation of peening with high energy densities as a type of peening with overlap.

In summary, the overall energy input an area receives depends on pulse energy, overlap and number of sequences. Each aspect is crucial for the resulting residual stress distribution.

4. Conclusions

Residual stresses in the AA2024-T3clad and AA2024-T3 specimens for different peening strategies were investigated in this study. The effects of the clad layer, Al foil, choice of advancing direction, overlap as well as peening with two sequences on residual stress profiles were studied. The residual stresses were measured using the incremental hole-drilling technique, and for two shot patterns EBSD measurements were performed in order to study the local orientation changes exemplarily. Furthermore, the influence of one single shot with different energy impacts on the affected surface area was investigated. The conclusions of the study can be summarized as follows:

- The clad layer influenced the stresses only within the clad layer and near the interface.
- In this study, the use of an aluminum foil had no significant impact on the residual stresses. However, the effect of the LP process on the material surface is less severe for peening with Al foil.

- Peening with one sequence led to a non-equibiaxial stress profile. The higher stresses were measured orthogonal to the chosen advancing direction. The residual stress profiles for an unidirectional peening strategy were not significantly different compared to bidirectional peening.
- The EBSD analysis showed that the crystal orientation is different after the LP treatment with one peening sequence in comparison to the unpeened material. Depending on the choice of advancing direction, different crystal orientations were intensified. This finding indicates that the material reaction is different depending on the advancing direction. This observation possibly correlates with different residual stress profiles.
- A shot overlap of 50 percent in one direction increased the residual stresses orthogonal to the advancing direction, leading to a more pronounced non-equibiaxial residual stress profile.
- Peening with two sequences with different advancing directions chosen for each sequence can lead to an equibiaxial residual stress profile.
- The investigation of single shots showed that the apparent influenced area on the surface of the specimens is significantly larger than the chosen focus size. Increasing laser energy led to an increase in the size of the affected surface area, which is more circular at high energy densities. If the energy input visually affects a larger area than the chosen focus size, a pressure overlap might occur and needs to be considered.

Author Contributions: Conceptualization, S.K. and B.K.; methodology, S.K.; investigation, Z.K.; EBSD investigation, V.V.; resources, N.K.; writing—original draft preparation, Z.K.; writing—review and editing, Z.K., S.K., V.V., N.K. and B.K.; visualization, Z.K.; supervision, B.K.

Funding: This research received no external funding

Acknowledgments: The authors would like to thank R. Dinse and F. Dorn for the preparation of the specimens.

Conflicts of Interest: The authors declare no conflict of interest.

References

1. Dursun, T.; Soutis, C. Recent developments in advanced aircraft aluminium alloys. *Mater. Des.* **2014**, *56*, 862–871. [CrossRef]
2. Reid, L. *Sustaining an Aging Aircraft Fleet With Practical Life Enhancement Methods*; Fatigue Technology: Tukwila, WA, USA, 2003.
3. Ding, K.; Ye, L. (Eds.) *Laser Shock Peening: Performance and Process Simulations*; Woodhead Publishing in Materials; Woodhead: Cambridge, UK, 2006.
4. Keller, S.; Horstmann, M.; Kashaev, N.; Klusemann, B. Experimentally validated multi-step simulation strategy to predict the fatigue crack propagation rate in residual stress fields after laser shock peening. *Int. J. Fatigue* **2019**, *124*, 265–276. [CrossRef]
5. Clauer, A.H. Laser Shock Peening for fatigue resistance. In Proceedings of the Surface Performance of Titanium, Cincinnati, OH, USA, 7–9 October 1996; pp. 217–230.
6. Warren, A.; Guo, Y.; Chen, S. Massive parallel laser shock peening: Simulation, analysis, and validation. *Int. J. Fatigue* **2008**, *30*, 188–197. [CrossRef]
7. Sticchi, M.; Staron, P.; Sano, Y.; Meixer, M.; Klaus, M.; Rebelo-Kornmeier, J.; Huber, N.; Kashaev, N. A parametric study of laser spot size and coverage on the laser shock peening induced residual stress in thin aluminium samples. *J. Eng.* **2015**, *2015*, 97–105. [CrossRef]
8. Hu, Y.; Gong, C.; Yao, Z.; Hu, J. Investigation on the non-homogeneity of residual stress field induced by laser shock peening. *Surf. Coat. Technol.* **2009**, *203*, 3503–3508. [CrossRef]
9. Montross, C. Laser shock processing and its effects on microstructure and properties of metal alloys: A review. *Int. J. Fatigue* **2002**, *24*, 1021–1036. [CrossRef]
10. Dorman, M.; Toparli, M.B.; Smyth, N.; Cini, A.; Fitzpatrick, M.E.; Irving, P.E. Effect of laser shock peening on residual stress and fatigue life of clad 2024 aluminium sheet containing scribe defects. *Mater. Sci. Eng. A* **2012**, *548*, 142–151. [CrossRef]

11. Keller, S.; Chupakhin, S.; Staron, P.; Maawad, E.; Kashaev, N.; Klusemann, B. Experimental and numerical investigation of residual stresses in laser shock peened AA2198. *J. Mater. Process. Technol.* **2018**, *255*, 294–307. [CrossRef]

12. Hu, Y.; Yao, Z.; Hu, J. 3-D FEM simulation of laser shock processing. *Surf. Coat. Technol.* **2006**, *201*, 1426–1435. [CrossRef]

13. Toparli, M.B.; Fitzpatrick, M.E. Effect of Overlapping of Peen Spots on Residual Stresses in Laser-Peened Aluminium Sheets. *Metall. Mater. Trans. A* **2019**, *50*, 1109–1112. [CrossRef]

14. Correa, C.; Ruiz de Lara, L.; Díaz, M.; Gil-Santos, A.; Porro, J.A.; Ocaña, J.L. Effect of advancing direction on fatigue life of 316L stainless steel specimens treated by double-sided laser shock peening. *Int. J. Fatigue* **2015**, *79*, 1–9. [CrossRef]

15. Xu, G.; Luo, K.Y.; Dai, F.Z.; Lu, J.Z. Effects of scanning path and overlapping rate on residual stress of 316L stainless steel blade subjected to massive laser shock peening treatment with square spots. *Appl. Surf. Sci.* **2019**, *481*, 1053–1063. [CrossRef]

16. Salimianrizi, A.; Foroozmehr, E.; Badrossamay, M.; Farrokhpour, H. Effect of Laser Shock Peening on surface properties and residual stress of Al6061-T6. *Opt. Lasers Eng.* **2016**, *77*, 112–117. [CrossRef]

17. Correa, C.; Ruiz de Lara, L.; Díaz, M.; Porro, J.A.; García-Beltrán, A.; Ocaña, J.L. Influence of pulse sequence and edge material effect on fatigue life of Al2024-T351 specimens treated by laser shock processing. *Int. J. Fatigue* **2015**, *70*, 196–204. [CrossRef]

18. Correa, C.; Peral, D.; Porro, J.A.; Díaz, M.; Ruiz de Lara, L.; García-Beltrán, A.; Ocaña, J.L. Random-type scanning patterns in laser shock peening without absorbing coating in 2024-T351 Al alloy: A solution to reduce residual stress anisotropy. *Opt. Laser Technol.* **2015**, *73*, 179–187. [CrossRef]

19. Schajer, G.S. Advances in Hole-Drilling Residual Stress Measurements. *Exp. Mech.* **2010**, *50*, 159–168. [CrossRef]

20. Steinzig, M.; Ponslet, E. Residual stress measurement using the hole drilling method and laser speckle interferometry Part I. *Exp. Tech.* **2003**, *27*, 59–63. [CrossRef]

21. Ponslet, E.; Steinzig, M. Residual stress measurement using the hole drilling method and laser speckle interferometry Part III: Analysis technique. *Exp. Tech.* **2003**, *27*, 17–21. [CrossRef]

22. Chupakhin, S.; Kashaev, N.; Klusemann, B.; Huber, N. Artificial neural network for correction of effects of plasticity in equibiaxial residual stress profiles measured by hole drilling. *J. Strain Anal. Eng.* **2017**, *52*, 137–151. [CrossRef]

23. Peyre, P.; Fabbro, R. Laser shock processing: A review of the physics and applications. *Opt. Quant. Electron.* **1995**, *27*, 1213–1229.

24. Xu, Y.Y.; Ren, X.D.; Zhang, Y.K.; Zhou, J.Z.; Zhang, X.Q. Coating Influence on Residual Stress in Laser Shock Processing. *Key Eng. Mater.* **2007**, *353–358*, 1753–1756. [CrossRef]

25. Kashaev, N.; Ventzke, V.; Horstmann, M.; Chupakhin, S.; Riekehr, S.; Falck, R.; Maawad, E.; Staron, P.; Schell, N.; Huber, N. Effects of laser shock peening on the microstructure and fatigue crack propagation behaviour of thin AA2024 specimens. *Int. J. Fatigue* **2017**, *98*, 223–233. [CrossRef]

26. Yang, J.M.; Her, Y.C.; Han, N.; Clauer, A.H. Laser shock peening on fatigue behavior of 2024-T3 Al alloy with fastener holes and stopholes. *Mater. Sci. Eng. A* **2001**, *298*, 296–299. [CrossRef]

Article

The Effect of Laser Peening without Coating on the Fatigue of a 6082-T6 Aluminum Alloy with a Curved Notch

Enrico Troiani and Nicola Zavatta *

Department of Industrial Engineering, University of Bologna, 47121 Forlì, Italy
* Correspondence: enrico.troiani@unibo.it

Received: 20 May 2019; Accepted: 26 June 2019; Published: 28 June 2019

Abstract: Laser shock peening has established itself as an effective surface treatment to enhance the fatigue properties of metallic materials. Although a number of works have dealt with the formation of residual stresses, and their consequent effects on the fatigue behavior, the influence of material geometry on the peening process has not been widely addressed. In this paper, Laser Peening without Coating (LPwC) is applied at the surface of a notch in specimens made of a 6082-T6 aluminum alloy. The treated specimens are tested by three-point bending fatigue tests, and their fatigue life is compared to that of untreated samples with an identical geometry. The fatigue life of the treated specimens is found to be 1.7 to 3.3 times longer. Brinell hardness measurements evidence an increase in the surface hardness of about 50%, following the treatment. The material response to peening is modelled by a finite element model, and the compressive residual stresses are computed accordingly. Stresses as high as −210 MPa are present at the notch. The ratio between the notch curvature and the laser spot radius is proposed as a parameter to evaluate the influence of the notch.

Keywords: laser shock peening; fatigue; notch; aluminium alloys; finite element method

---

1. Introduction

Laser Shock Peening (LSP) is a technology that makes use of shock waves induced by a laser to improve the mechanical properties of a metallic component. Short laser pulses (1–50 ns) with a high-power intensity are shot at the surface of the component. The laser beam vaporizes a superficial layer of the treated material, with the local formation of high-pressure plasma, as noted by the authors of [1]. Fabbro et al. [2] studied the use of a transparent overlay as an effective method to confine the generated plasma. This results in the formation of intense shock waves, which induce high residual stresses in the surrounding material, as shown by Sano et al. [3].

Conventional laser peening usually employs an ablative layer (i.e., a coating applied on the surface of the material) to prevent damage to the metal surface. Another technique also exists, called Laser Peening without Coating (LPwC), in which no ablative layer is used and the treated specimen is immersed in water during exposure to laser pulses. Compared with conventional laser peening, LPwC does not require a specific surface preparation and can be performed by commercial Nd:YAG lasers, which makes it particularly appealing for a number of applications.

The effects of the LSP-induced residual stresses on the fatigue of aluminum alloys are well documented in the literature. A review by Montross et al. [4] reported an improved fatigue life of Al 2024 and Al 7075 specimens treated with laser peening, while the authors of [5] show an effective reduction in the fatigue crack growth rate in a laser-peened 6061-T6 aluminum alloy. Gao [6] noted that the superior performances of laser shock peening compared with mechanical shot peening are due to deeper compressive stresses and a better surface finish.

Similar improvements of the fatigue behavior have been reported for laser peening without coating. In 2006, Sano et al. [7] observed a substantial prolongation of the fatigue life of an LPwC-treated AC4CH Al alloy, even though an increase of the surface roughness was found. Similar results were obtained by the authors of [8] for AA7075-T73 open-hole specimens. A recent review [9] reports enhanced fatigue properties for a wide class of metals treated with LPwC, including 6061 and 6082 Al alloys.

Geometry is known to play a role in the efficacy of LSP, as shown in the literature [10], where the effect of LSP on thin AA2024 panels typical of aeronautical applications were studied, and retarded crack propagation was observed. A work by Troiani et al. [11] highlights the potential drawbacks of the LSP of thin panels, depending on the selected peening path. Of particular interest is the influence of geometric discontinuities on the peening process. Yang [12] reported an improved fatigue life of Al 2024 specimens with pre-existing holes treated with LSP; on the other hand, Ivetic et al. [13] showed that the interaction between laser peening and an open hole in aluminum panels could potentially result in a decreased fatigue life. Dorman [14] addressed the presence of scribe defects and their effects on the fatigue life of LSP-treated 2024 aluminum alloys.

Although curved notches are often critical for the nucleation of fatigue cracks, only a few works have addressed the effects of laser shock peening on a curved surface. Notably, Peyre [15] studied the effects of laser shock peening in specimens with a curved notch. The authors reported extended fatigue lives of laser-peened specimens compared with mechanically shot-peened and untreated ones. They also measured high compressive stresses around the notch. Vasu and Grandhi [16] analyzed the residual stress field induced by laser peening on a curved surface by means of a finite element model. The compressive stress was found to increase in a concave geometry, as compared with a flat or convex surface. The increase was related to the radius of the curvature, in that the smaller the radius, the higher the compressive stress.

This work aims to study the effects of laser peening without coating applied at a circular notch, as shown in Figure 1. In particular, the effects of the process on the fatigue behavior are investigated by three-point bending specimens, while the residual stresses at the notch are computed by means of a finite element model.

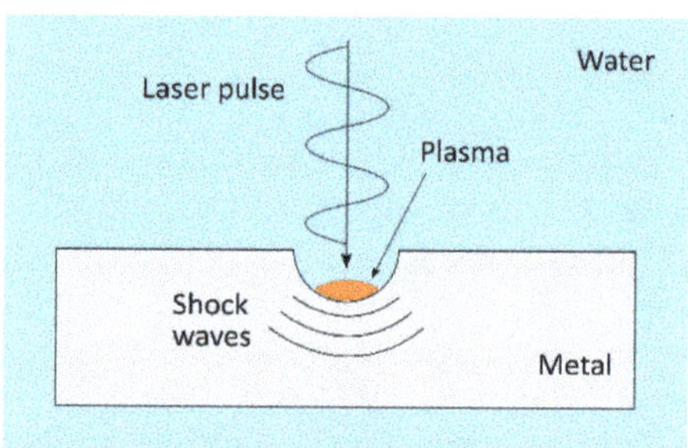

Figure 1. Application of the Laser Peening without Coating (LPwC) process to a notched specimen.

2. Materials and Methods

2.1. Characteristics of the Baseline Material

The specimens used for the analysis were made of aluminum alloy 6082-T6, with the geometry given in Figure 2. The specimens were machined out of a $400 \times 260 \times 20$ mm^3 plate, with a circular notch milled at the center. A Mitutoyo Surface Roughness Tester (Mitutoyo Italiana, Lainate, Italy) was used to measure the surface roughness at the notch—five measurements were taken on each specimen and the nominal roughness was found to be equal to 3.2 μm. The thickness of the specimen was high enough, so that it did not warp during the peening process. The geometry of the specimen was similar to that reported in the literature [15] for three-point bending fatigue tests, with the central notch acting as a stress concentration for the easier localization of the fatigue crack nuclei.

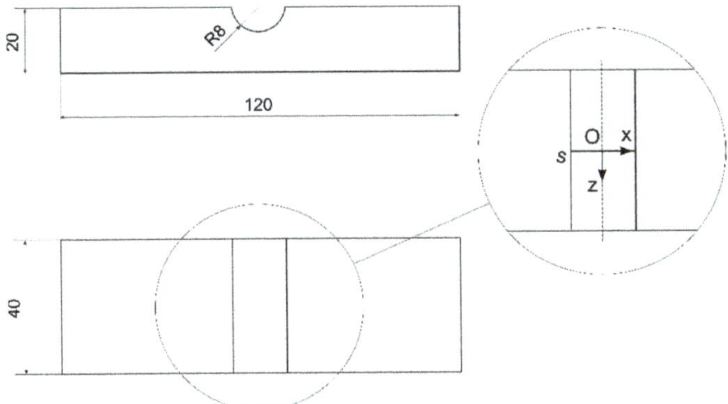

Figure 2. Specimen geometry. The midline path (s) and the local coordinates are shown in the detailed view of the notch.

2.2. Laser Peening Treatment

The LPwC treatment was performed in the Department of Industrial Engineering Laboratories of the University of Bologna (Bologna, Italy). The setup used for the laser shock peening is shown in Figure 3.

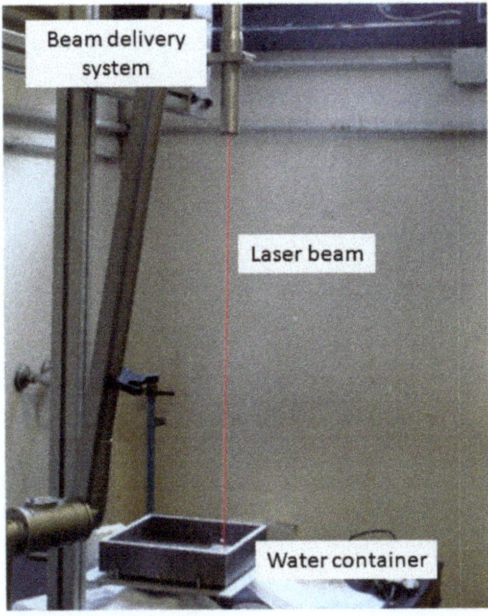

Figure 3. Setup used for the laser shock peening without coating treatment.

The laser source was provided by a Nd:YAG pulsed laser produced by Quanta Systems (Samarate, Italy). Its settings are as follows: wavelength of 1064 nm, nominal pulse energy of 1.8 J, and a pulse duration equal to 8 ns. The nominal laser power density was 7 GW/cm^2.

The peening was applied in the central part of the notch on a rectangular area of 40×4 mm^2, as shown in Figure 4a. A circular laser spot size of 2 mm in diameter was used, with a distance of 0.25 mm between two adjacent spot-centers. A spot overlapping of 75%, as shown in Figure 4b, was chosen so as to attain complete coverage of the treated area on a single layer, as discussed in the literature [13]. This resulted in a pulse density equal to 1600 pulses/cm^2.

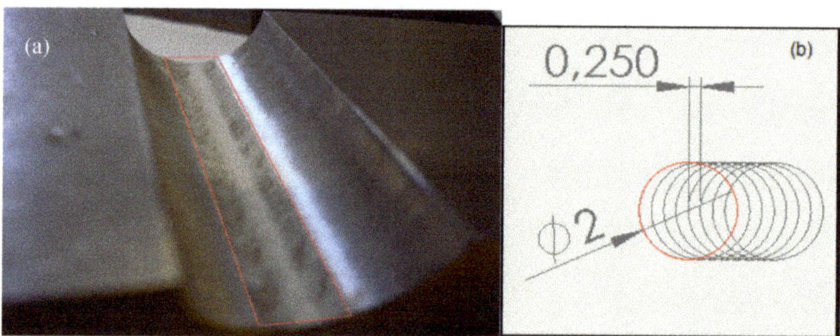

Figure 4. (a) Detailed view of the peened zone. (b) The treated area is bounded in the red box.

2.3. Fatigue Testing

The three-point bending fatigue tests were performed at the MaSTeR Lab laboratories in Forlì (University of Bologna), with the use of a 100 kN servo-hydraulic testing machine (Italsigma, Forlì, Italy). The experimental setup is illustrated in Figure 5. The tests were conducted under force-controlled conditions at four different stress levels, with a constant R-ratio of R = 0.1. The stress concentration factor for the given specimen configuration was Kt = 1.3, which resulted in maximum stresses at the notch ranging from 220 to 260 MPa.

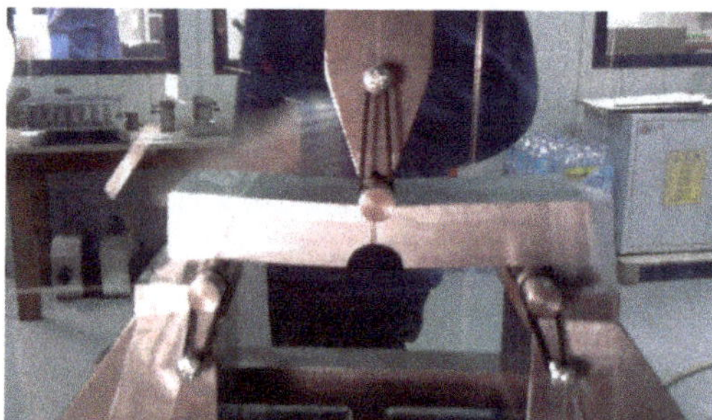

Figure 5. Experimental setup for three-point bending fatigue tests.

2.4. Hardness Measurement

The hardness measurements were performed in order to quantify the effect of the LPwC treatment on the surface hardness of the treated specimens. The measurements were conducted according to the E10-18 standard test procedure for Brinell hardness [17]. The sample used for testing was taken from the LPwC-treated specimens, and was additionally machined in order to accommodate the hardness testing machine. Five measurements were performed, two in the central, peened section of the specimen (Figure 6), and three on the untreated part of the sample (Figure 7). As the diameter of the indentation (e.g., about 1 mm) is small compared with the radius of the notch, the effect of the curvature was not considered.

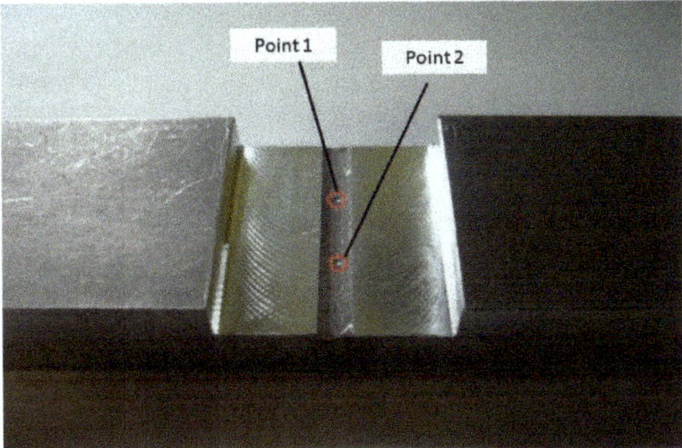

Figure 6. Central hardness measurement points. The two locations used for the Brinell hardness measurement in the region treated with LPwC are shown.

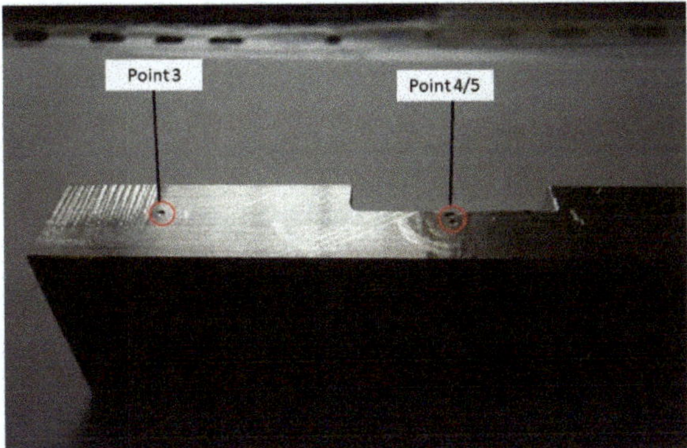

Figure 7. Side hardness measurement points. Locations considered for hardness measurement of the untreated material.

2.5. Numerical Model

The residual stresses induced by laser peening were computed by finite element analyses in the commercial software Abaqus/Explicit v6.12 (Dassault Systemes, Paris, France). Several models can be found in the literature to characterize the material response in the context of laser shock peening. The Johnson–Cook constitutive model considers the effect of large strains and high strain rates on material behavior, and was applied in an explicit/implicit finite element model by the authors of [18]. Ivetic et al. [8] optimized the non-linear response through full-explicit numerical analyses, taking into account the actual overlapping of the peening shots. Nevertheless, the material response to the shock waves propagation can be thought of as a condition of high-strain rate cyclic loading. The effect of the cyclic deformation induced by laser peening was already studied by Angulo et al. [19]. In this regard, a nonlinear isotropic/kinematic hardening plastic model, which also includes the change of properties due to the Bauschinger effect, can be used to simulate the material behavior. This model consists of isotropic hardening (Equation (1)) plus a nonlinear kinematic hardening component. The back stresses in kinematic hardening can be computed by Equation (2), according to Chaboche [20].

$$\sigma_Y = \sigma_0 + Q_\infty\left[1 - \exp\left(-b\varepsilon^{pl}\right)\right] \tag{1}$$

$$d\alpha = Cd\varepsilon^{pl} - \gamma\alpha\left|d\varepsilon^{pl}\right| \tag{2}$$

The material properties of Al 6082-T6 were selected according to the works by Chen et al. [21] and Chen et al. [22], which present thorough measurements of the hardening parameters, including testing at high strain rates via a split-Hopkinson tension bar. The maximum strain rate measured in the tests was around 3400 s^{-1}, which is lower than that occurring during laser peening (i.e., in the order of 10^6 s^{-1}), according to the authors of [15]. However, Al 6082-T6 is reported to have a low sensitivity on the strain rate. Moreover, a study by Langer et al. [23] investigated the effects of using conventional test data to model the laser peening process; they concluded that consistent results could be obtained with strain rates in the order of 10^3 s^{-1} (i.e., similar to those encountered in the literature of [21,22]). The hardening parameters input into the material model are summarized in Table 1.

Table 1. Kinematic hardening parameters for Al 6082-T6.

σ_0 [MPa]	C [MPa s]	γ [s]	Q_∞ [MPa]	b [–]
319	3211.7	25	64.5	24.3

Instead of direct modelling of the laser and the plasma layer, the equivalent pressure of the plasma was considered, following the approach described by the authors [18]. As a result of the low energy of the laser pulses and the short pulse duration, the thermal effect on the material is negligible compared to the effect of shock waves, as reported by the authors of [24]. The simulation is split into the following two steps: first, pressure is applied in the peened area and the waves propagate elastically in the material; second, the load and constraints are released and residual stresses develop, following material relaxation. An excessive computational time would have been required to model multiple laser shots. As a result, only one single shot was modelled, with a spot size equivalent to the treated area of the specimens.

The pressure exerted by the plasma depends on the laser parameters, namely, wavelength, power density, and pulse duration. The laser spot was simulated by applying a pressure load with a triangular temporal profile and uniform spatial distribution. The triangular temporal profile approximates the results reported in the literature [25] for a water confined plasma, with a maximum pressure of 4 GPa attained after 25 ns, which decays to zero at 50 ns.

The geometry of the finite element model reproduces that of the real coupons, cut on the two sides so that the total width is 48 mm. The size of the peened region is 40 × 4 mm^2. The mesh size is not homogeneous, as shown in Figure 8, as thin elements are needed in the thickness direction to capture the residual stress gradient at the notch; the minimum element is 0.13 × 0.20 × 0.03 mm^3. Clamped boundary conditions are applied at the edges of the specimen, and the bottom surface is constrained against displacements.

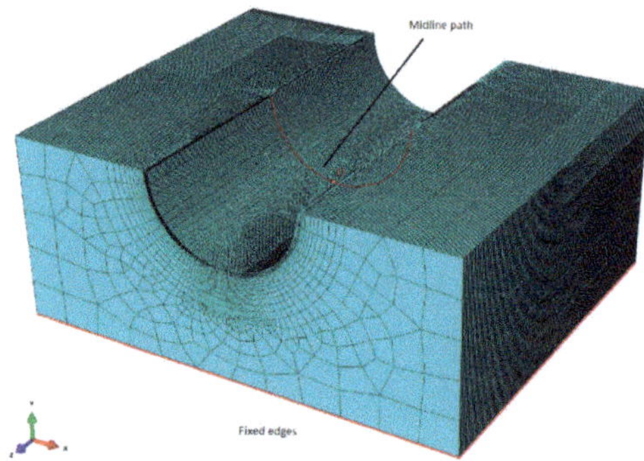

Figure 8. Finite element model of the specimen. Built-in constraints at the bottom edges are shown. The midline path is highlighted in red and its origin is denoted by the letter O.

3. Results

3.1. Fatigue Tests

A total of eight specimens were tested—four samples (denoted as 2, 3, 4, and 5) were LPwC-treated, as described above; the other four (6B, 7B, 8B, and 9B) were untreated and served as a baseline for comparison. The Maximum Stress–Number of cycles (S–N) results of the two cases are plotted in Figure 9. Two distinct S–N curves are fitted through the data of the peened and unpeened samples using the least-squares method. Even if four specimens are not sufficient for constructing a statistically significant S–N curve, the trends obtained from fatigue testing are very clear. The fatigue life of the treated specimens was found to be longer for all load levels, with a maximum increase equal to 3.34 times the baseline value.

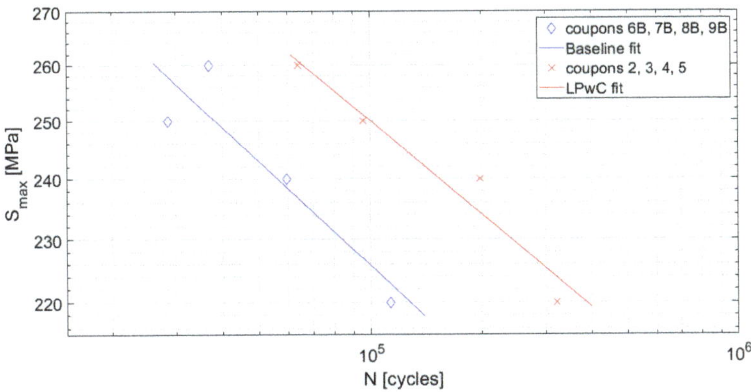

Figure 9. S–N curves from the fatigue tests. The maximum nominal stresses vs fatigue lives of the tested specimens are plotted. Least-square fittings of the untreated and peened samples are shown in red and blue, respectively.

3.2. Hardness Measurement

The measured hardness values are collected in Table 2. The average hardness measured in the unpeened region, around 120 HB, is 34% higher than the nominal hardness reported by the manufacturer, which is a result of the milling operation used to cut the specimens. Comparing this to the measurements at the notch, it shows a hardness increase of about 50% after LPwC.

Table 2. Brinell hardness measurement results. LSP—laser shock peening.

Measurement Point	Surface State	Brinell Hardness
1	LSP-treated	178 HB
2	LSP-treated	185 HB
3	As machined	123 HB
4	As machined	117 HB
5	As machined	120 HB

3.3. FEM Results

The distribution of the normal stress σ_{xx} n the peened area is shown in Figure 10. The treated area extends 2 mm left and right from the origin (O), also shown in the figure. Compressive residual stresses are present in this region, with compression peaks as high as 210 MPa. The σ_{xx} component shown here acts in the same direction of the bending stress in the tests. The finite elements analysis also provides information on the in-depth distribution of the stresses. The in-depth residual stress σ_{xx} in Figure 11 is computed at location O, perpendicularly to the surface of the notch. A compressive peak is found at a depth of about 0.2 mm, while a strong stress gradient is observed close to the surface. At depths greater than 0.7 mm, a tractive residual stress field is present as a consequence of equilibrium.

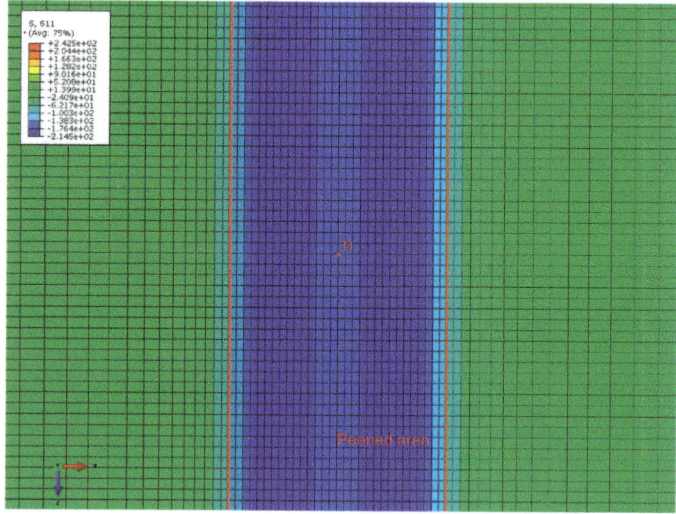

Figure 10. Distribution of residual stress σ_{xx} in the peened strip, bounded by the red lines.

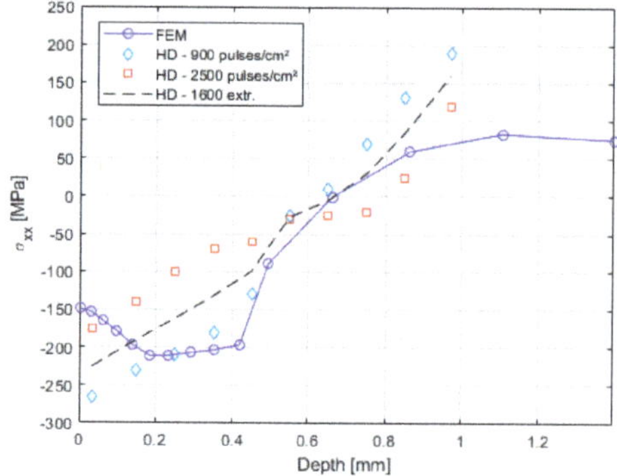

Figure 11. Residual stress distribution in the in-depth direction. The FEM results are plotted together with the hole drilling measurements in the literature [22] (blue and red dots) and to their extrapolation (dashed line).

4. Discussion

The tests conducted on the untreated and peened specimens show a clear improvement of the fatigue behavior after laser peening. For all of the applied stress levels, a net increase of the fatigue life was observed. The fatigue life of the baseline samples (i.e., 110 kcycles at 220 MPa) is in line with previous findings by the authors of [26], who reported about 120 kcycles at 230 MPa in four-point bending tests of Al 6082 specimens. The fatigue life of the LPwC-treated samples shows an improvement from 1.73 to 3.34 times the baseline value. These findings can be compared to the results in the literature [15], which include the three-point bending of notched specimens treated with conventional laser peening with an ablative layer. The authors reported a net improvement following LSP, with the fatigue lives enhanced by a factor of 10 compared to the baseline; albeit, it should be noted that a different material, namely Al 7075, was tested. For a better characterization of the scatter in the S–N curves of Al 6082, additional testing would be recommended, which falls beyond the scope of this work.

The Brinell hardness measurements evidence a surface hardness increase in the order of 50% in the peened region. Trdan et al. [24] already reported a micro-hardness increase for Al 6082-T651 after LPwC. A further study [27] of the microstructural evolution of Al 6082-T651 evidenced an increase of the dislocation density after LPwC, with the production of ultra-fine and nano-grains, and related it to the induced residual stresses. In light of this, the increase of dislocation density can be seen as the prime driver of the enhanced surface hardness, as well as being involved in the plastic deformation, which results in the formation of residual stresses.

The improved fatigue life of the peened samples is ascribed mainly to the compressive residual stresses induced by the process. The numerical results confirm that compressive stresses are present at the notch surface along the treated strip, as shown in Figure 10; in particular, compressive residual stresses σ_{xx} are computed at the notch. The three-point bending tests produce a tensile stress σ_{xx} at the notch, which tends to propagate a crack, normal to the x-direction. The compressive residual stress σ_{xx} contributes to closing the crack by decreasing the effective stress intensity factor range at the crack tip. For metallic materials, this is known to generate a reduction of the crack growth rate, thus postponing the specimen's failure. This mechanism could explain the prolonged fatigue life of the treated samples observed in Figure 9.

The computed in-depth residual stress field is shown in Figure 11. Also plotted are the experimental findings by the auhors of [24] obtained by hole drilling (HD) measurements on flat Al 6082 specimens treated with LPwC. As two different pulse densities were used in the literature [25], namely 900 and 2500 pulses/cm^2, a direct comparison with the numerical results presented here for a density of 1600 pulses/cm^2 is somewhat biased. To this end, an extrapolation of the measurements to a density equal to 1600 pulses/cm^2 is also shown; this is obtained by linearly interpolating the measurements for the upper and lower pulse density. The maximum compressive residual stresses calculated by FEM are about −210 MPa, which is 16% higher than the value of −180 MPa extrapolated from the measurements at the same depth.

The model predicts a compressive stress gradient close to the surface, which is not observed in the measurements. However, the first measurement by the authors of [25] was taken at a depth of 0.1 mm, and the points at 0.03 mm are the result of the method used for post-processing, as explicitly pointed out by the authors themselves. Remarkably, a similar stress gradient was found by the authors of [28] in a numerical study of a flat Al 2050 specimen. When validating the numerical results against X-ray diffraction measurements, the authors reported on the difficulty of measuring the residual stress gradient. In light of this, the steep stress gradient observed in the numerical results should not be regarded as an effect strictly related to the curvature at the notch. Interestingly, the depth of the compressive region seems to be basically unaffected by the presence of the notch, with compressive residual stresses extending to about 0.7 mm deep, similarly to the case of the unnotched specimen.

The effect of the notch on the residual stress can be estimated as follows. Vasu and Grandhi [16] reported that the effect of the surface curvature on the residual stresses induced by peening strongly depends on the radius of curvature (RC). Their results can be normalized, taking the ratio of the RC to the laser spot radius (SR); this shows that, for 2.4 < RC/SR < 5, the increase of the maximum compressive stress compared to a flat surface is between 17% and 8%. Assuming a linear variation between these two extremes, the value of RC/SR = 4 used in the present work would result in an increase of about 14% compared to the flat case. A comparison with the results discussed above (i.e., an increment equal to 16%) shows a good consistency for this estimate.

5. Conclusions

Laser peening without coating is a promising treatment for aluminum alloys. The outcomes of this work, focusing on LPwC applied at a notch, could be of interest for those applications in which the structural integrity is severely affected by the presence of holes and notches, such as aircraft components.

In summarizing, the following can be drawn from our observations:

- The fatigue life of the peened samples shows an improvement over that of the untreated ones, suggesting that laser peening without coating could enhance the fatigue behavior of notched components, similarly to conventional laser shock peening.
- The FEM results indicate that deep compressive residual stresses are induced in the treated area after LPwC, which are known to obstruct fatigue crack growth, and are deemed to be the main reason for the improved fatigue life observed in the tests.
- The Brinell tests confirmed the ability of LPwC to increase surface hardness, possibly improving the wear performances of the peened specimens, which, together with the enhanced fatigue properties, could make this treatment appealing for applications such as lugs and bolted joints.

What has been presented here suggests that laser peening without coating has the potential to improve the fatigue behavior of aluminium components, even when it is applied directly at a notch. In this regard, the LPwC technology could emerge as an alternative to conventional LSP.

Author Contributions: Conceptualization, E.T. and N.Z.; data curation, E.T. and N.Z.; formal analysis, E.T. and N.Z.; investigation, E.T. and N.Z.; methodology, E.T. and N.Z.; software, N.Z.; supervision, E.T.; validation, N.Z.; writing (original draft), E.T.; writing (review and editing), E.T. and N.Z.

Funding: This research received no external funding.

Acknowledgments: The authors are thankful to Paolo Proli for technical support, Goran Ivetic and Ivan Meneghin for the help in the tests.

Conflicts of Interest: The authors declare no conflict of interest.

References

1. Fairand, B.P.; Clauer, A.H.; Jung, R.G.; Wilcox, B.A. Quantitative assessment of laser-induced stress waves generated at confined surfaces. *Appl. Phys. Lett.* **1974**, *25*, 431–433. [CrossRef]
2. Fabbro, R.; Fournier, J.; Ballard, P.; Devaux, D.; Virmont, J. Physical study of laser-produced plasma in confined geometry. *J. Appl. Phys.* **1990**, *68*, 775–784. [CrossRef]
3. Sano, Y.; Mukai, N.; Okazaki, K.; Obata, M. Residual stress improvement in metal surface by underwater laser irradiation. *Nucl. Instrum. Methods Phys. Res. Sect. B Beam Interact. Mater. Atoms.* **1997**, *121*, 432–436. [CrossRef]
4. Montross, C.S.; Wei, T.; Ye, L.; Clark, G.; Mai, Y.-W. Laser shock processing and its effects on microstructure and properties of metal alloys: A review. *Int. J. Fatigue* **2002**, *24*, 1021–1036. [CrossRef]
5. Rubio-González, C.; Ocaña, J.L.; Gomez-Rosas, G.; Molpeceres, C.; Paredes, M.; Banderas, A.; Porro, J.; Morales, M. Effect of laser shock processing on fatigue crack growth and fracture toughness of 6061-T6 aluminium alloy. *Mater. Sci. Eng. A* **2004**, *386*, 291–295. [CrossRef]
6. Gao, Y.K. Improvement of fatigue property in 7050-T7451 aluminium alloy by laser peening and shot peening. *Mater. Sci. Eng. A* **2011**, *528*, 3823–3828. [CrossRef]
7. Sano, Y.; Masaki, K.; Ochi, Y.; Altenberger, I.; Scholtes, B. Laser Peening without Coating as a Surface Enhancement Technology. *J. Laser Micro Nanoeng.* **2006**, *1*, 161–166. [CrossRef]
8. Ivetic, G.; Meneghin, I.; Troiani, E.; Molinari, G.; Lanciotti, A.; Ristori, V.; Ocaña, J.L.; Morales, M.; Porro, J.A.; Polese, C.; et al. Characterisation of Fatigue and Crack Propagation in Laser Shock Peened Open Hole 7075-T73 Aluminium Specimens. In *ICAF 2011 Structural Integrity: Influence of Efficiency and Green Imperatives, Proceedings of the 26th Symposium of the International Committee on Aeronautical Fatigue, Montreal, QC, Canada, 1–3 June 2011*; Komorowski, J., Ed.; Springer: Dordrecht, The Netherlands, 2011; pp. 855–866.
9. Karthik, D.; Swaroop, S. Laser peening without coating—An advanced surface treatment: A review. *Mater. Manuf. Process.* **2017**, *32*, 1565–1572. [CrossRef]
10. Kashaev, N.; Ventzke, V.; Horstmann, M.; Chupakhin, S.; Riekehr, S.; Falck, R.; Maawad, E.; Staron, P.; Schell, N.; Huber, N. Effects of laser shock peening on the microstructure and fatigue crack propagation behaviour of thin AA2024 specimens. *Int. J. Fatigue* **2017**, *98*, 223–233. [CrossRef]

11. Troiani, E.; Taddia, S.; Meneghin, I.; Molinari, G. Fatigue Crack Growth in Laser Shock Peened Thin Metallic Panels. *Adv. Mater. Res.* **2014**, *996*, 775–781. [CrossRef]
12. Yang, J.-M.; Her, Y.C.; Han, N.; Clauer, A. Laser shock peening on fatigue behaviour of 2024-T3 Al alloy with fastener holes and stopholes. *Mater. Sci. Eng. A* **2001**, *298*, 296–299. [CrossRef]
13. Ivetic, G.; Meneghin, I.; Troiani, E.; Molinari, G.; Ocaña, J.; Morales, M.; Porro, J.; Lanciotti, A.; Ristori, V.; Polese, C.; et al. Fatigue in laser shock peened open-hole thin aluminium specimens. *Mater. Sci. Eng. A* **2012**, *534*, 573–579. [CrossRef]
14. Dorman, M.; Toparli, M.B.; Smyth, N.; Cini, A.; Fitzpatrick, M.E.; Irving, P.E. Effect of laser shock peening on residual stress and fatigue life of clad 2024 aluminium sheet containing scribe defects. *Mater. Sci. Eng. A* **2012**, *548*, 142–151. [CrossRef]
15. Peyre, P.; Fabbro, R.; Merrien, P.; Lieurade, H.P. Laser shock processing of aluminium alloys. Application to high cycle fatigue behaviour. *Mater. Sci. Eng. A* **1996**, *210*, 102–113. [CrossRef]
16. Vasu, A.; Grandhi, R.V. Effects of curved geometry on residual stress in laser peening. *Surf. Coat. Technol.* **2013**, *218*, 71–79. [CrossRef]
17. ASTM International. *ASTM E10-18 Standard Test Method for Brinell Hardness of Metallic Materials*; ASTM International: West Conshohocken, PA, USA, 2018. [CrossRef]
18. Peyre, P.; Sollier, A.; Chaieb, I.; Berthe, L.; Bartnicki, E.; Braham, C.; Fabbro, R. FEM simulation of residual stresses induced by laser Peening. *Eur. Phys. J. Appl. Phys.* **2003**, *23*, 83–88. [CrossRef]
19. Angulo, I.; Cordovilla, F.; García-Beltrán, A.; Smyth, N.S.; Langer, K.; Fitzpatrick, M.E.; Ocaña, J.L. The effect of material cyclic deformation properties on residual stress generation by laser shock processing. *Int. J. Mech. Sci.* **2019**, *156*, 370–381. [CrossRef]
20. Chaboche, J.L. Time-independent constitutive theories for cyclic plasticity. *Int. J. Plast.* **1986**, *2*, 149–188. [CrossRef]
21. Chen, Y.; Clausen, A.H.; Hopperstad, O.S.; Langseth, M. Stress–strain behaviour of aluminium alloys at a wide range of strain rates. *Int. J. Solids Struct.* **2009**, *46*, 3825–3835. [CrossRef]
22. Chen, X.; Peng, Y.; Peng, S.; Yao, S.; Chen, C.; Xu, P. Flow and fracture behaviour of aluminium alloy 6082-T6 at different tensile strain rates and triaxialities. *PLoS ONE* **2017**, *12*, e0181983. [CrossRef]
23. Langer, K.; Olson, S.; Brockman, R.; Braisted, W.; Spradlin, T.; Fitzpatrick, M.E. High strain-rate material model validation for laser peening simulation. *J. Eng.* **2015**, *13*, 150–157. [CrossRef]
24. Trdan, U.; Porro, J.A.; Ocaña, J.L.; Grum, J. Laser shock peening without absorbent coating (LSPwC) effect on 3D surface topography and mechanical properties of 6082-T651 Al alloy. *Surf. Coat. Technol.* **2012**, *208*, 109–116. [CrossRef]
25. Berthe, L.; Fabbro, R.; Peyre, P.; Tollier, L.; Bartnicki, E. Shock waves from a water-confined laser-generated plasma. *J. Appl. Phys.* **1997**, *82*, 2826–2832. [CrossRef]
26. Benedetti, M.; Bortolamedi, T.; Fontanari, V.; Frendo, F. Bending fatigue behaviour of differently shot peened Al 6082 T5 alloy. *Int. J. Fatigue* **2004**, *26*, 889–897. [CrossRef]
27. Trdan, U.; Skarba, M.; Ocaña, J.L. Laser shock peening effect on the dislocation transitions and grain refinement of Al–Mg–Si alloy. *Mater. Character.* **2014**, *97*, 57–68. [CrossRef]
28. Hfaiedh, N.; Peyre, P.; Song, H.; Popa, I.; Ji, V.; Vignal, V. Finite element analysis of laser shock peening of 2050-T8 aluminium alloy. *Int. J. Fatigue* **2015**, *70*, 480–489. [CrossRef]

Article

Fatigue Properties of Maraging Steel after Laser Peening

Luca Petan [1], Janez Grum [1,*], Juan Antonio Porro [2], José Luis Ocaña [2] and Roman Šturm [1]

[1]   Faculty of Mechanical Engineering, University of Ljubljana, 1000 Ljubljana, Slovenia;
      luca.petan@gmail.com (L.P.); roman.sturm@fs.uni-lj.si (R.Š.)

[2]   UPM Laser Centre, Polytechnical University of Madrid, 28031 Madrid, Spain; japorro@etsii.upm.es (J.A.P.);
      jlocana@etsii.upm.es (J.L.O.)

*   Correspondence: janez.grum@fs.uni-lj.si; Tel.: +386-41-725-520

Received: 30 October 2019; Accepted: 24 November 2019; Published: 28 November 2019

Abstract: Maraging steels are precipitation hardening steels used for highly loaded components in aeronautical and tooling industry. They are subjected to thermomechanical loads and wear, which significantly shorten their service life. Improvements of their surface mechanical properties to overcome such phenomena are of great interest. The purpose of our research was to investigate the influence of pulse density and spot size of a laser shock peening (LSP) process on the surface integrity with the fatigue resistance of X2NiCoMo18-9-5 maraging steel. Surface integrity was analyzed through roughness, residual stress, and microhardness measurements. The tests performed on resonant testing machine confirmed LSP is a promising process for increasing fatigue resistance of a component. Fatigue crack occurs, when the resonance frequency decreases. This moment, when the fatigue crack initiation phase ends and the fatigue crack propagation phase starts, was chosen as the moment of failure. We have proved LSP is a successful method in improving fatigue resistance of maraging steel by appropriate combination of laser spot size and pulse density tested in our research.

Keywords: microhardness; residual stresses; resonant fatigue resistance; roughness

---

1. Introduction

Ultrahigh-strength maraging steels achieve exceptional mechanical properties through precipitation strengthening [1]. Upon cooling from a solution annealing temperature, a nickel-rich austenite matrix with a virtual absence of carbon transforms to a soft and fully martensitic structure with a high dislocation density [2]. Solution annealing followed by an artificial aging process at a temperature of around 480 °C [3] involves the precipitation of intermetallic compounds at dislocation sites, thus contributing to the achievement of an excellent combination of strength and toughness. Commercially available maraging steels can reach yield strengths of over two GPa. Other characteristics of maraging steels are dimensional stability, good machinability and weldability, high fracture toughness, good thermal conductivity, and significantly high resistance to crack propagation and thermal fatigue [4]. Properties, such as good machinability, dimensional stability during heat treatment, and significantly high resistance to thermal fatigue, are needed in tooling applications, such as plastic molds and die casting dies for magnesium and aluminum alloys.

When used as a structural steel, maraging components can be exposed to various forms of detrimental phenomena, such as high-cycle mechanical fatigue [5], synergetic effects of stress and corrosion [6], wear, and thermal fatigue [7], which causes heat checks and stress cracks.

Mechanical properties and fatigue behavior of highly stressed metallic components can be significantly improved by generating compressive residual stresses (RSs) in the surface layer of a material using peening techniques [8], such as shot peening (SP) and laser shock peening (LSP). LSP can produce highly compressive RSs up to 1 mm in depth, which is about four times deeper than with a traditional SP process [9]. LSP is an innovative surface treatment [10], during which the surface of a treated component, usually covered with an absorbent coating and a transparent confining medium, is exposed to nanosecond long laser pulses of intense energy [11–13]. The irradiated zone vaporizes and transforms into plasma by ionization. Rapid expansion of the high-temperature plasma generates pressure, which is transmitted into the metal through shock waves. The movement of the shock waves from the surface to the depth of the material causes in-plane expansion of the material. When the stresses, caused by the shock waves, exceed the dynamic yield strength of the material, plastic deformation occurs. Those changes in the material generate compressive RSs.

A lot of research work was done on conventional steels, but very little is known about the effects of LSP on the fatigue strength of maraging steels. Banas et al. [14], who exposed maraging steel weldments to high-power Nd:YAG laser pulses, presented one of the early research papers in this context. The mechanical effect of shock waves increased the dislocation density in a heat-affected zone (HAZ) that led to a 17% increase in fatigue strength after the LSP. Grum et al. [15] analyzed the effects of LSP on a die casting maraging steel, i.e., X2NiCoMo12-8-8. They found out the LSP generated highly compressive RSs in the component surface layer. Petan et al. [16] have also shown that LSP with relatively low pulse energy can generate compressive RSs in maraging steel at a level of 500 MPa. When they increased laser pulse density (PD), the surface roughness increased. Similarly, Lavender et al. [17] found out compressive RSs in pilger dies, made of A2 tool steel, were produced by the effects of LSP. The RSs reached a depth of 1.5 mm with the maximum surface values up to −1050 MPa. This influenced an increase of the fatigue life of the pilger dies by 300%. Studies [18–21] on other high-strength tool steels have shown that the implementation of peening techniques can increase wear and thermal fatigue resistance by generating compressive RSs and inducing strain hardening in a surface layer.

The purpose of our research was to investigate the influence of laser processing parameters on the effects of the surface properties with fatigue resistance of X2NiCoMo18-9-5 maraging steel. Surface integrity was analyzed with surface roughness, residual stress, microhardness measurements, and resonant fatigue tests, while the influence of each processing parameter and their interactions was statistically evaluated using the analysis of variance (ANOVA) [22].

2. Materials and Methods

2.1. Material Properties

Experimental work was conducted on the 300-grade X2NiCoMo18-9-5 maraging steel, which was provided by Deutsche Edelstahlwerke. The given chemical composition of the maraging steel is presented in Table 1. Specimens prepared for fatigue tests were cut out of a delivered maraging rolled plate with a thickness of 9.5 mm (the width of the fatigue test specimen) and then heat-treated. The specimens were first solution annealed for 1 h at a temperature of 820 °C, followed by quenching, and artificially ageing for 3 h at 480 °C, cooled in air (this process was called maraging precipitation hardening (MPH)). The mechanical properties of the precipitation-hardened maraging steel are listed in Table 2. Before performing the LSP, all the specimens were ground and polished to ensure surface roughness uniformity.

Table 1. Nominal chemical composition of the X2NiCoMo18-9-5 maraging steel (unit in wt. %).

Fe	Ni	Co	Mo	Ti	Al	C
Bal.	17–19	8–10	4.5–5.5	0.5–0.8	0.05–0.15	≤0.03

Table 2. Mechanical properties of the X2NiCoMo18-9-5 maraging steel after the maraging precipitation hardening (MPH) process.

Tensile Strength [MPa]	Young's Modulus [GPa]	Density [kg/m³]	Elongation in 50 mm [%]	Reduction in Area [%]	Fracture Toughness [MPa·m$^{1/2}$]	Rockwell Hardness [Hardness Rockwell Scale C (HRC)]
1800–2100	195	8100	8–9	40–53	67–80	52–56

2.2. Laser Shock Peening

The heat-treated and polished specimens of maraging steel (l = 80 mm × w = 20 mm × t = 9.5 mm) were exposed to LSP in a confined mode using water as a transparent overlay and without an absorbent coating. In comparison to the conventional LSP with an absorbent coating, this technique uses lower laser pulse energy in order to avoid surface melting. This permits laser treatment to be performed without a sacrificial layer and allows for higher overlapping rates between laser spots, in our case between 78% and 92%. The LSP path mode on the specimens is presented in Figure 1. The direction of the laser processing was parallel to the rolling direction. Several specimens were exposed to LSP simultaneously. The width of the LSP area on each specimen was 16 mm. We used different combinations of laser pulse parameters. We changed the laser PD in a range from 900 to 2500 cm^{-2} with a step of 100 cm^{-2} at three different laser spot diameters (SDs; i.e., 1.5, 2.0, and 2.5 mm). The LSP processing parameters used in the experiment are presented in Table 3. The Laser pulse energy and the laser pulse duration were constant, i.e., 2.8 J and 10 ns, respectively. Therefore, the laser hit the specimen surface with a power density in a range between 5.7 and 15.8 GW·cm^{-2}. We used a Q-switched Nd: YAG laser (model: Brilliant B, wavelength: 1064 nm, Gaussian spatial distribution; Quantel, Lannion, France) operating at 10 Hz.

Table 3. Research design: laser shock peening (LSP) process parameters and plan of measurements.

Laser Spot Diameter (SD) [mm]	Laser Power Density [GW·cm^{-2}]	Laser Pulse Density (PD) [cm^{-2}]	Pulse Overlapping Rate [%]	Measurements of Fatigue Resistance
1.5	15.8	900, 1600, and 2500	78–87	Surface roughness
2.0	8.9	900, 1600, and 2500	83–90	Hardness
2.5	5.7	900, 1600, and 2500	87–92	Residual stresses

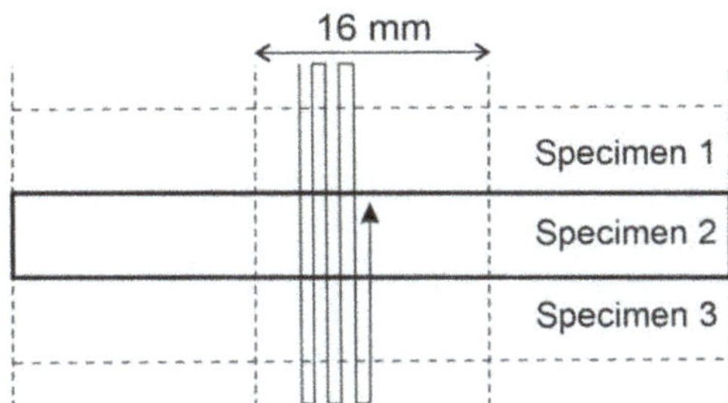

Figure 1. Principle of the LSP path mode on the specimens.

2.3. Resonant Fatigue Tests

In order to obtain insight into the effects of LSP on the mechanical fatigue resistance of the chosen maraging steel, fatigue tests were carried out using a resonant testing machine, CRACKTRONIC (RUMUL, Neuhausen am Rheinfall, The Switzerland), which is a table model for dynamic bending load applications with testing frequencies between 40 and 300 Hz. The kinematic conditions allows for pure bending between the gripping heads. An electromagnetically driven resonator, built as a rotary oscillator, creates appropriate bending moments. The elastic modulus of a material and the specimen geometry have effects on the resonance frequency of the machine. The crack initiation and propagation in this unit reduce its cross-sectional area, which affects stiffness reduction and consequently resonance frequency, as stated also in [23]. This machine measures resonance frequencies with a resolution of 0.01 Hz. In this research, the bending moment M was applied in a sinusoidal wave form at a stress ratio R of 0.1. We chose this stress ratio to provide and keep the upper surface of the specimen, where we expected crack initiation, in tensile conditions during fatigue tests. The bending frequency was around 114 Hz. We further reduced the middle of the specimen to ensure higher semicircular bending stresses in this area (Figure 2). In our resonant tests, we applied bending moments in a range between 60 and 78 N·m. The parameters of fatigue loading used in the experiment are presented in Table 4. According to our simulation, the maximum bending moment within a fatigue cycle generated a load of 1082 MPa in the most critical point of the specimen. The stress map during the bending is represented in Figure 3. Stress analysis was simulated with a software package SolidWorks (v23, 2015) according to the finite elements method (FEM) (3DEXPERIENCE, Vélizy-Villacoublay, France).

Table 4. Fatigue loading parameters.

Laser SD [mm]	Laser PD [cm^{-2}]	Maximum Bending Moment [N·m]	Maximum Bending Stress [MPa]
1.5	900 and 2500	60, 66, 72, and 78	833, 916, 999, and 1082
2.0	1600	60, 66, 72, and 78	833, 916, 999, and 1082
2.5	900 and 2500	60 and 78	833 and 1082
Base metal (MPH)	-	60, 66, 72, and 78	833, 916, 999, and 1082

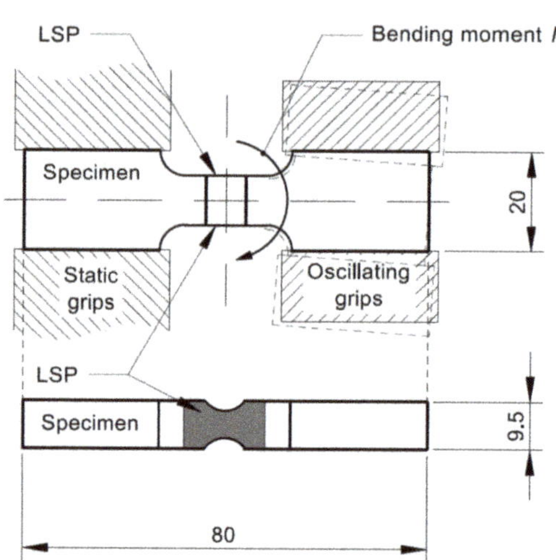

Figure 2. Principle of mechanical fatigue testing. The bending moment (M) range: 60–78 N·m.

Figure 4 shows the change of the resonant frequency in dependence of the number of fatigue cycles, where N_i represents an initiation period and N_p indicates a propagation period. The resonant frequency, conditioned by the specimen's geometry, began to decrease, when the fatigue crack occurred. This event, which also separated the fatigue crack initiation phase and the fatigue crack propagation phase, was chosen as the moment of failure. We did not measure specimens' temperatures during the fatigue testing, and we did not notice any temperature increase in the specimens during and after the test. The CRACKTRONIC is a compact testing device, so heat generated in the specimen during the fatigue loading could possibly be transferred to the resonant device.

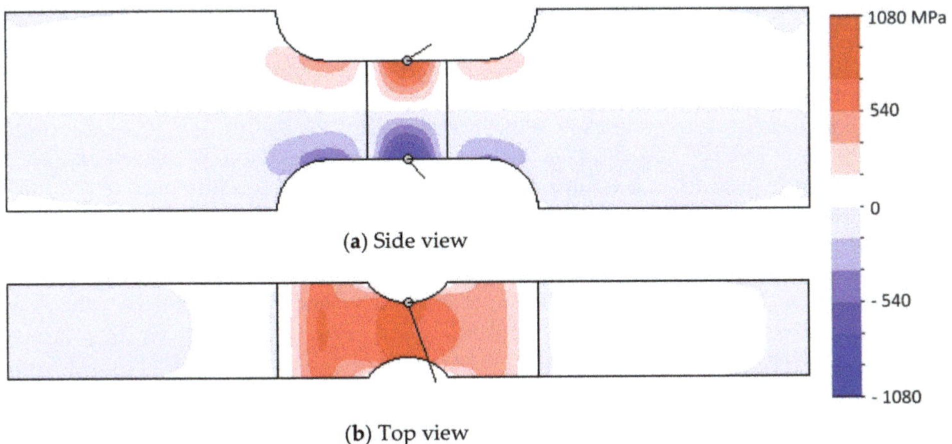

Figure 3. Stress maps during the bending within the fatigue specimen: (**a**) side-view stress map; (**b**) top-view stress map. The maximum stress indicated in red was approximately 1000 MPa. The bending moment was 78 N·m.

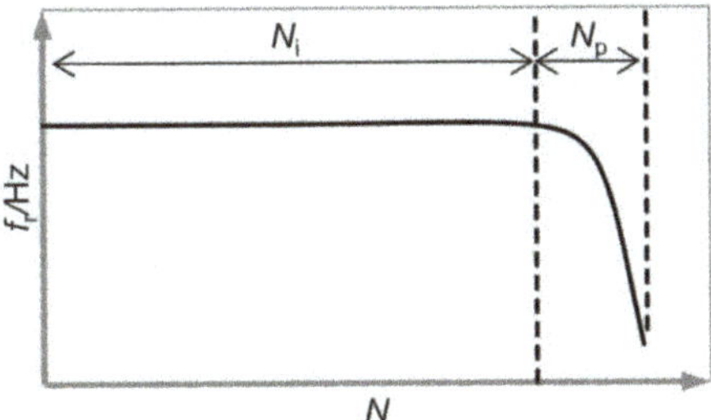

Figure 4. Resonant frequency as a function of the number of fatigue cycles during the fatigue test.

3. Results and Discussion

3.1. Roughness Measurements

Surface roughness measurements (Ra—the arithmetic average of the absolute values of the profile heights over the evaluation length, and Rz—the average value of the absolute values of the heights of five highest-profile peaks and the depths of five deepest valleys within the evaluation length) were performed with a Surtronic 3+ contact profilometer (Taylor Hobson Ltd, Leicester, UK). Before the LSP was applied, 10 individual surface roughness measurements were carried out, with five in the rolling direction and five in the transversal direction. The number of measurements after the LSP was doubled. The characteristic surface roughness was calculated as an average of the results from longitudinal and transversal measurements.

Before the LSP, the surface roughness was at a level of Ra = 0.2 μm and Rz = 1.2 μm. After the LSP, the surface roughness increased to Ra = 0.6–1.2 μm and Rz = 3.9–7.4 μm. The minimum values of roughness were measured on the specimen treated with the smallest laser spots and the lowest PD, while the maximum values were measured on the specimen treated with the largest laser spots and the highest PD.

To obtain an overview of the effects of laser PD and laser spot size on the surface integrity evaluated by roughness, hardness, and RS, a general factorial design was carried out. The laser parameters were examined using the ANOVA and the response surface methodology (RSM), where the influence of individual factor was considered to be statistically significant for $P < 0.05$. According to the statistical analysis (ANOVA), we found out that interactions of PD and SD also had a significant influence on surface roughness, RS, and hardness ($P < 0.0001$). We tested several polynomial models on statistical characteristics (F-value (statistical characteristics used to test the significance of adding new model terms to those terms already in the model), R^2 (reports the strength of the relationship between the set of independent variables and the dependent variable)) to fit the response to the measured values. We found out, according to the F-values and the R^2 values, the most suitable model for surface roughness, hardness, and RS is a quadratic model. Therefore, according to the mentioned measurements at laser SDs of 1.5, 2.0, and 2.5 mm, the results are presented as contour plots.

As can be observed in the contour plots in Figures 5 and 6, laser PD had the main influence on surface roughness. Higher PD influenced higher surface roughness, at all the diameters of the laser spot. When processing hard materials, it is difficult to detect a clear relation between LSP processing parameters and surface roughness due to moderate increases of Ra and Rz. At a PD of 900 cm^{-2}, the minimum roughness values were obtained when using an SD of 1.9 mm. At a PD of 1600 cm^{-2}, the minimum roughness values were obtained when using an SD of 1.7 mm. At a PD of 2500 cm^{-2}, the minimum roughness values were obtained when using an SD of 1.5 mm. That happened partially at the expense of decreasing the overlapping of laser spots during the LSP. At the highest PD, we can reduce surface roughness by decreasing laser spot size. At a lower PD, the optimal laser SD was between 1.8 and 2.0 mm.

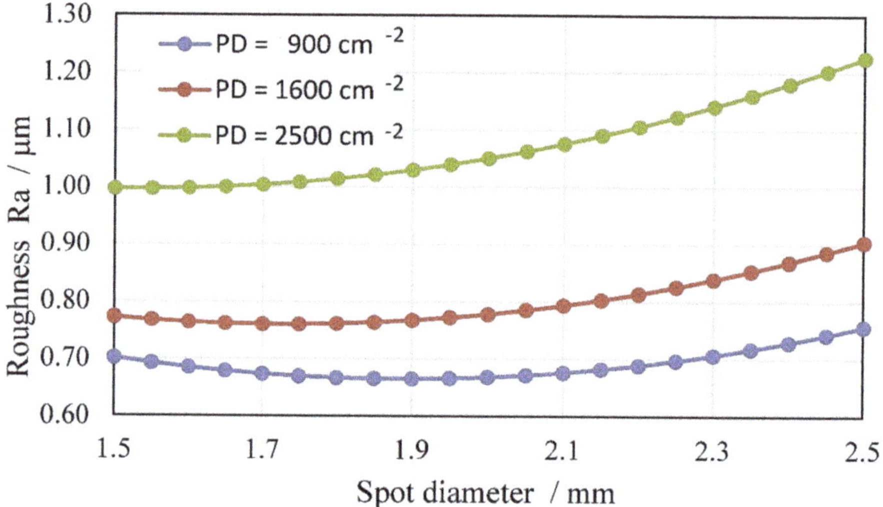

Figure 5. Roughness (Ra) as a function of laser SD and PD.

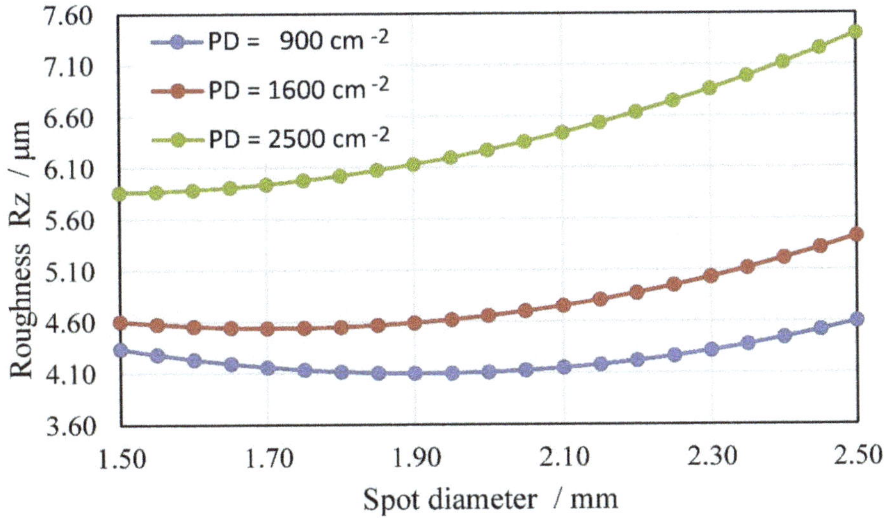

Figure 6. Roughness (*Rz*) as a function of laser SD and PD.

The ablative nature of a laser process, because of the absence of an absorbent coating, combined with mechanical effects of laser pulse pressure, leads to profile deepening.

Profile depth (Pt) is a kind of important information for planning additional process operations after LSP, like grinding and polishing. It shows us the height difference between the untreated and LSP-treated surfaces (inserted in Figure 7). We can find out that increasing PD influences the increase of profile depth and it is more distinct at a large diameter of a laser spot. Profile depth increases linearly in conjunction with increasing PD. The line is tilt more greatly in the case of bigger SDs. Especially the trend of Pt results is very unfavorable when applying LSP with a laser SD of 2.5 mm (Figure 7). The surface profile was lowered by almost 100 µm in the case of PD = 2500 cm^{-2}.

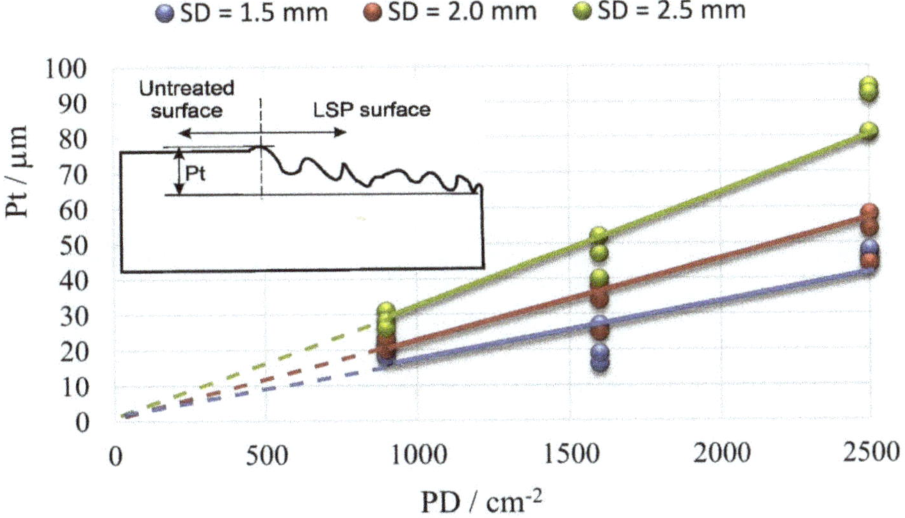

Figure 7. Profile depth (Pt) as a function of laser PD and SD.

3.2. Residual Stress Measurements

RSs were measured with a standard hole-drilling method. We used a milling cutter with a diameter of 1.6 mm. Deformation of the specimen caused during the drilling was measured with 06-062-UM strain gage rosettes, which were connected to the LabVIEW software. Blind holes in the surface layer were incrementally drilled to a depth of 1 mm with a 0.1 mm increment, where the drilling process was temporarily interrupted, enabling deformations resulting from material relaxation to fully occur and stabilize. The final RS profiles were calculated with the H-Drill software, where an integral method with automatic smoothing was applied.

In the base metal, after the heat treatment, RSs were low, ranging between −13 and 49 MPa. After the LSP, at different process parameters, compressive RSs arose in the surface layer. The maximum compressive RSs were at the surface in a range between −1000 and −350 MPa. At depths between 0.5 and 1.0 mm, the transition from a compressive state to a tensile state occurred. Some typical RS distributions in depth are presented in Figure 8. According to the statistical analysis (ANOVA), we found out that the interactions of PD and SD have a significant influence on the RS ($P < 0.0001$). We fitted the measured values with the quadratic model. Therefore, according to the mentioned measurements at laser SDs of 1.5, 2.0, and 2.5 mm and at laser PDs of 900, 1600, and 2500 cm^{-2}, the results are presented as contour plots for the whole range of laser SD from 1.5 to 2.5 mm.

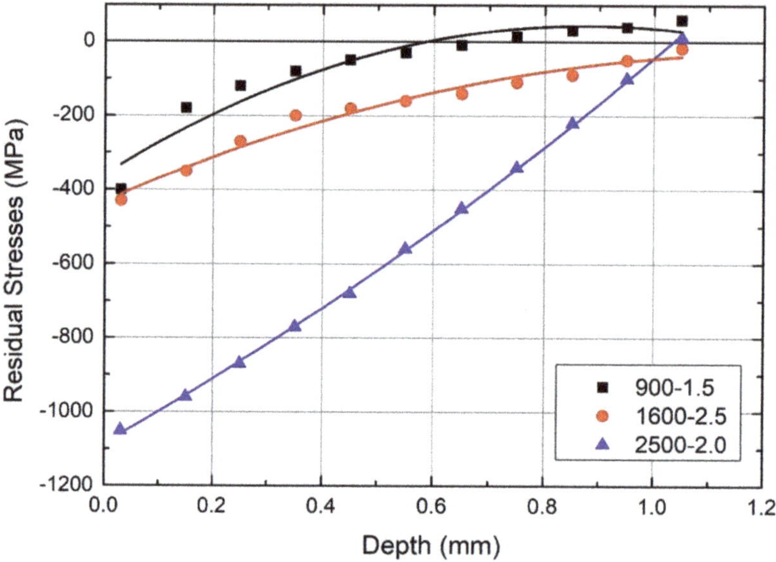

Figure 8. Residual stress distributions in depth for different PDs (i.e., 900, 1600, and 2500 cm^{-2}) and different SDs (i.e., 1.5, 2.0, and 2.5 mm).

The plots of the average RSs at specified depths, shown in Figures 9 and 10, suggest that the maximum compressive RSs were achieved with a 2.0 mm-diameter laser spot, both at the surface and at a depth of 1.0 mm. This could not be directly connected with the pulse power density (PPD) only, which is the highest with a 1.5 mm-diameter laser spot. The interaction between laser beam size and material affected the propagation nature of the shock waves. Smaller-diameter shock waves probably expand like spheres with a higher attenuation rate than larger-diameter shock waves, which behave like planar fronts [15,16,24,25]. This phenomenon, together with a higher overlapping rate between laser spots, may explain our findings. The RS profiles are almost the same for different PDs at a depth of 1.0 mm. The shift from big compressive RSs at the surface towards tensile RSs at a depth of 1.0 mm is between 400 and 600 MPa. In Figure 11, we can see a comparison of the RS profiles at the surface and at a depth of 1.0 mm for PD = 1600 cm^{-2}.

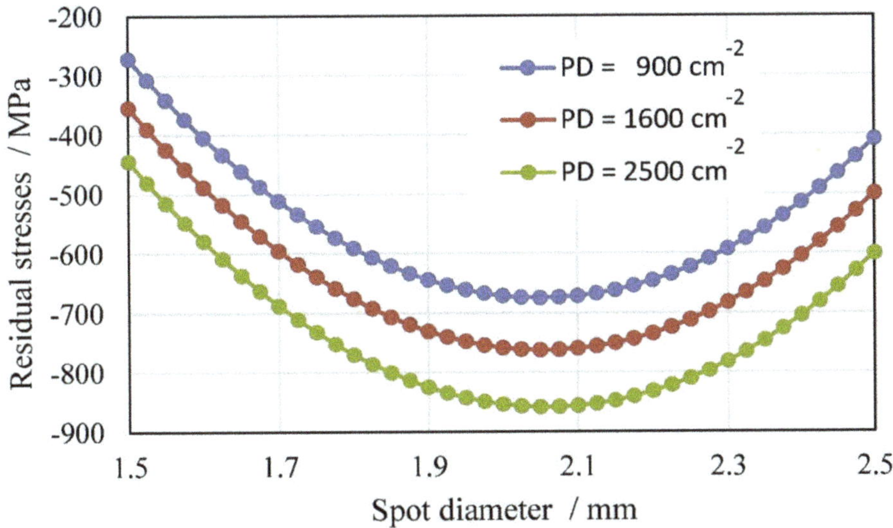

Figure 9. Residual stresses at the surface as a function of laser SD and PD.

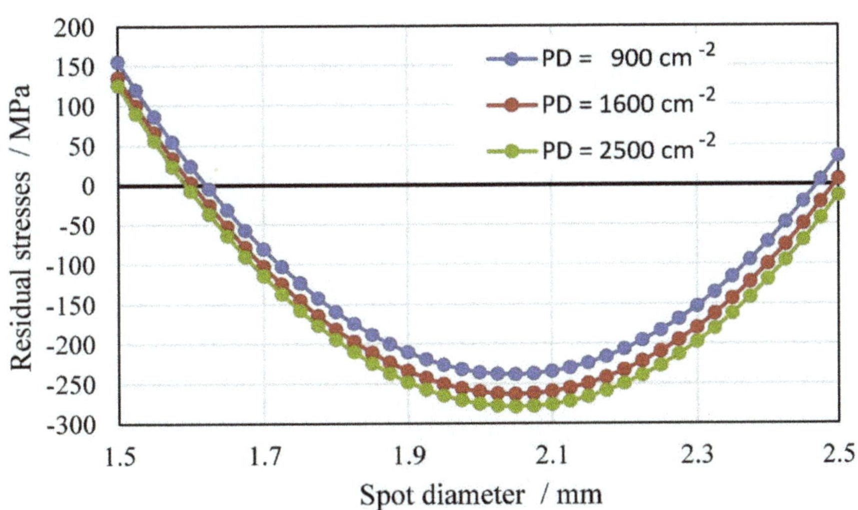

Figure 10. Residual stresses at a depth of 1.0 mm as a function of laser SD and PD.

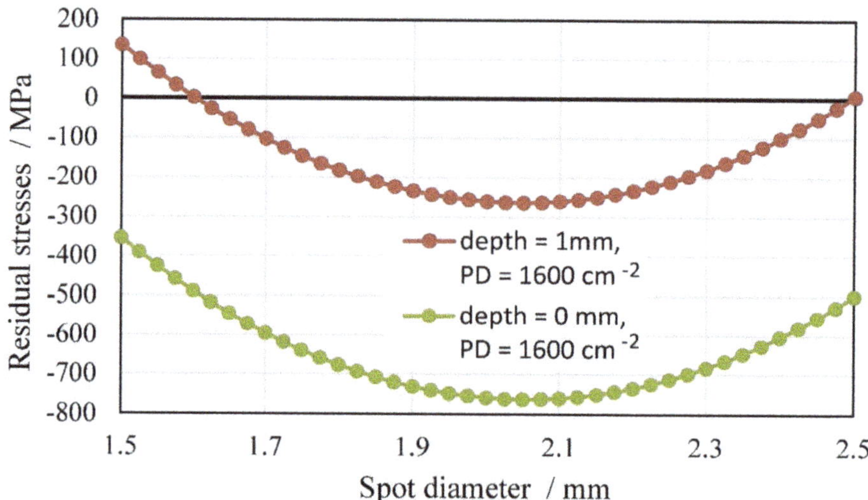

Figure 11. Residual stresses at the surface and at a depth of 1.0 mm as a function of laser SD for PD = 1600 cm^{-2}.

3.3. Microhardness Measurements

In-depth microhardness distributions were obtained with the standard Vickers test method using a Leitz Wetzlar hardness tester with a 200 g load and a 15 s load time. Vickers microhardness was measured at a depth of 2 mm from the LSP-treated surface (Figure 12). The microhardness analysis of the surface layer indicated that, after the LSP, strain hardening occurred. Strain hardening was detected as an increase in microhardness. The microhardness of the heat-treated maraging steel before the LSP was 667 HV$_{0.2}$. The highest microhardness value after the LSP was measured just below the surface in a range between 730 and 740 HV$_{0.2}$. The increased hardness was detected in the surface layer. The thickness of the strain-hardened surface was between 0.6 and 1.6 mm, depending on laser PD and laser SD.

Interactions of laser PD and SD have a significant influence on the hardness of an LSP-treated surface, found by a statistical analysis (ANOVA) ($P < 0.0001$) [25]. We fitted the measured values with the quadratic model. The results are presented as contour plots for the whole range of laser SD from 1.5 to 2.5 mm. The contour plots, shown in Figures 13 and 14, indicate that, in general, the maximum microhardness was achieved with a 2.0 mm-diameter laser spot, both at the surface and at a depth of 1.0 mm. Once again, this could not be directly connected with PPD. In the case of a 1.5 mm-diameter laser spot, the PPD was higher than in the case of a 2.0 or 2.5 mm-diameter laser spot. These findings confirm that, for our laser source, with a 2.0 mm-diameter laser spot, the most pronounced mechanical effect was obtained.

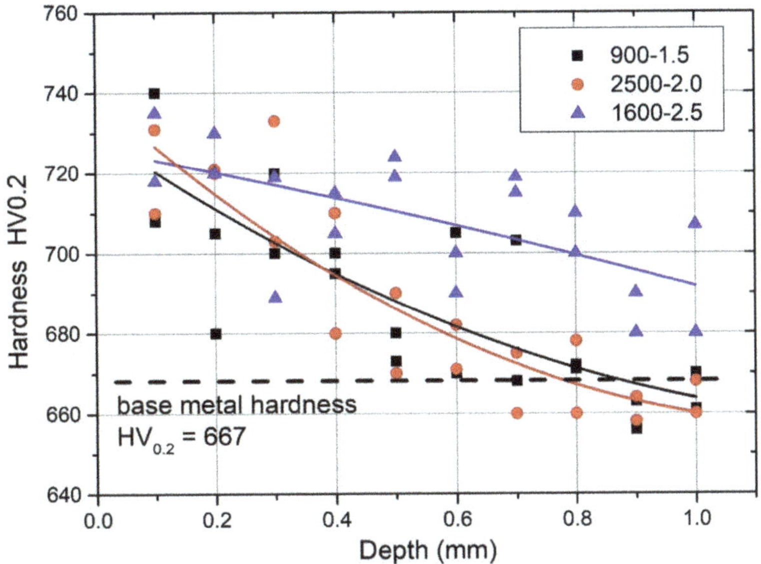

Figure 12. Micro-hardness measurements in depth for different PDs (i.e., 900, 1600, and 2500 cm^{-2}) and different SDs (i.e., 1.5, 2.0, and 2.5 mm).

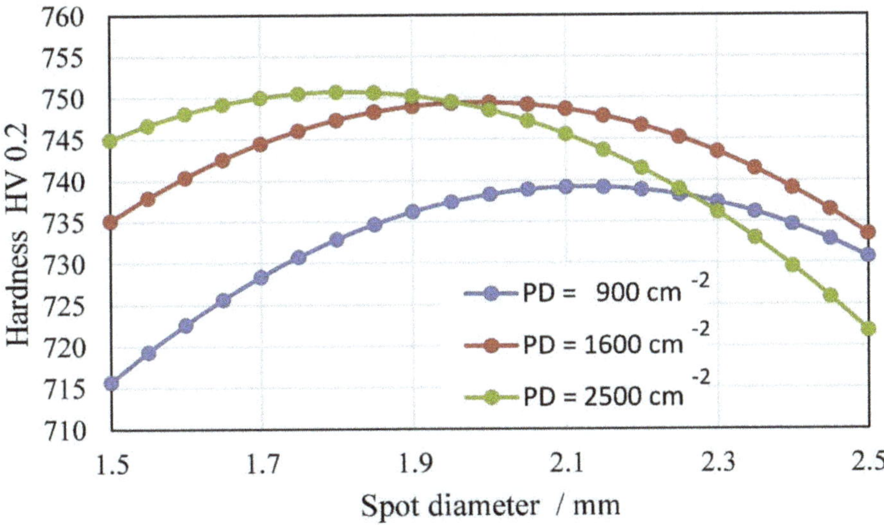

Figure 13. Microhardness at the surface as a function of laser SD for different PDs (i.e., 900, 1600, and 2500 cm^{-2}).

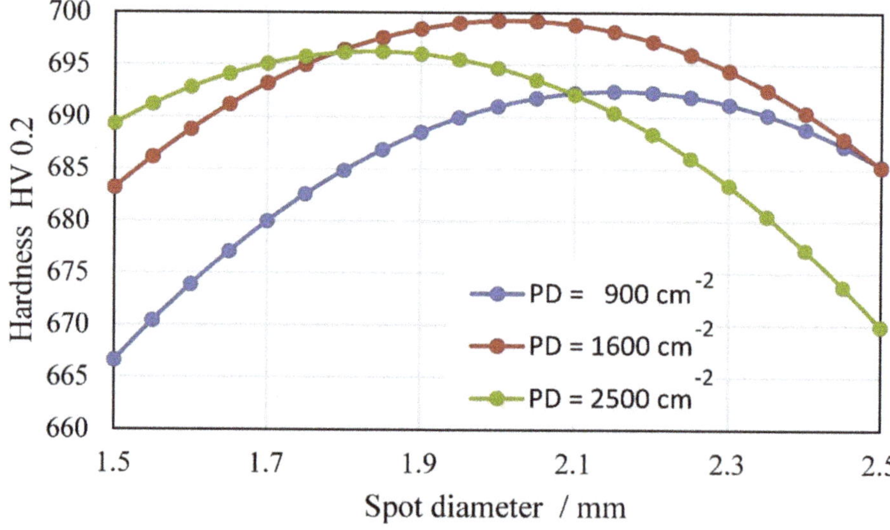

Figure 14. Microhardness at a depth of 1.0 mm as a function of laser SD for different PDs (i.e., 900, 1600, and 2500 cm^{-2}).

Additionally, other researchers found out that higher overlapping rates and multiple LSP impacts tend to cause a hardness increase [4,9]. We have previously shown that hardness increases after LSP is mainly due to the presence of compressive RSs [15,22]. With different LSP parameters, the microhardness decrease from the surface to a depth of 1.0 mm is almost linear. The relationship between microhardness and laser SD for PD = 1600 cm^{-2} is shown in Figure 15.

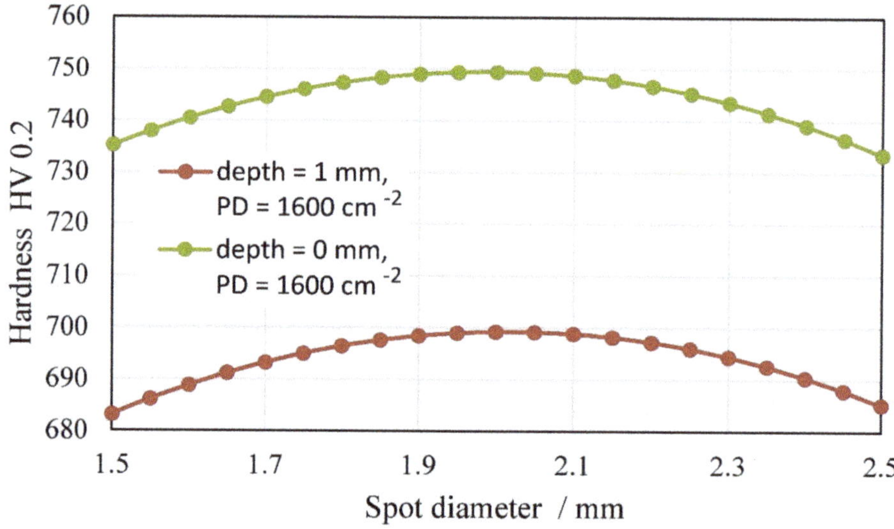

Figure 15. Microhardness at the surface and at a depth of 1.0 mm as a function of laser SD for PD = 1600 cm^{-2}.

3.4. Resonant Fatigue Resistance

The presence of compressive RS itself does not insure an improvement in fatigue strength. The increase in surface roughness must be considered, because surface defects may promote initialization and propagation of cracks during dynamic loading. The fatigue test results are shown in Figure 16. It was found out that the maraging steel in a delivered state (maraging solution annealing, with hardness values of 32–37 hardness Rockwell scale C (HRC) and an *Rm* value of 1200 MPa) withstood the same number of fatigue cycles for crack initiation, which was precipitation-hardened, when the fatigue testing was performed at a bending moment of 60 N·m and a bending stress of 833 Mpa. When the bending stress was increased to 1082 Mpa, the precipitation-hardened maraging steel showed a 43% increase in crack initiation time (blue line in Figure 16). As can be observed, the LSP was proven to be successful in improving the fatigue resistance of precipitation-hardened maraging steel. The fatigue resistance was improved after every combination of laser SD and PD, which was tested within our research. Therefore, the negative effect of the increased surface roughness did not overcome the positive effect of the compressive RSs, present in the thin surface layer of the laser-peened material. According to our statistical analysis, we discovered that the selection of laser SD between 1.5 and 2.5 mm in combination with a PD of 900 or even 2500 cm^{-2} did not have a statistically significant effect on the fatigue resistance when comparing the LSP-treated specimens. The selection of laser processing parameters had a statistically significant effect on fatigue life only, when a laser SD of 2.0 mm was used ($P < 0.0001$). The LSP with an SD of 2.0 mm and a PD of 1600 cm^{-2} increased the fatigue life of the precipitation-hardened maraging steel specimens from a range of 2–4 × 10^4 cycles to a range of 5–9 × 10^5 cycles. Therefore, the number of fatigue cycles, necessary for fatigue crack initiation, increased by 25 times. In some cases, we stopped the fatigue test after 10^7 cycles without crack initiation.

The implementation of the LSP technique can increase fatigue resistance, by generating compressive RSs and inducing strain hardening in most loaded surface areas, which consequently reduces the need for tool repair. Hence, when we succeed in increasing tool maintenance intervals (repair or change), the production cost is reduced.

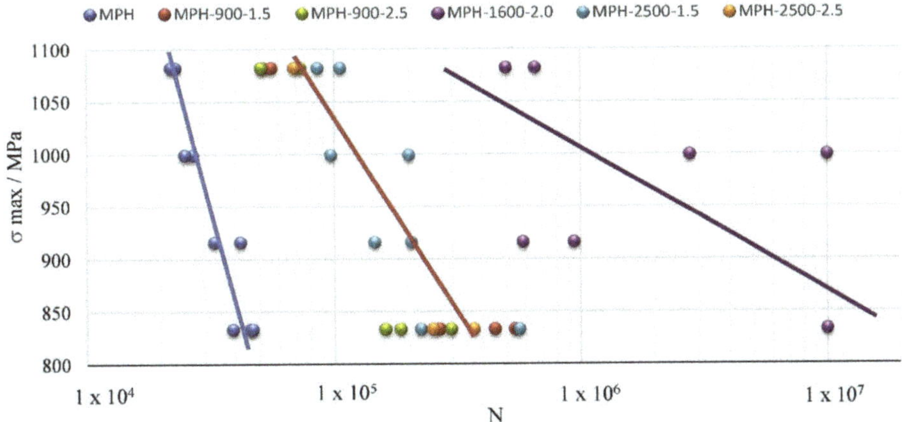

Figure 16. Fatigue test results.

Further, the cyclic loading after the crack initiation phase causes crack propagation and leads to a continuous decrease of the resonant frequency until final fracture. Other researchers have also studied the effect of LSP on crack propagation behavior [24]. Within our research, we compared the behavior of the resonant frequency between the untreated and laser-peened specimens during the fatigue loading.

As can be observed in Figure 17, the resonant frequency was decreasing more slowly, when the surface was treated with LSP. This finding indicated that laser peening not only extended the fatigue crack initiation time, but also reduced the crack propagation rate.

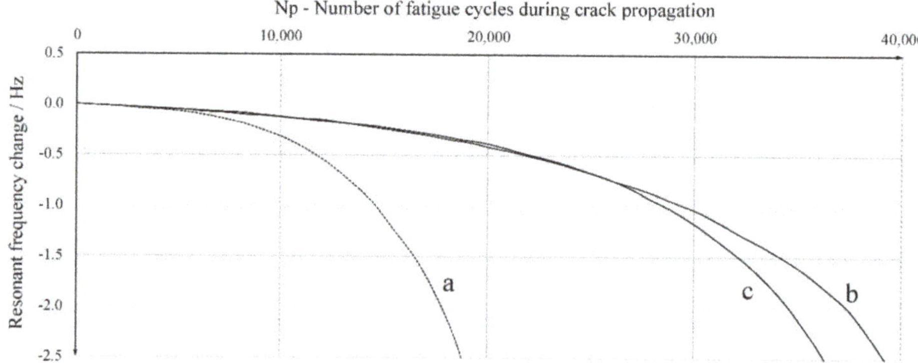

Figure 17. Resonant frequency behavior within the crack propagation phase during the fatigue bending: (a) MPH; (b) MPH-900-2.5; and (c) MPH-1600-2.0. The maximum bending moment was 78 N·m.

Figure 18a presents the failure locations on the specimens without LSP. The crack grew at the surface, because the LSP process slowed down thanks to the presence of the compressive RSs.

By laser peening with a 2.0 mm-diameter laser spot, the generated compressive RSs were in a range of 800–900 MPa, such that their positive effect could overcome the critical stress on the narrowed site in the middle of the fatigue specimen. It can be observed that the fatigue crack initiated at the edge of the LSP area (Figure 18c) and was not in the narrowest position in the middle of the specimen (Figure 18b).

When choosing the best laser peening parameters for increasing the fatigue life of a component, we must take into account a combination of the highest hardness and the highest compressive RS at a great depth (in our case, it is 1 mm). Laser spot size, in combination with laser power density and overlapping degree of laser pulses, affects the propagation of shock waves [26]. Small-diameter shock waves have different attenuation rates from those of large-diameter shock waves, when they go deep. In our case, we obtained the best combination of surface quality and material properties with the following LSP parameters: laser PPD = 8.9 GW·cm^{-2}, laser PD = 1600 cm^{-2}, laser SD = 2 mm, laser spot overlapping degree = 87%.

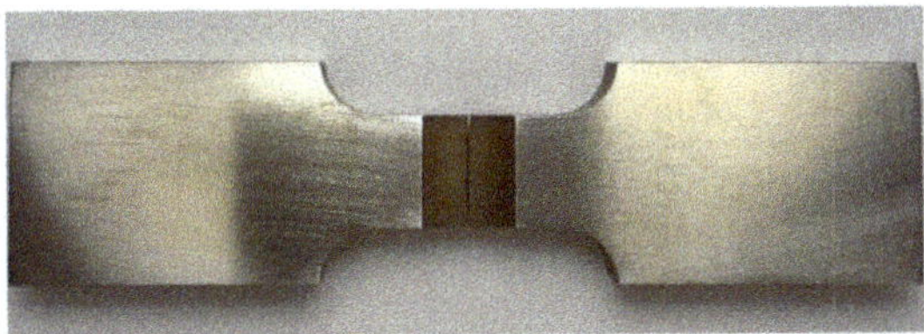

(**a**) Without LSP

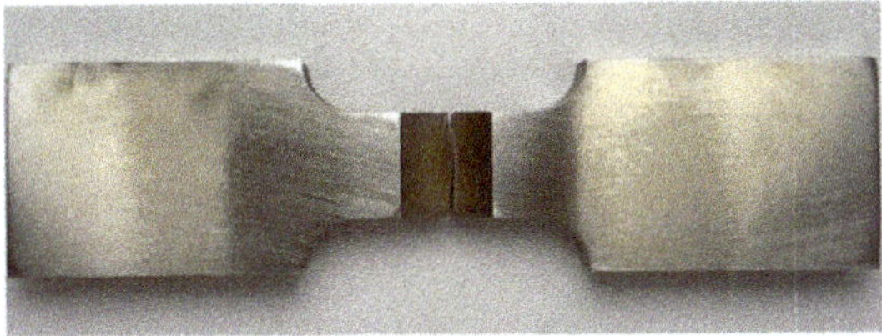

(**b**) With LSP, PD = 900 cm⁻², SD = 2.5 mm, crack in the middle of the specimen

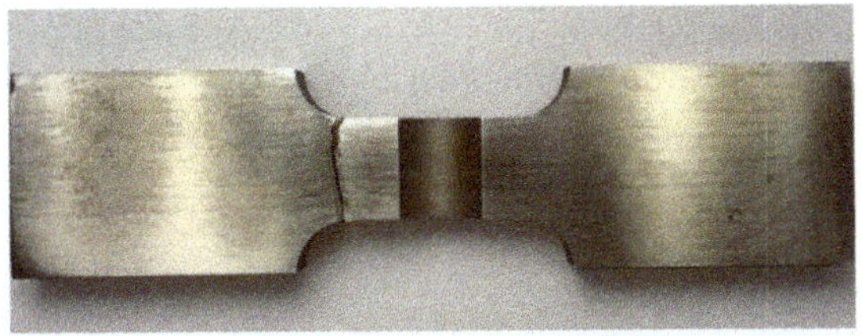

(**c**) With LSP, PD = 1600 cm⁻², SD = 2.0 mm, crack at the edge of LSP area

Figure 18. Fatigue crack and failure location on the specimen without LSP (**a**), the specimen with LSP for PD = 900 cm^{-2} and SD = 2.5 mm (**b**), and the specimen with LSP for PD = 1600 cm^{-2} and SD = 2.0 mm (**c**).

4. Conclusions

In our research, we have investigated the effects of LSP treatment on the surface properties of the maraging steel X2NiCoMo18-9-5 in a precipitation-hardened state. We have used different combinations of PD and spot size on the surface. The LSP was performed in a confined mode without a protective coating using constant laser pulse energy and duration. The conclusions could be summarized as following:

- After the LSP treatment, the maximum compressive RS of 1050 MPa was generated at the surface. At a depth of 1.0 mm, they still could be at a level of −300 MPa when using the optimal LSP parameters. Otherwise, the transition from a compressive state to a tensile state occurred at depths between 0.5 and 1.0 mm.
- Before the LSP, the surface roughness was Ra = 0.2 μm and Rz = 1.2 μm. After the LSP, the surface roughness increased to Ra = 0.6–1.2 μm and Rz = 3.9–7.4 μm. Higher roughness was obtained, when we increased the laser SD and PD.
- With increasing PD, profile depth Pt was increased. This phenomenon is more distinct at a large diameter of a laser spot. In the case of SD = 2.5 mm and PD = 2500 cm^{-2}, the profile depth was almost 100 μm.
- Strain hardening was detected as an increase in microhardness. The microhardness of the precipitation-hardened maraging steel before the LSP was 667 HV$_{0.2}$. The highest microhardness

value at the surface after the LSP is around 750 $HV_{0.2}$ and was registered on the specimen treated with a 2.0 mm-diameter laser spot and laser PDs of 1600 and of 2500 cm^{-2}.

- We have obtained the best combination of mechanical properties of the modified surface layer using a 2.0 mm-diameter laser spot.
- The LSP successfully improved the mechanical fatigue resistance of the maraging steel.
- The negative effect of the increased surface roughness did not overcome the positive effect of compressive RSs generated by LSP.
- Considering the chosen testing parameters, the number of fatigue cycles, necessary for fatigue crack initiation, was increased by 25 times.
- By analyzing the resonant frequency during the fatigue testing, we have discovered that the rate of the decrease of the resonant frequency was lower after the LSP, which indicated that laser peening not only extended the fatigue initiation time, but also reduced the crack propagation rate.

Author Contributions: J.G. and J.L.O. conceived, designed and performed the experiments; J.A.P. performed LSP experiments, R.Š. and L.P. performed the microstructural, roughness, fatigue, and RS characterizations; L.P. and J.G. analyzed and evaluated the data; L.P. and R.Š. wrote the manuscript. J.G. supervised the study. All the authors reviewed the final paper.

Funding: This research received no external funding.

Acknowledgments: This paper is part of the research work within the National Research Program (Nr. P2-0270) financed by the Slovene Ministry of Education, Science and Sport. The authors are very grateful for the financial support.

Conflicts of Interest: The authors declare no conflict of interest.

References

1. Guo, Z.; Li, D.; Sha, W. Quantification of precipitate fraction in maraging steels by X-ray diffraction analysis. *Mater. Sci. Technol.* **2004**, *20*, 126–130. [CrossRef]
2. Santos, L.P.M.; Béreš, M.; Bastos, I.N.; Tavares, S.S.M.; Abreu, H.F.G.; Gomes da Silva, M.J. Hydrogen embrittlement of ultrahigh strength 300 grade maraging steel. *Corros. Sci.* **2015**, *101*, 12–18. [CrossRef]
3. Rohrbach, K.; Schmidt, M. Maraging steels. In *ASM Handbook—Properties and Selection: Irons Steels and High Performance Alloys*; Davis, J.R., Ed.; ASM International: Materials Park, OH, USA, 1990; pp. 793–800.
4. Decker, R.F.; Floreen, S. Maraging Steel—The First 30 Years. In *Maraging Steels Recent Developments and Applications, Proceedings of the Symposium TMS Meeting, Phoenix, AZ, USA, 25–29 January 1988*; TMS: Pittsburgh, PA, USA; pp. 1–38.
5. Mayer, H.; Schuller, R.; Fitzka, M.; Tran, D.; Pennings, B. Very high cycle fatigue of nitride 18Ni maraging steel sheet. *Int. J. Fatigue* **2014**, *64*, 140–146. [CrossRef]
6. Mitterer, C.; Holler, F.; Üstel, F.; Heim, D. Application of hard coatings in aluminium die casting—Soldering, erosion and thermal fatigue behavior. *Surf. Coat. Technol.* **2000**, *125*, 233–239. [CrossRef]
7. Shivpuri, R.; Semiatin, S.L. Friction and wear of dies and die materials. In *ASM Handbook—Friction, Lubrication and Wear Technology*; Henry, S.D., Ed.; ASM International: Materials Park, OH, USA, 1992; pp. 621–648.
8. Nalla, R.K.; Altenberger, I.; Noster, U.; Liu, G.Y.; Scholtes, B.; Ritchie, R.O. On the influence of mechanical surface treatments—Deep rolling and laser shock peening—on the fatigue behavior of Ti-6Al-4V at ambient and elevated temperatures. *Mater. Sci. A* **2003**, *355*, 216–230. [CrossRef]
9. Clauer, A.H. Laser shock peening for fatigue resistance. *Surf. Perform. Titan.* **1996**, *217*, 230.
10. Ding, K.; Ye, L. *Laser Shock Peening: Performance and Process Simulation*; Woodhead Publishing Limited: Boca Raton, FL, USA, 2006.
11. Bolger, J.A.; Montross, C.S.; Rode, A.V. Shock waves in basalt rock generated with highpowered lasers in a confined geometry. *J. Appl. Phys.* **1999**, *86*, 5461–5466. [CrossRef]
12. Kalainathan, S.; Prabhakaran, S. Recent development and future perspectives of low energy laser shock peening. *Opt. Laser Technol.* **2016**, *81*, 137–144. [CrossRef]

13. Peyre, P.; Berthe, L.; Scherpereel, X.; Fabbro, R. Laser-shock processing of aluminiumcoated 55C1 steel in water-confinement regime, characterization and application to high-cycle fatigue behavior. *J. Mater. Sci.* **1998**, *33*, 1421–1429. [CrossRef]

14. Banas, G.; Elsayed-Ali, H.E.; Lawrence, F.V.; Rigsbee, J.M. Laser shock-induced mechanical and microstructural modification of welded maraging steel. *J. Appl. Phys.* **1990**, *67*, 2380–2384. [CrossRef]

15. Grum, J.; Zupančič, M.; Ocana, J.L. Laser shock processing of the maraging steel surface. *Mater. Sci. Forum* **2007**, *537–838*, 655–662. [CrossRef]

16. Petan, L.; Ocaña, J.L.; Grum, J. Effects of LSP on surface integrity of 18% Ni maraging steel. *J. Mech. Eng.* **2016**, *62*, 291–298. [CrossRef]

17. Lavender, C.A.; Hong, S.-T.; Smith, M.T.; Johnson, R.T.; Lahrman, D. The effect of laser shock peening on the life and failure mode of a cold pilger die. *J. Mater. Process. Technol.* **2008**, *204*, 486–491. [CrossRef]

18. Chang, S.-H.; Tang, T.-P.; Tai, F.-C. Enhancement of thermal cracking and mechanical properties of H13 tool steel by shot peening treatment. *Surf. Eng.* **2011**, *27*, 581–586. [CrossRef]

19. Cho, K.T.; Song, K.; Oh, S.H.; Lee, Y.-K.; Lee, W.B. Surface hardening of shot peened H13 steel by enhanced nitrogen diffusion. *Surf. Coat. Technol.* **2013**, *232*, 912–919. [CrossRef]

20. Harada, Y.; Fukaura, K.; Haga, S. Influence of microshot peening on surface layer characteristics of structural steel. *J. Mater. Process. Technol.* **2007**, *191*, 297–301. [CrossRef]

21. Sawada, T.; Yanagitani, A. Properties of cold work tool steel shot peened by 1200 HV-class Fe-Cr-B gas atomized powder as shot peening media. *Mater. Transact.* **2010**, *51*, 735–739. [CrossRef]

22. Trdan, U.; Porro, H.A.; Ocaña, J.L.; Grum, J. Laser shock peening without absorbent coating (LSPwC) effect on 3D surface topography and mechanical properties of 6082-T651 Al alloy. *Surf. Coat. Technol.* **2012**, *208*, 109–116. [CrossRef]

23. Lorenzino, P.; Navarro, A. The variation of resonance frequency in fatigue tests as a tool for in-situ identification of crack initiation and propagation, and fort the determination of cracked areas. *Int. J. Fatigue* **2015**, *70*, 374–382. [CrossRef]

24. Zhao, Y.; Dong, Y.; Ye, C. Laser shock peening induced residual stresses and the effect on crack propagation behavior. *Int. J. Fatigue* **2017**, *100*, 407–417. [CrossRef]

25. Petan, L.; Ocaña, J.L.; Grum, J. Influence of laser shock peening pulse density and spot size on the surface integrity of X2NiCoMo 18-9-5 maraging steel. *Surf. Coat. Technol.* **2016**, *307*, 262–270. [CrossRef]

26. Montross, C.S.; Wei, T.; Ye, L.; Clark, G.; Mai, Y.W. Laser shock processing and its effects on microstructure and properties of metal alloys: A review. *Int. J. Fatigue* **2002**, *24*, 1021–1036. [CrossRef]

Article

Laser Shock Peening: Toward the Use of Pliable Solid Polymers for Confinement

Corentin Le Bras [1,2,*], Alexandre Rondepierre [1,3], Raoudha Seddik [1], Marine Scius-Bertrand [1,4], Yann Rouchausse [1], Laurent Videau [4], Bruno Fayolle [1,*], Matthieu Gervais [1,*], Leo Morin [1], Stéphane Valadon [2], Romain Ecault [2], Domenico Furfari [5] and Laurent Berthe [1,*]

[1] Laboratoire PIMM, UMR 8006, ENSAM, CNRS, CNAM, HESAM, 151 boulevard de l'Hôpital, 75013 Paris, France
[2] Airbus Operation S.A.S, 316 route de Bayonne-B.P. D4101, CEDEX 9, F-31060 Toulouse, France
[3] Thales Las France, 78990 Elancourt, France
[4] CEA, DAM, DIF, 91297 Arpajon, France
[5] Airbus Operations GmbH, Kreetslag 10, 21129 Hamburg, Germany
*   Correspondence: corentin.le-bras@airbus.com (C.L.B.); bruno.fayolle@ensam.eu (B.F.); matthieu.gervais@lecnam.net (M.G.); laurent.berthe@ensam.eu (L.B.); Tel.: +33-171-936-533 (C.L.B.)

Received: 20 May 2019; Accepted: 11 July 2019; Published: 17 July 2019

Abstract: This paper presents the first extensive study of the performances of solid polymers used as confinement materials for laser shock applications such as laser shock peening (LSP) as opposed to the exclusively used water-confined regime up to now. The use of this new confinement approach allows the treatment of metal pieces needing fatigue behavior enhancement but located in areas which are sensitive to water. Accurate pressure determination in the polymer confinement regime was performed by coupling finite element simulation and experimental measurements of rear free-surface velocity using the velocity interferometer system for any reflector (VISAR). Pressure could reach 7.6 and 4.6 GPa for acrylate-based polymer and cross-linked polydimethylsiloxane (PDMS), respectively. At 7 and 4.7 GW/cm^2, respectively, detrimental laser breakdown limited pressure for acrylate and PDMS. These results show that the pressures produced were also as high as in water confinement, attaining values allowing the treatment of all types of metals with LSP and laying the groundwork for future determination of the fatigue behavior exhibited by this type of treated materials.

Keywords: laser shock peening; polymers; solid confinement; VISAR measurement; finite element method

---

1. Introduction

The first discoveries leading to the development of modern-day laser shock peening (LSP) started in the 1960s with the spread of pulsed laser technology. The study of laser interaction with different materials led to pressure measurements on a surface ablated by a pulsed laser [1]. A major breakthrough occurred in 1970 when Anderholm discovered that the pressure delivered through a laser shock could be greatly improved by confining the plasma produced by placing a quartz overlay, transparent to the laser beam, on the target [2]. At the beginning of the 1970s, studies on the effect of LSP applied on metallic targets began at the Battelle institute in Columbus, Ohio and demonstrated an improvement of mechanical properties in the treated zone [3].

From there, LSP started becoming a competitor of the conventional shot peening for the improvement of the fatigue performance of treated components, and laser peening is now one of the most effective surface pre-stressing treatments used to enhance the fatigue strength of metallic structures. The laser-shock-peening-induced compressive residual stresses offer better fatigue life properties [4] by retarding the crack growth and inhibiting the crack initiation [5–7]. In addition, LSP can significantly enhance the resistance of the treated components to stress corrosion [8].

Compared to the conventional shot peening, the affected depth is much larger for laser shock peening—up to ≈1.5 mm compared to ≈300 μm for shot peening [9] of Al alloy materials. In addition, the LSP-induced work hardening is generally limited (about +10% to +30%) compared to conventional shot peening [10]. This can be explained by the fact that the loading duration is very short, which consequently does not allow the activation of all the sliding systems of the material and thus generates fewer cross dislocations. Only cyclic hardening materials such as 304L and 316L have their hardness and their level of residual stresses increase with impact repetition. Nevertheless, it is not possible to draw conclusions regarding the amplitude of the induced compressive residual stresses, given that contradictions exist in the literature. Therefore, conventional shot peening may be unfavorable over a material's lifetime due to controlled deformation loading (high cycle thermal loading), reducing the beneficial effects of compressive residual stresses. For laser shock, such a risk can be avoided because of the low work hardening.

Today, LSP at an industrial scale is mainly applied in aeronautics for the treatment of sensitive areas on certain parts to increase their lifespan. Water is the usual confining material because it is cheap, transparent to the laser, and ensures contact with surfaces. Other areas are developing quickly toward industrial applications such as the treatment of parts produced by additive manufacturing. LSP treatment of these types of materials allows more shaping and forming possibilities as well as shape correction treatment due to the highly controlled nature of the process compared to conventional shot peening [10,11]. Coupled with the deeper levels of residual stress produced, it has shown to be highly cost effective despite its higher operating cost compared to conventional shot peening.

One of the obstacles to extending LSP's applications is the impossibility of using water in a reactive atmosphere or near electronic devices. A solution to this issue should be a solid material, as demonstrated by the pioneering work on laser shock [2]. However, the use of rigid glasses for the treatment of pieces presenting complex geometries such as the ones encountered in the aeronautics industry is impossible. In contrast, a soft polymer confinement, with its adaptability, shaping possibilities, and wide range of formulations and properties, is an ideal candidate for this type of need. Laser shock peening with polymer confinement has been studied only by Hong et al. [12]. Authors evaluated only the influence of the confining medium used on mechanical impedance, without carrying out a complete investigation of the performances exhibited by these materials.

This paper presents a study concerning the use of polymers for laser shock applications. It focuses on the capability of materials to be transparent to a laser beam at high intensity and to generate a high-pressure plasma when used for the laser shock peening process in a confined regime and without a thermal coating. Plasma pressure is the driving force of the process for evaluating the range of materials which can be processed and the boundary limit for simulation. In the case of confinement in specific conditions, only experimental approaches can determine this key parameter of LSP. We present optical transmission and the characterization of plasma pressures produced by the laser in a confined regime with a choice of polymer confinements (i.e., acrylate and polydimethylsiloxane). Experiments were performed from rear free-surface velocity measurements using the velocity interferometer system for any reflector (VISAR) on pure aluminum foils coupled with numerical simulations. Results are compared to the pressure produced with water confinement interaction. The first part of this paper presents the experimental setup and methodology, while the second part is dedicated to results and discussion.

2. Materials and Methods

Figure 1 presents the experimental setup and methods used to determine plasma pressure from the velocity profile measured by VISAR.

The velocity profile was reproduced using a finite element (FE) model in which the plasma pressure was adjusted as an input condition. A direct correlation between rear free-surface velocity and the maximum pressure of the plasma was also extracted to provide fast signal analysis.

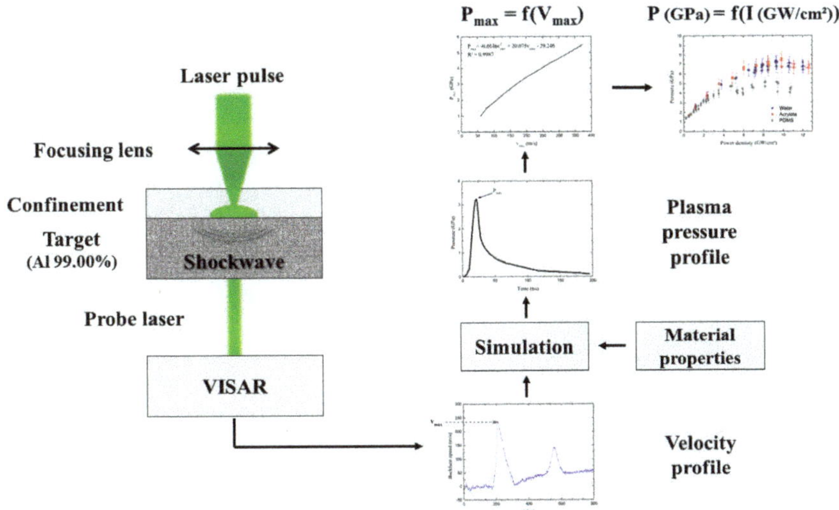

Figure 1. Setup used in the Hephaïstos platform to realize the pressure measurements. VISAR: velocity interferometer system for any reflector.

2.1. Material

The choice of the polymers was based on a few criteria in order for them to be able to deliver good results when used for LSP; namely, good optical transmission to the laser wavelength at high laser intensity, and the capability to generate sufficient pressures. Two polymers fulfilling these conditions were chosen for use as confining materials for the study. The first one was an acrylate-based tape with a thickness of 1 mm . The second one was a cross-linked PDMS with a thickness of 2 mm. Both can be defined as pressure-sensitive adhesives [13], allowing an easy bonding and debonding of the confining material to the target.

For pressure determination, the targets chosen were pure aluminum foils 99.00% from Goodfellows company (AL000645, Coraopolis, PA, USA) with a thickness of 1 mm. Aluminum is a material of choice for its well-defined properties. The thickness was chosen as a way to separate the first peak from the velocity profile from the second peak caused by the shock wave going back and forth inside the material. To ensure contact between the polymer and the aluminum target, every potential bubble was pushed away by applying low pressure on the adhesive. For the water-confined regime, a water droplet was apposed on the surface of the aluminum target. For each confinement and each laser intensity used, three samples were tested to minimize potential error and ensure reproducibility of the results.

2.2. Laser

The laser shots were performed with the Hephaïstos facility at PIMM (Laboratory for Processes and Engineering in Materials and Mechanics, Paris, France). The platform is equipped with a Gaïa HP laser from THALES (Elancourt, France). It is composed of two Nd:YAG synchronized or delayed lasers, both emitting at 532 nm. It delivers up to 14 J with a Gaussian temporal profile with a full width half-maximum (FWHM) of 9 ns. The laser spot diameter is 3 mm through a diffractive optical element (DOE) and lens with a focal length of 148 mm [14,15]. The parameters chosen and the range of laser intensity covered by this study are standard parameters for laser peening applications such as fatigue behavior improvement and the reduction of crack propagation by the induction of compressive residual stress. Energy and pulse duration were measured shot by shot. The spatial profile was measured every day.

Figure 2 shows a typical image and corresponding profile of the laser spot at the location of the target, demonstrating a uniformity of intensity below 6%. With these laser parameters, the power density reached up to 13 GW/cm^2 to achieve the breakdown threshold for all materials, which is the physical limit of the process.

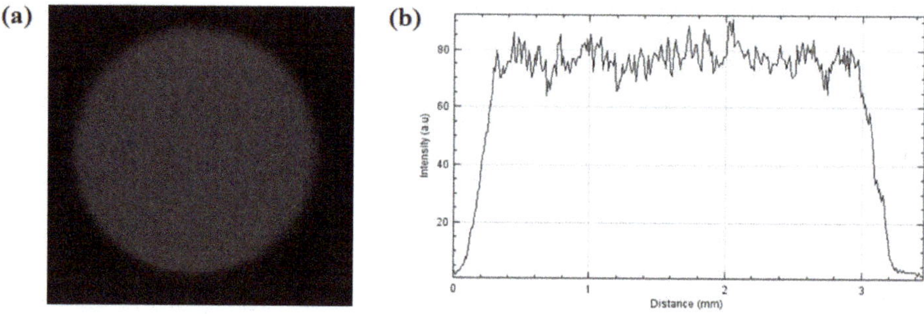

Figure 2. (**a**) Spatial shape and (**b**) intensity profile for a 3-mm laser spot with a diffractive optical element (DOE) at the Hephaïstos facility.

2.3. Transmission Measurements

Optical transmission measurements of the polymer were performed using a calorimeter (QE50LP-H-MB-QED, Gentec, Québec, QC, Canada) located on the back of the sample holder used for the laser shock. The transmission was calculated from the difference of energy with or without material. The incident energy was chosen for a power density below the laser threshold (0.2 GW/cm^2).

2.4. VISAR

The rear free-surface velocity of pure aluminum foils was measured with VISAR [16]. It is an optical diagnostic tool allowing the characterization of shock waves via the obtention of the back face velocity of the sample. It is a non-contact acquisition method which can be used with water or solid confinement. The device is a Michelson-type interferometer measuring the Doppler shift in the wavelength of a probe laser reflected on the rear free surface of a target which is moved as a result of the induced shock wave.

The PIMM VISAR is composed of two main parts: the laser probe, which is a continuous-wave laser (532 nm, 5 W) from the Coherent company (Santa Clara, CA, USA), and a home-made interferometer. The laser probe is focused on the back face of the target on the center of the laser spot of the incident laser (Figure 1). The wavelength shift induced by the rear free-surface deformation with the shock-wave propagation is then transmitted to the interferometer part, where the signal interferes with itself. This device has already been used and described for plasma characterization in the direct regime [17] and in the water-confined regime [18].

2.5. Simulation

To determine plasma pressure profiles, experimental rear free-velocity profiles were predicted using simulation. The laser shock propagation simulation was carried out using the commercial FE code ABAQUS 6.14 [19]. A 2D axisymmetric model was developed. Since laser shock is a high-speed process, an explicit solver is used while considering the dynamic effects.

2.5.1. Target Geometry and Boundary Conditions

Figures 3 and 4 respectively show the target configuration and the mesh used for simulation. The target was modeled as a planar shell with a 1 mm width and a 7.5 mm length. It was meshed by means of CAX4R elements (Continuum, 4-node bilinear axisymmetric, quadrilateral, reduced integration, hourglass control). To improve the accuracy of the FE solutions, a refinement mesh was used in the treated area with the use of a BIAS function in the X direction (Figure 4). As the element dimensions decreased, the obtained results became more stable. A decrease between experimental and numerical rear free-velocity profiles as observed. The treated region was refined with small elements of 1 µm × 1 µm dimensions. For the boundary condition, one node of the bottom surface was fixed (Figure 3).

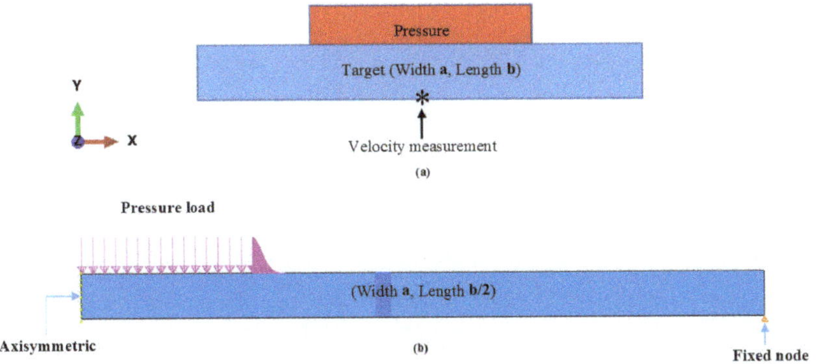

Figure 3. Target used for the laser shock peening (LSP) finite element (FE) modeling: (**a**) geometry and (**b**) 2D axisymmetric FE model.

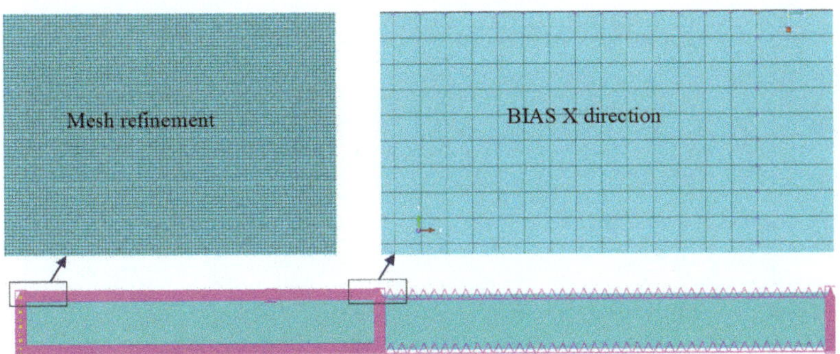

Figure 4. Mesh refinement used for the LSP simulation.

2.5.2. Constitutive Material Model

For shock produced by laser-induced high strain rate (i.e., up to 10^6 s^{-1}), the appropriate behavior law of the treated target must consider the effect of the high strain rate. In this context, the Johnson–Cook constitutive model [20] was used. In this model, the Von Mises yield criterion is defined as:

$$\sigma = \left(\sigma_y + K\epsilon_p^n\right)\left(1 + C ln\left(\frac{\dot{\epsilon}}{\dot{\epsilon}_0}\right)\right)\left(1 - \left(\frac{T - T_0}{T_{melt} - T_0}\right)\right). \tag{1}$$

The first part describes the strain-hardening effect. The second part characterizes the strain rate effect. The last part of the Johnson–Cook constitutive law associates the stress with material temperature evolution during the plastic deformation. σ_y is the yield stress, B is the hardening modulus. It encompasses the second member of the above mentioned equation, ϵ_p is the equivalent plastic deformation, n is the hardening coefficient, C is the strain rate sensitivity parameter, $\dot{\epsilon}$ is the strain rate during the process, $\dot{\epsilon}_0$ is the reference strain rate, T_{melt} is the fusion temperature, T_0 is the room temperature, E is Young's modulus, and v is Poisson's ratio.

In the present study, a preliminary simulation using CAX4RT elements (Continuum, A 4-node thermally coupled axisymmetric quadrilateral, bilinear displacement and temperature, reduced integration, hourglass control) confirmed that the thermal softening effect could be neglected (for the case of laser intensity $I = 2.7$ GW/cm^2, plasma pressure FWHM equal to 18 ns, and spot size of 3 mm). In fact, the local temperature increases due to the plastic deformation and the wave propagation did not have a significant influence on the rear free velocity. For this reason, the thermal part in Johnson–Cook's constitutive law was neglected [21]. For pure aluminum, Table 1 [22] gives the Johnson–Cook parameters used for the simulation.

Table 1. Parameters used for the Johnson–Cook material model with an aluminum target (99.00%) [22].

Material	σ_y(GPa)	B (GPa)	n	C	$\dot{\epsilon}_0$	E (GPa)	v
Aluminum	0.129	0.2	0.45	0.03	0.01	69	0.33

2.5.3. Spatial and Temporal Pressure Profiles

To generate spatial and temporal pressure profiles $P = f(x, y, t)$, a VDOLAD subroutine was used. The $P(t)$ profiles given in Figure 5 [23] were adjusted to provide coherence between the experimental and the numerical results. The $P(x, y)$ distribution was obtained (Figure 6) from beam analysis. The intensity profile from Figure 2b, obtained through a camera, was used to generate the $P(x, y)$ distribution. Previous work on the subject showed the equivalence between maximum pressure and intensity [18,24]. Preliminary simulation showed that modulation had no effect on the rear free-velocity profile.

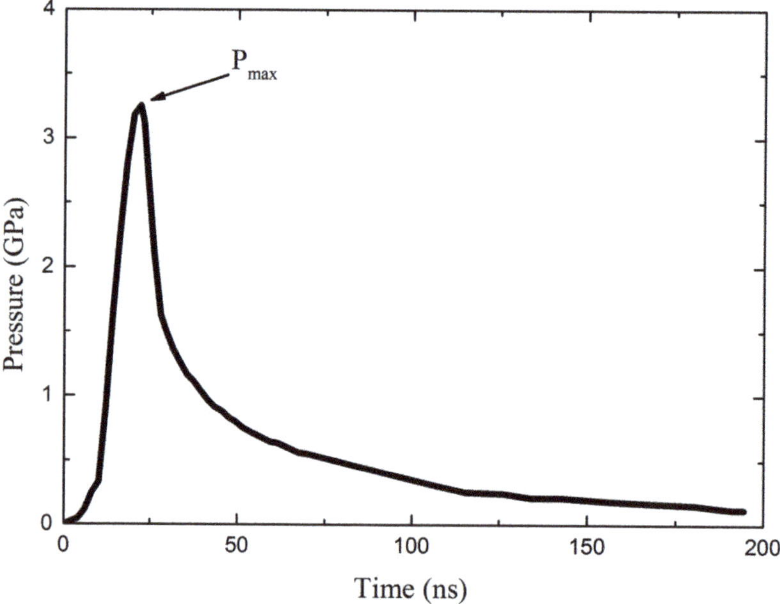

Figure 5. $P(t)$ used for the simulation.

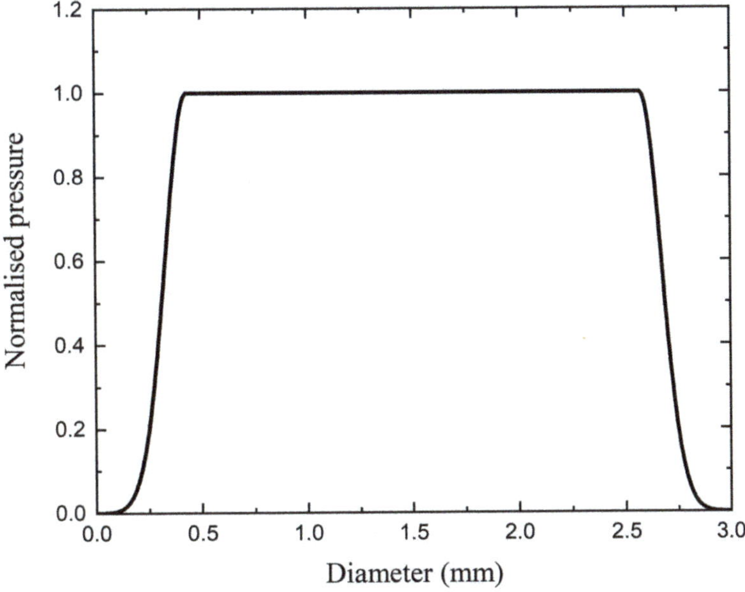

Figure 6. $P(x, y)$ distribution simulation obtained from the intensity profile from Figure 2b.

3. Results and Discussion

3.1. Transmission

Table 2 presents the optical transmission of acrylate and PDMS used for the confined laser interaction. It shows that acrylate and PDMS exhibited very good transmissions of 0.92 and 0.82, respectively. This confirms that they should be good candidates for applications. The corresponding absorption can be accounted for by following the incident power density. For the water-confined regime, no incident energy loss was considered since water is transparent at the wavelength used [25].

Table 2. Optical transmission of acrylate and PDMS at 532 nm.

Confinements	Transmission
Acrylate	0.92
PDMS	0.82

3.2. Pressure Measurement

Figure 7 presents a comparison of the rear free-velocity profile of the aluminum target in the water-confined regime at 2.7 GW/cm^2 measured with VISAR with the values obtained by simulation. The velocity profile presented two peaks corresponding respectively to the first and second occurrences of the shock wave at the rear surface of the target.

On the front shock, an inflection of the velocity corresponded to the elastic precursor at 38 m/s which equates to a Hugoniot limit P_H of 0.312 GPa. We could calculate the dynamic yield stress using the following formula [26]:

$$\sigma_0^{Dyn} = P_H \frac{1 - 2\nu}{1 - \nu}. \tag{2}$$

This calculation yields 0.179 GPa, which is in agreement with value previously determined in Table 1 [22]. The simulation agreed very well with the experiments for both peaks. The corresponding maximum plasma pressure, denoted as P_{max}, was 3.252 GPa. Simulation showed a negative peak (25 m/s) at the end of release at 300 ns. This corresponds to effects in relation with the edge of the spatial profile [27], which are very sensitive to pressure gradients.

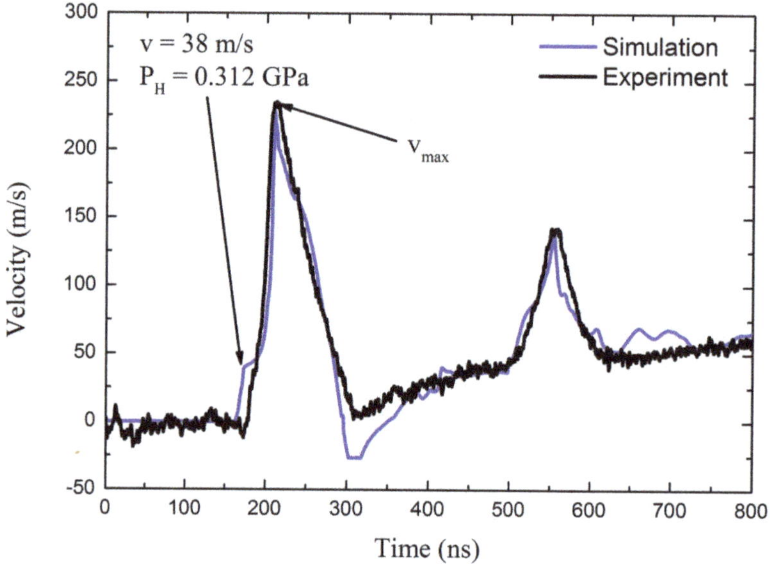

Figure 7. Simulated and experimental rear free-velocity profiles of a water-confined laser shot on 1-mm-thick aluminum at 2.7 GW/cm^2 with a laser spot diameter of 3 mm.

From the simulation, the maximum pressure of plasma P_{max} could be related to the maximum velocity of the first peak (v_{max}) measured by VISAR, as shown in Figure 8:

$$P_{max}(\text{GPa}) = -0.01456 v_{max}(\text{m/s})^2 - 20.075 v_{max}(\text{m/s}) - 29.246. \tag{3}$$

Compared to previous results [18], this includes shock-wave attenuation depending on the incident pressure and target thickness.

Figure 9 presents typical rear free-velocity profiles at 2.7 GW/cm^2 for water as well as acrylate and PDMS polymeric confining materials. Figure 9 focuses on the first occurrence of the laser shock wave. The global shape was similar for the three materials: both the inflection on the shock front due to elastic precursor and the FWHM (55 ns) were alike. Some differences were highlighted in the end of the relaxation in relation to edge effects. However, maximum velocities v_{max} were different, showing 237 m/s, 217 m/s, and 187 m/s for water, acrylate, and PDMS respectively. From rear free-velocity profiles, corresponding P_{max} values were calculated from the incident power densities ranging from 0.22 GW/cm^2 to 12.50 GW/cm^2. They are gathered in Figure 10 for water, acrylate, and PDMS.

For the three confining materials, the data can be separated in two parts. In the first part, pressure increased with the incident power density, and in the second part the pressure saturated due to the detrimental effect of plasma breakdown of the material confinement [28]. The power density for which the breakdown limits the plasma pressure of a confining material is called the breakdown threshold. These are reported with maximum pressure in Table 3.

The acrylate and water confinements exhibited the same pressures up to 2.3 GW/cm^2. At higher intensities, the acrylate produced slightly higher pressures, respectively 6.5 GPa at 6.02 GW/cm^2 and 6.4 GPa at 6.53 GW/cm^2. The breakdown threshold was 7 GW/cm^2 for both confinements, and saturating pressures were produced at 7.6 GPa for the acrylate confinement and 7 GPa for water.

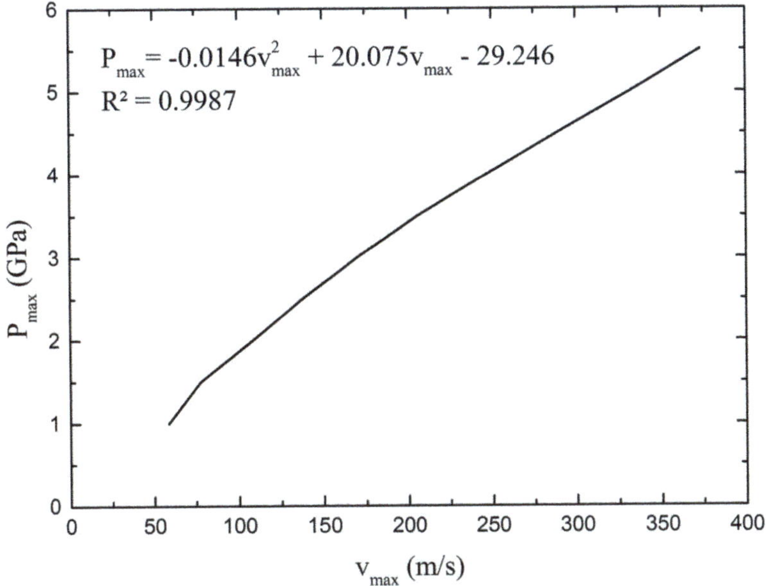

Figure 8. Maximum pressure of the plasma extracted from simulation as a function of the maximum velocity developed by the laser shock process depending on the rear free-surface velocity obtained with VISAR measurement.

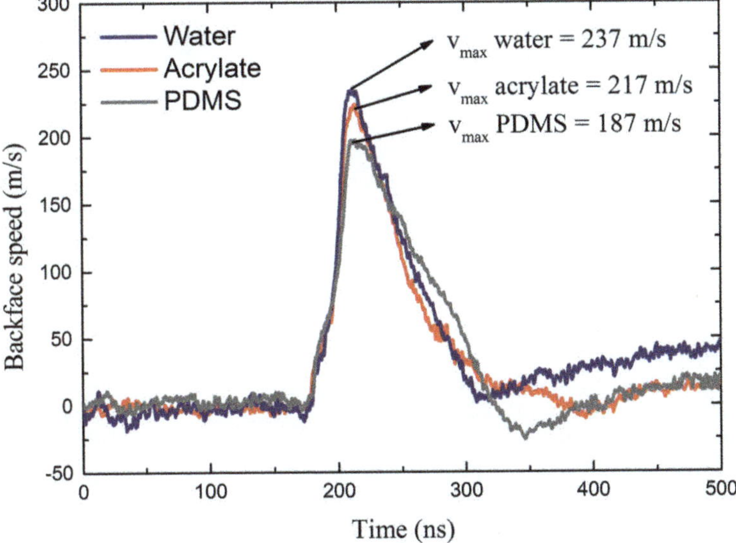

Figure 9. Velocity profiles with a laser focal diameter of 3 mm, an incident power density of 2.7 GW/cm^2, and a 1-mm-thick aluminum target for water, acrylate, and PDMS confinements.

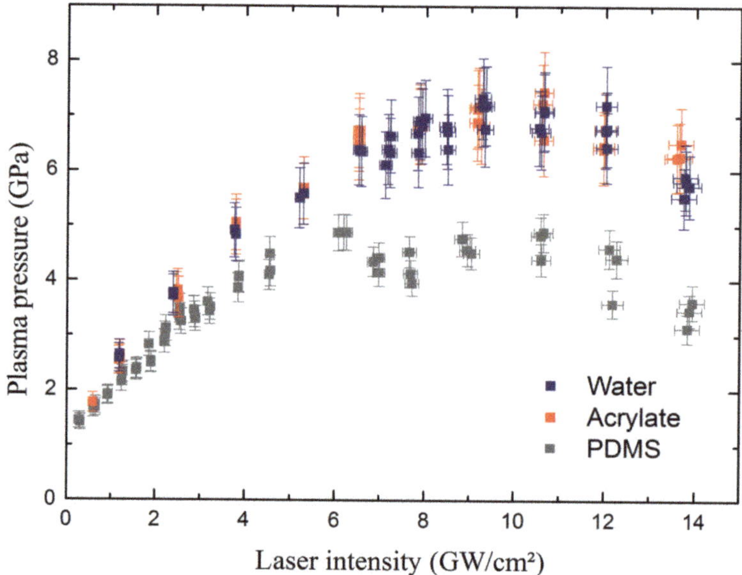

Figure 10. Pressure measurements obtained from the rear free-surface velocity measured with VISAR for the three different confinements: water, acrylate, and PDMS.

As expected with its optical transmission, the PDMS confinement exhibited pressures that were slightly lower than water and acrylate confinements, respectively 4.25 GPa at 3.50 GW/cm^2, 4.90 GPa at 3.75 GW/cm^2, and 4.98 GPa at 3.50 GW/cm^2. The breakdown threshold was also lower at 4.7 GW/cm^2, producing a maximum pressure of 4.6 GPa.

The chemical composition of the polymers can influence the breakdown threshold. Following this, the higher optical absorption of the PDMS favored material damage initiation compared to the acrylate confining medium. In fact, polymer damage was observed regardless of the polymer confinement used when high laser intensities were applied, typically when going higher than 3 GW/cm^2 for the acrylate confinement, which seemed to be perfectly transparent except for the laser energy loss at the interfaces. With growing damages, plasma breakdown can initiate on defects, thus lowering the breakdown threshold. The confined plasma composition is a mix, coming from the contribution of the aluminum target and the polymer confinement material through ablation and plasma heating, respectively [29]. Maximum plasma pressures were also higher than previously published by Berthe in [18,30]. The difference could be due to the better laser spatial profile obtained by using a DOE and the simulation of rear free-surface velocity on the range of incident power density instead of using a constant attenuation. The present results clearly indicate the capability of polymers to reach pressures equivalent to water confinement. However, efforts must be made concerning the adhesion and mechanical properties of polymers for LSP configuration using high repetition rate shot without coatings by varying chemical composition and manufacturing parameters.

Table 3. Breakdown thresholds and maximum pressures extracted from Figure 10.

Confinement		Water	Acrylate	PDMS
Breakdown threshold	(GW/cm^2)	7	7	4.7
Maximum pressure	(GPa)	7	7.6	4.6

4. Conclusions

This papers presents for the first time pressure measurements in a polymer confinement regime for 9 ns pulse duration at 532 nm wavelength up to 12.5 GW/cm^2 corresponding to the current laser parameters of LSP. Pressures were determined by coupling FE simulation and experimental measurements of rear free-velocity performed with VISAR.

- Simulation and experiment coupling allowed accurate pressure measurements and full time and spatial determination of boundary limits for process modeling.
- The pressures generated with acrylate were as high as those produced in the water confinement regime, up to 7 GPa, while reaching 4.6 GPa for PDMS.
- A breakdown phenomenon occurred at 7 GW/cm^2 for the water and acrylate confinements while it happened at around 4.7 GW/cm^2 for the PDMS confinement.
- Results demonstrate that LSP with confinement polymer allows the treatment of all types of metals, from low to high elastic limit material. [31].

Therefore, solid polymer confinements are an alternative to water for laser shock applications such as LSP. Further works will concern:

- The influence of polymer chemical composition on plasma breakdown (extended to infrared wavelength)
- Plasma physics at the metal/polymer interface as well as the plasma composition.
- Optimization of polymer mechanical and adhesive properties for high-repetition-rate processing.

Author Contributions: This work was supervised by L.B. The experiments on Hephaistos' facility were conducted by C.L.B., A.R., M.S.-B., and L.B., with Y.R.'s technical support. The design and manufacture of the polymers were done by B.F., M.G., and C.L.B. The simulations were made by R.S. under the supervision of L.M., L.V., R.E., and S.V., and D.F. participated in the exploitation and design of the Hephaistos experiments.

Funding: This study was supported by the Authors' affiliate companies (Airbus and Thales) and institutions (CEA, CNRS, and ENSAM) and Agence pour la Recherche Forge Laser.

Conflicts of Interest: Authors declare no conflict of interest.

Reference

1. Askar'yan, G.A.; Moroz, E.M. Pressure on Evaporation of Matter in a Radiation Beam. *Sov. J. Exp. Theor. Phys.* **1963**, *16*, 1638.
2. Anderholm, N. Laser-generated stress waves. *Appl. Phys. Lett.* **1970**, *16*, 113–115. [CrossRef]
3. Fairand, B.P.; Wilcox, B.A.; Gallagher, W.J.; Williams, D.N. Laser shock-induced microstructural and mechanical property changes in 7075 aluminum. *J. Appl. Phys.* **1972**, *43*, 3893–3895.10.1063/1.1661837. [CrossRef]
4. Peyre, P.; Fabbro, R.; Merrien, P.; Lieurade, H. Laser shock processing of aluminium alloys. Application to high cycle fatigue behaviour. *Mater. Sci. Eng. A* **1996**, *210*, 102–113, doi:10.1016/0921-5093(95)10084-9. [CrossRef]
5. Pavan, M.; Furfari, D.; Ahmad, B.; Gharghouri, M.; Fitzpatrick, M. Fatigue crack growth in a laser shock peened residual stress field. *Int. J. Fatigue* **2019**, *123*, 157–167, doi:10.1016/j.ijfatigue.2019.01.020. [CrossRef]
6. Dhakal, B.; Swaroop, S. Review: Laser shock peening as post welding treatment technique. *J. Manuf. Process.* **2018**, *32*, 721–733, doi:10.1016/j.jmapro.2018.04.006. [CrossRef]
7. Sun, R.; Li, L.; Guo, W.; Peng, P.; Zhai, T.; Che, Z.; Li, B.; Guo, C.; Zhu, Y. Laser shock peening induced fatigue crack retardation in Ti-17 titanium alloy. *Mater. Sci. Eng. A* **2018**, *737*, 94–104, doi:10.1016/j.msea.2018.09.016. [CrossRef]
8. Peyre, P.; Scherpereel, X.; Berthe, L.; Carboni, C.; Fabbro, R.; Béranger, G.; Lemaitre, C. Surface modifications induced in 316L steel by laser peening and shot-peening. Influence on pitting corrosion resistance. *Mater. Sci. Eng. A* **2000**, *280*, 294–302, doi:10.1016/S0921-5093(99)00698-X. [CrossRef]

9. Peyre, P.; Fabbro, R. Laser shock processing: a review of the physics and applications. *Opt. Quantum Electron.* **1995**, *27*, 1213–1229, doi:10.1007/BF00326477. [CrossRef]

10. Hackel, L.; Rankin, J.R.; Rubenchik, A.; King, W.E.; Matthews, M. Laser peening: A tool for additive manufacturing post-processing. *Addit. Manuf.* **2018**, *24*, 67–75. [CrossRef]

11. Maamoun, A.; Elbestawi, M.; Veldhuis, S. Influence of shot peening on AlSi10Mg parts fabricated by additive manufacturing. *J. Manuf. Mater. Process.* **2018**, *2*, 40. [CrossRef]

12. Hong, X.; Wang, S.; Guo, D.; Wu, H.; Wang, J.; Dai, Y.; Xia, X.; Xie, Y. Confining medium and absorptive overlay: Their effects on a laser-induced shock wave. *Opt. Lasers Eng.* **1998**, *29*, 447–455, doi:10.1016/S0143-8166(98)80012-2. [CrossRef]

13. Creton, C. Pressure-sensitive adhesives: an introductory course. *MRS Bull.* **2003**, *28*, 434–439. [CrossRef]

14. Bardy, S.; Aubert, B.; Berthe, L.; Combis, P.; Hébert, D.; Lescoute, E.; Rullier, J.L.; Videau, L. Numerical study of laser ablation on aluminum for shock-wave applications: development of a suitable model by comparison with recent experiments. *Opt. Eng.* **2016**, *56*, 011014. [CrossRef]

15. Sagnard, M.; Berthe, L.; Ecault, R.; Touchard, F.; Boustie, M. Development of the symmetrical laser shock test for weak bond inspection. *Opt. Laser Technol.* **2019**, *111*, 644-652. [CrossRef]

16. Barker, L.; Hollenbach, R. Laser interferometer for measuring high velocities of any reflecting surface. *J. Appl. Phys.* **1972**, *43*, 4669–4675. [CrossRef]

17. Tollier, L.; Fabbro, R.; Bartnicki, E. Study of the laser-driven spallation process by the velocity interferometer system for any reflector interferometry technique. I. Laser-shock characterization. *J. Appl. Phys.* **1998**, *83*, 1224–1230, doi:10.1063/1.366819. [CrossRef]

18. Berthe, L.; Fabbro, R.; Peyre, P.; Tollier, L.; Bartnicki, E. Shock waves from a water-confined laser-generated plasma. *J. Appl. Phys.* **1997**, *82*, 2826–2832, doi:10.1063/1.366113. [CrossRef]

19. Hibbitt, Karlsson, & Sorensen. *ABAQUS/Explicit: User's Manual*; Hibbitt, Karlsson and Sorenson Incorporated: New York, NY, USA, 1998; Volume 1.

20. Johnson, G.R. A constitutive model and data for materials subjected to large strains, high strain rates, and high temperatures. In Proceedings of the 7th International Symposium on Ballistics, The Hague, The Netherlands, 19–21 April 1983; pp. 541–547.

21. Hfaiedh, N.; Peyre, P.; Song, H.; Popa, I.; Ji, V.; Vignal, V. Finite element analysis of laser shock peening of 2050-T8 aluminum alloy. *Int. J. Fatigue* **2015**, *70*, 480–489. [CrossRef]

22. Cuq-Lelandais, J.P. Etude du Comportement Dynamique de Matériaux Sous Choc Laser Subpicoseconde. Ph.D. Thesis, ISAE-ENSMA Ecole Nationale Supérieure de Mécanique et d'Aérotechique, Poitiers, France, 2010.

23. Fabbro, R.; Fournier, J.; Ballard, P.; Devaux, D.; Virmont, J. Physical study of laser-produced plasma in confined geometry. *J. Appl. Phys.* **1990**, *68*, 775–784, doi:10.1063/1.346783. [CrossRef]

24. Peyre, P.; Berthe, L.; Scherpereel, X.; Fabbro, R.; Bartnicki, E. Experimental study of laser-driven shock waves in stainless steels. *J. Appl. Phys.* **1998**, *84*, 5985–5992. [CrossRef]

25. Pope, R.M.; Fry, E.S. Absorption spectrum (380–700 nm) of pure water. II. Integrating cavity measurements. *Appl. Opt.* **1997**, *36*, 8710–8723. [CrossRef] [PubMed]

26. Peyre, P.; Berthe, L.; Vignal, V.; Popa, I.; Baudin, T. Analysis of laser shock waves and resulting surface deformations in an Al–Cu–Li aluminum alloy. *J. Phys. D Appl. Phys.* **2012**, *45*, 335304. [CrossRef]

27. Boustie, M.; Cuq-Lelandais, J.; Bolis, C.; Berthe, L.; Barradas, S.; Arrigoni, M.; De Resseguier, T.; Jeandin, M. Study of damage phenomena induced by edge effects into materials under laser driven shocks. *J. Phys. D Appl. Phys.* **2007**, *40*, 7103. [CrossRef]

28. Sollier, A.; Berthe, L.; Fabbro, R. Numerical modeling of the transmission of breakdown plasma generated in water during laser shock processing. *EPJ Appl. Phys.* **2001**, *16*, 131–139, doi:10.1051/epjap:2001202. [CrossRef]

29. Wu, B.; Shin, Y.C. A self-closed thermal model for laser shock peening under the water confinement regime configuration and comparisons to experiments. *J. Appl. Phys.* **2005**, *97*, 113517. [CrossRef]

30. Berthe, L.; Courapied, D.; El Karnighi, S.; Peyre, P.; Gorny, C.; Rouchausse, Y. Study of laser interaction in water flow confinement at high repetition rate. *J. Laser Appl.* **2017**, *29*, 042006, doi:10.2351/1.5007947. [CrossRef]

31. Peyre, P.; Fabbro, R.; Berthe, L.; Scherpereel, X.; Bartnicki, E. Laser-shock processing of materials and related measurements. In Proceedings of the High-Power Laser Ablation, Santa Fe, NM, USA, 14 September 1998; Volume 3343, pp. 183–194.

 *metals*

Article

Improving Fatigue Performance of Laser-Welded 2024-T3 Aluminum Alloy Using Dry Laser Peening

Tomokazu Sano [1,*], Takayuki Eimura [1], Akio Hirose [1], Yosuke Kawahito [2], Seiji Katayama [2], Kazuto Arakawa [3], Kiyotaka Masaki [4], Ayumi Shiro [5], Takahisa Shobu [6] and Yuji Sano [7,8]

[1] Division of Materials and Manufacturing Science, Graduate School of Engineering, Osaka University, 2-1 Yamada-oka, Suita 565-0871, Japan; toyahi31@gmail.com (T.E.); hirose@mapse.eng.osaka-u.ac.jp (A.H.)

[2] Joining and Welding Research Institute, Osaka University, 11-1 Mihogaoka, Ibaraki 567-0047, Japan; kawahito@jwri.osaka-u.ac.jp (Y.K.); katayama@jwri.osaka-u.ac.jp (S.K.)

[3] Interdisciplinary Faculty of Science and Engineering, Shimane University, 1060 Nishikawatsu-cho, Matsue 690-8504, Japan; arakawa@riko.shimane-u.ac.jp

[4] National Institute of Technology, Okinawa College, 905 Henoko, Nago 905-2192, Japan; masaki-k@okinawa-ct.ac.jp

[5] Quantum Beam Science Research Directorate, National Institute for Quantum and Radiological Science and Technology, Kouto, Sayo 679-5148, Japan; shiro.ayumi@qst.go.jp

[6] Materials Sciences Research Center, Japan Atomic Energy Agency, Kouto, Sayo 679-5148, Japan; shobu@spring8.or.jp

[7] Division of Research Innovation and Collaboration, Institute for Molecular Science, National Institutes of Natural Sciences, 38 Nishigo-Naka, Myodaiji, Okazaki 444-8585, Japan; yuji-sano@ims.ac.jp

[8] Department of Quantum Beam Physics, Institute of Scientific and Industrial Research, Osaka University, 8-1 Mihogaoka 567-0047, Japan

* Correspondence: sano@mapse.eng.osaka-u.ac.jp

Received: 23 September 2019; Accepted: 4 November 2019; Published: 6 November 2019

Abstract: The purpose of the present study was to verify the effectiveness of dry laser peening (DryLP), which is the peening technique without a sacrificial overlay under atmospheric conditions using femtosecond laser pulses on the mechanical properties such as hardness, residual stress, and fatigue performance of laser-welded 2024 aluminum alloy containing welding defects such as undercuts and blowholes. After DryLP treatment of the laser-welded 2024 aluminum alloy, the softened weld metal recovered to the original hardness of base metal, while residual tensile stress in the weld metal and heat-affected zone changed to compressive stresses. As a result, DryLP treatment improved the fatigue performances of welded specimens with and without the weld reinforcement almost equally. The fatigue life almost doubled at a stress amplitude of 180 MPa and increased by a factor of more than 50 at 120 MPa. DryLP was found to be more effective for improving the fatigue performance of laser-welded aluminum specimens with welding defects at lower stress amplitudes, as stress concentration at the defects did not significantly influence the fatigue performance.

Keywords: dry laser peening; femtosecond laser; shock wave; laser welding; 2024 aluminum alloy

---

1. Introduction

Laser peening (LP), or laser shock peening (LSP), is a surface treatment method used to improve mechanical properties, such as fatigue performance and corrosion resistance, by hardening the material and adding compressive residual stresses on the surfaces via a laser-driven shock wave that causes plastic deformation of the material [1–5]. In recent years, the application of LP has been extended to the aerospace, nuclear, automotive, and biomedical industries [6]. The LP process can efficiently induce plasticity because of the high-strain-rate deformation caused by shock compression. Unlike other peening processes, such as shot peening, hammer peening, and ultrasonic peening, LP is a noncontact process that does not contaminate the sample.

During LP processing, the material surface is irradiated with a laser pulse passed through a transparent plasma confinement medium, such as water or glass, resulting in explosive ablation of the material. The material is plastically deformed by a shock wave formed by the recoil force of plasma expansion during ablation, which propagates into the material. An opaque overlay, such as paint, black tape, or metal foils, is conventionally applied to the material surface to avoid thermal effects, such as excessive temperature increase or melting, due to the high energy of the laser itself or laser-generated plasma. A LP method that does not require such a coating has been developed using an irradiation source formed by overlapping low-energy pulses [7,8]. Furthermore, a dry LP (DryLP) technique has been developed that does not require a sacrificial overlay and is performed under ambient conditions, where the material is directly irradiated using femtosecond laser pulses [9–11]. Although femtosecond laser pulses have a small pulse energy, the electric field is so strong that the material is explosively ablated without requiring the opaque overlay or the plasma confinement medium [12,13]. This induces a plastic shock wave strong enough to cause shock effects in the material [14–19].

Here, we proposed the application of DryLP to improve the mechanical properties of laser-welded 2024 aluminum alloy. LP is generally effective for improving the fatigue performance of arc-welded [20] and friction stir-welded joints [21–23]. The fatigue performance of welded precipitation-strengthened aluminum alloys, such as the 2000, 6000, and 7000 series, were worse than the corresponding base material (BM) because of the softening of the weld metal (WM), heat-affected zone (HAZ), and residual tensile stress on the surface after welding [24,25]. Therefore, in recent years, friction stir welding (FSW) has been widely used to join precipitation-strengthened aluminum alloys because it results in only a small decrease in the strength of the weld joint and small distortion of the joint after welding [26–28], although the welding speed is relatively low. Laser welding is a high-speed welding method that has been used for achieving high-productivity welding of precipitation-strengthened aluminum alloys [29,30]. Although the weldability of 2024 aluminum alloy is generally low, fast full-penetration welding of this alloy using highly focused fiber laser achieved weld joints with smaller HAZ regions and no cracking [31]. However, areas of WM with reduced strength exist, and avoiding generation of blowholes in the laser-welded joints is difficult. Although the thickness with the compressive residual stress induced by DryLP process is almost one-tenth of conventional LP methods [9,10], this method was shown to be effective for FSW-processed 7075-T73 aluminum alloy, where the stir zone, thermo-mechanically affected zone, and HAZ were softened, but no welding defects occurred, confirming that the fatigue performance was better than that of the BM at lower stress amplitude after DryLP treatment. However, the effectiveness of DryLP on welded precipitation-strengthened aluminum alloy containing welding defects has never been investigated. Hence, the purpose of the present study is to verify the effectiveness of DryLP for laser-welded 2024 aluminum alloy containing welding defects by investigating the mechanical properties.

2. Experimental Method

A 2024-T3 aluminum alloy with thickness of 3 mm was used. The chemical composition of this alloy is shown in Table 1. The original alloy had a 0.2% proof stress of 334 MPa, tensile strength of 464 MPa, and elongation of 21.8%.

Table 1. Chemical composition of 2024-T3 aluminum alloy used in this study (mass%).

Si	Fe	Cu	Mn	Mg	Cr	Zn	Ti	Others	Al
0.02	0.05	4.4	0.55	1.4	0.00	0.02	0.01	0.01	Bal.

A single-mode fiber laser (IPG Photonics, YLS-2000-SM, Japan, wavelength: 1070 nm, CW) was used for full-penetration bead-on-plate welding of the aluminum alloy, as shown in Figure 1a. The fiber diameter was 14 μm and we used a laser power of 2.0 kW. The laser was focused on the alloy surface with a spot size of 54 μm. Ar was used for shielding gas with a flow rate of 30 L/min. A welding speed of 2.5 m/min was used. The top and bottom surfaces of the laser-welded specimens were observed using an optical digital microscopy (Hirox, KH-7700, Japan). The cross-section of the weld bead was observed using optical microscopy (Olympus, SZX7, Tokyo, Japan).

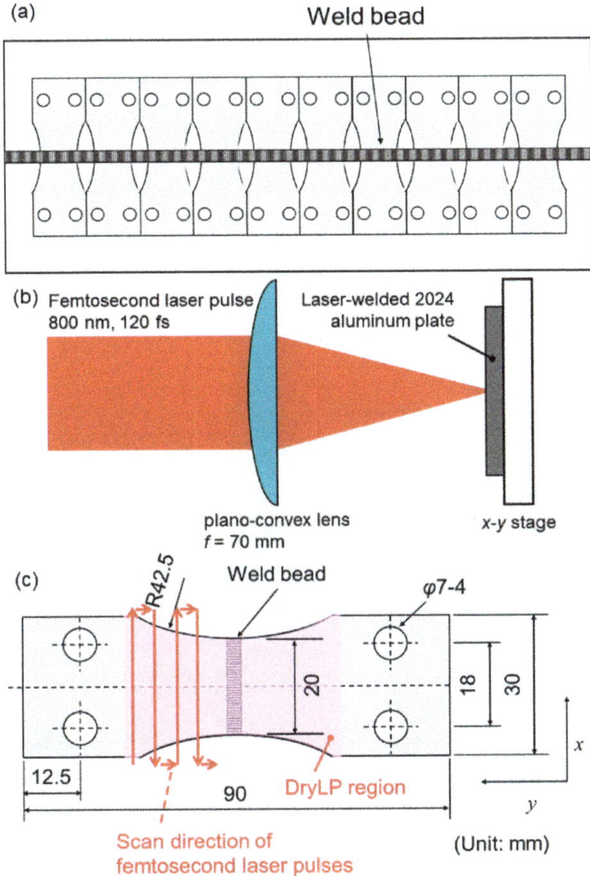

Figure 1. Schematic illustrations of the experimental procedures. (**a**) Full-penetration bead-on-plate laser welding and preparation of fatigue test specimens cut from the laser-welded plate. (**b**) Dry laser peening (DryLP) process using femtosecond laser pulses. (**c**) Geometry of the fatigue test specimens with the weld bead located in the center. DryLP was performed using an x–y automatic stage which sequentially moved the specimen in a serpentine pattern, as indicated by the red arrows.

Then, the laser-welded specimens were subjected to DryLP in air. The peening was performed 15 months after welding to allow the completion of natural aging. As shown in Figure 1b,c, femtosecond laser pulses with a wavelength of 800 nm, pulse duration of 130 fs, and pulse energy of 0.6 mJ (Spectra-Physics, Spitfire, Japan) were focused using a plano-convex lens with focal length of 70 mm onto the specimen. The laser pulses were overlapped, with a coverage of 692%, which was shown to be the most effective condition for DryLP of 2024-T3 aluminum alloy [9]. A detailed description of the DryLP process was provided in the previous study.

For the preparation of specimens for hardness tests, the weld reinforcement was removed and electropolished in 20% sulfuric acid-methanol electrolyte for 30 s to remove the work-strained layer before DryLP treatment. The hardness of the top surface was measured using a Vickers hardness tester (Mitsutoyo, HM-221, Kawasaki, Japan) with a load of 1.96 N and loading time of 15 s.

For the preparation of specimens for residual stress measurement, DryLP treatment was conducted on as-welded specimens without removing the weld reinforcement. Depth profiling of the residual stress which was normal to the weld bead in the specimens was conducted nondestructively using the BL22XU beamline at SPring-8 [32], using the strain scanning method [33] with monochromatic X-rays with a photon energy of 30.013 keV, as shown in Figure 2. A CdTe detector was used for the measurements. The residual stress σ was estimated using $\sigma = E(d - d_0)/d_0$, where E is the Young's modulus of 61.7 GPa, d is the d-spacing of the (311) plane of aluminum in the welded or DryLPed specimens, and d_0 is the d-spacing of the (311) plane in the BM of 0.12196 nm. The d-spacing of the (311) plane parallel to the weld bead in the gauge volume was measured, as shown in Figure 2a,b. The widths of both the incident and receiving slits were 0.2 mm. For the surface measurements, the heights of these slits were 50 μm. For depth profiling, the slit heights, which determine the depth resolution, were 10 μm from the surface to a depth of 40 μm, and the slit heights were 30 μm deeper than a depth of 40 μm. The d-spacings of the (311) plane of the WM, below the weld toe, and in the HAZ were measured, as shown in Figure 2c.

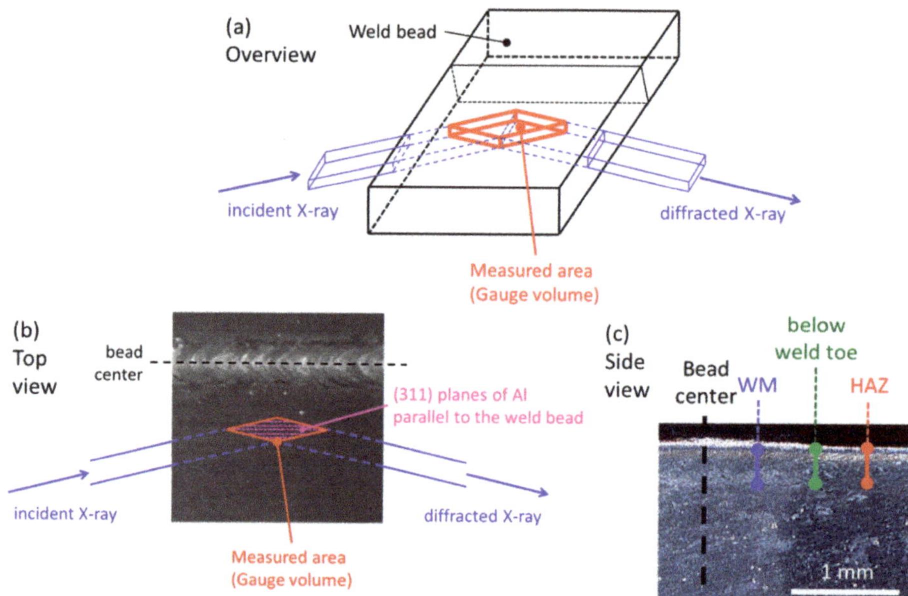

Figure 2. Schematic illustrations of (**a**) overview, (**b**) top, and (**c**) side views of residual stress measurements using the strain scanning method with synchrotron X-rays.

Four kinds of specimens for fatigue testing were prepared: (i) As-welded specimen; (ii) reinforcement-removed welded specimen; (iii) DryLPed welded specimen; and (iv) DryLPed reinforcement-removed welded specimen. The stress concentration influenced the fatigue properties of the as-welded specimen due to both reinforcements and undercuts. Hence, to investigate the stress concentration only influenced by the undercuts, the reinforcements were removed. The reinforcements were removed using diamond pastes with a particle size of 1 μm. These specimens were cut from the laser-welded specimen, as shown in Figure 1a. DryLP was conducted on both surfaces of the laser-welded specimen, as shown in Figure 1b. Plane bending fatigue tests (PBF-30, Tokyo Koki, Tokyo, Japan) were conducted at a cyclic speed of 1400 cycles/min with a constant strain amplitude and a stress ratio of $R = -1$ in air at room temperature based on Little's method [34]. The stress ratio of $R = -1$ was selected to indicate the effectiveness of the DryLP more clearly because both surfaces were treated. The fracture surfaces were observed using optical microscopy (Olympus, SZX7, Tokyo, Japan) and scanning electron microscopy (SEM; Hitachi, S-3000H, Tokyo, Japan).

The microstructures were observed to estimate dislocation densities in the specimens using a transmission electron microscopy (TEM; JEOL JEM-2010, Tokyo, Japan) with an acceleration voltage of 200 kV. For TEM observations, a small piece of the cross-section was thinned using a 30-keV-focused Ga-ion beam (Hitachi, FB-2000A, Tokyo, Japan).

3. Results and Discussion

3.1. Laser Welding

Optical microscopy images of the top and bottom rear surfaces of the laser-welded specimens are shown in Figure 3a,b. Although no cracks were observed on the surfaces, some pores existed on the top surface and some undercuts (indicated by yellow arrows in the figures) were found on both surfaces. The cross-section of the weld bead shows that full-penetration welding was achieved, where the reinforcement did not show any cracks (Figure 3c). The bead widths on the top and rear surfaces were around 2.0 mm and 1.2 mm, respectively. The optical microscopy image of the bottom surface of the reinforcement-removed welded specimen is shown in Figure 3d, where the yellow arrows indicate the undercuts.

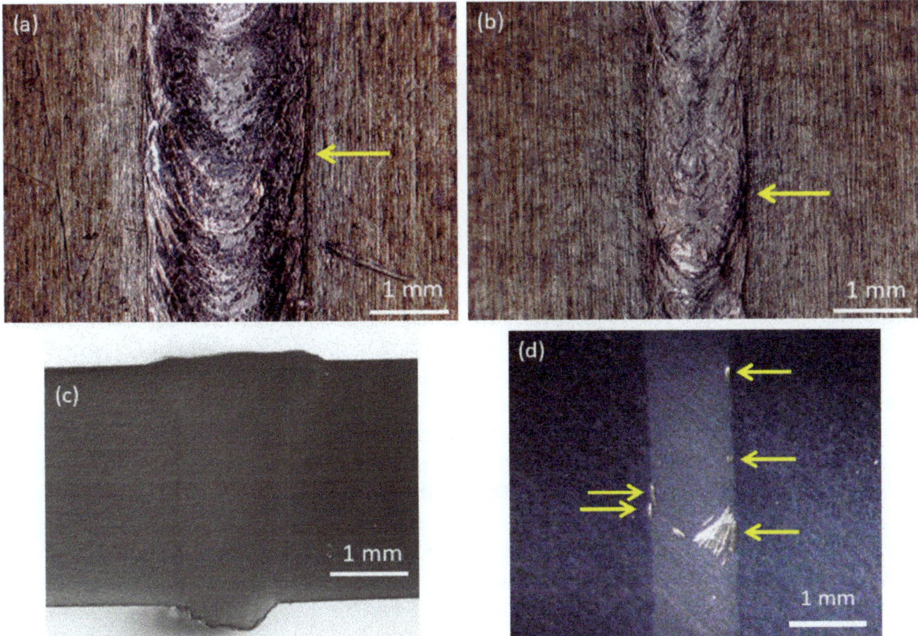

Figure 3. Optical microscopy images of (**a**) top surface, (**b**) bottom surface, (**c**) cross-section of the laser-weld bead, and (**d**) bottom surface of reinforcement-removed welded specimen.

3.2. Hardness

The results of the hardness tests for the samples with the reinforcement removed are shown in Figure 4. Before DryLP, the hardness of the BM was 138 HV, while that on the surface of the WM was ~100 HV. It was reported that this decrease in hardness is due to i) the segregation of the strengthening elements such as magnesium, copper, and their intermetallic compounds; ii) formation and growth of non-strengthening coarse precipitates; iii) dissolution of strengthening precipitates; iv) uniform re-distribution of precipitating elements; and v) vaporization of low boiling point magnesium during heating and the following freezing due to the fast cooling rates [35–37], resulting in fewer precipitates being formed, even after natural aging for 15 months. The hardness of the HAZ in this specimen was around 130 HV (similar to the BM) because of the dissolution of precipitates and overaging [35,38]. After DryLP, the hardness of all areas of the sample increased compared to that of the as-welded sample. The hardness of the WM was similar to that of the BM before peening, while the hardness of the HAZ and BM after DryLP was around 178 HV.

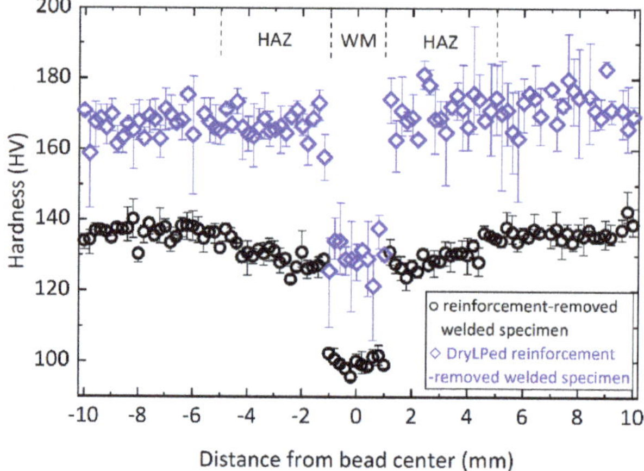

Figure 4. Hardness distributions over the surfaces of welded samples with the reinforcement removed, before, and after DryLP treatment. Error bar indicates the maximum and minimum values.

3.3. Residual Stress

Residual stress curves of the top surface before and after DryLP treatment of the laser-welded specimens are shown in Figure 5a. The residual stress in the WM and HAZ areas of the as-welded specimen were tensile, while other areas had compressive stresses, which is a typical residual stress distribution for welded joints. This tensile residual surface stress in the WM and HAZ areas changed to compressive stress after DryLP treatment, while the magnitude of the compressive residual stresses outside these areas increased. The depth profiles of the residual stress in the WM, below the weld toe, and in the HAZ before and after DryLP treatment are shown in Figure 5b–d. The tensile residual stresses in the WM, below the weld toe, and in the HAZ were observed to a depth of ~300 μm from the weld center in the as-welded specimen. These tensile residual stresses inside the material between the surface and a depth of ~100 μm changed to compressive stresses after DryLP, which is comparable to the thickness of the compressive layer in the DryLPed BM [9].

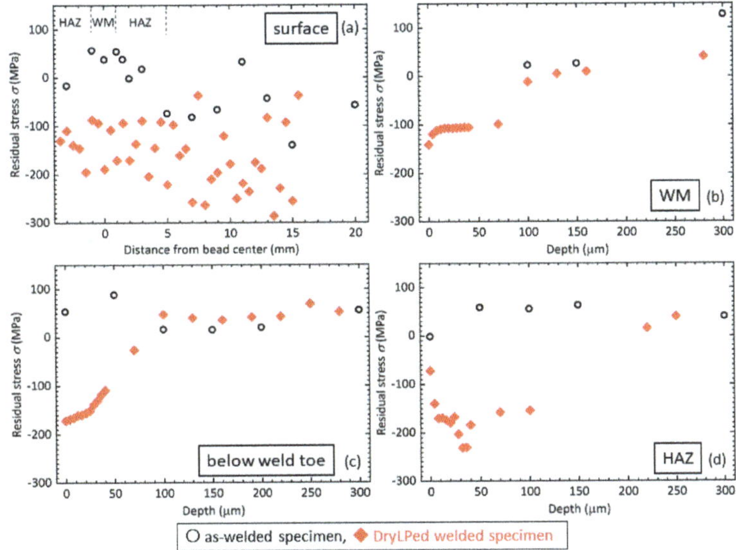

Figure 5. (**a**) Residual stress distributions along the surface (up to 50-μm depth). Depth profiles of the residual stress in the (**b**) WM, (**c**) below the weld toe, and (**d**) in the HAZ of laser-welded specimens before and after DryLP.

3.4. Fatigue Performance

The results of the fatigue tests are shown in Figure 6. The fitted curves for each specimen were obtained using Stromeyer's expression, $\log(\sigma - a) = -b \log N + c$, where σ is the stress amplitude, N is the number of cycles to failure, and a, b, and c are the fitting parameters. The fatigue performances of the as-welded specimens with and without reinforcement were worse than that of the BM. Although the fatigue lives of these specimens at a stress amplitude of 180 MPa were almost the same, that of the reinforcement-removed welded specimen was shorter than that of the as-welded specimen at 120 MPa. After DryLP treatment, the fatigue performances of the specimens with and without reinforcement were enhanced to a similar degree. The fatigue life increased by a factor of almost two at a stress amplitude of 180 MPa and more than 50 times at 120 MPa, which indicates that the DryLP treatment is more effective at lower stress amplitudes.

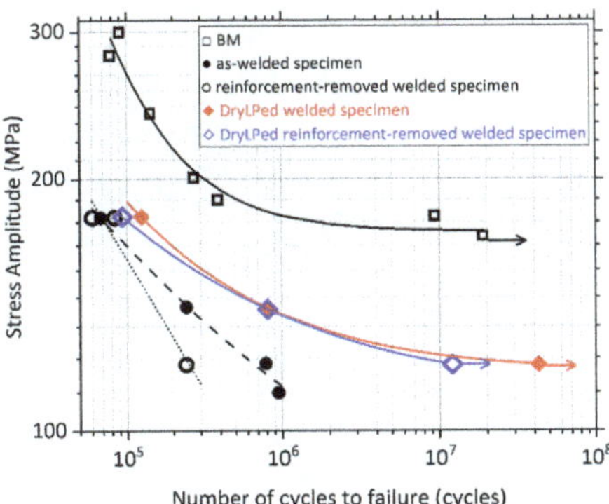

Figure 6. Results of fatigue tests for the base material (BM) and as-welded specimens (with and without reinforcement) before and after DryLP treatment.

The fracture surfaces of samples broken at 120 MPa and 180 MPa are shown in Figure 7, where the red arrows indicate crack initiation sites. The crack initiation sites for any specimens, such as as-welded and reinforcement-removed specimens before and after DryLP treatment, are undercuts, not blowholes. The fractures initiated at the boundary between the WM and HAZ for all specimens, regardless of DryLP treatment or the existence of weld reinforcement. Cracks initiated at undercuts are shown in the magnified views of the surfaces in Figure 7g.

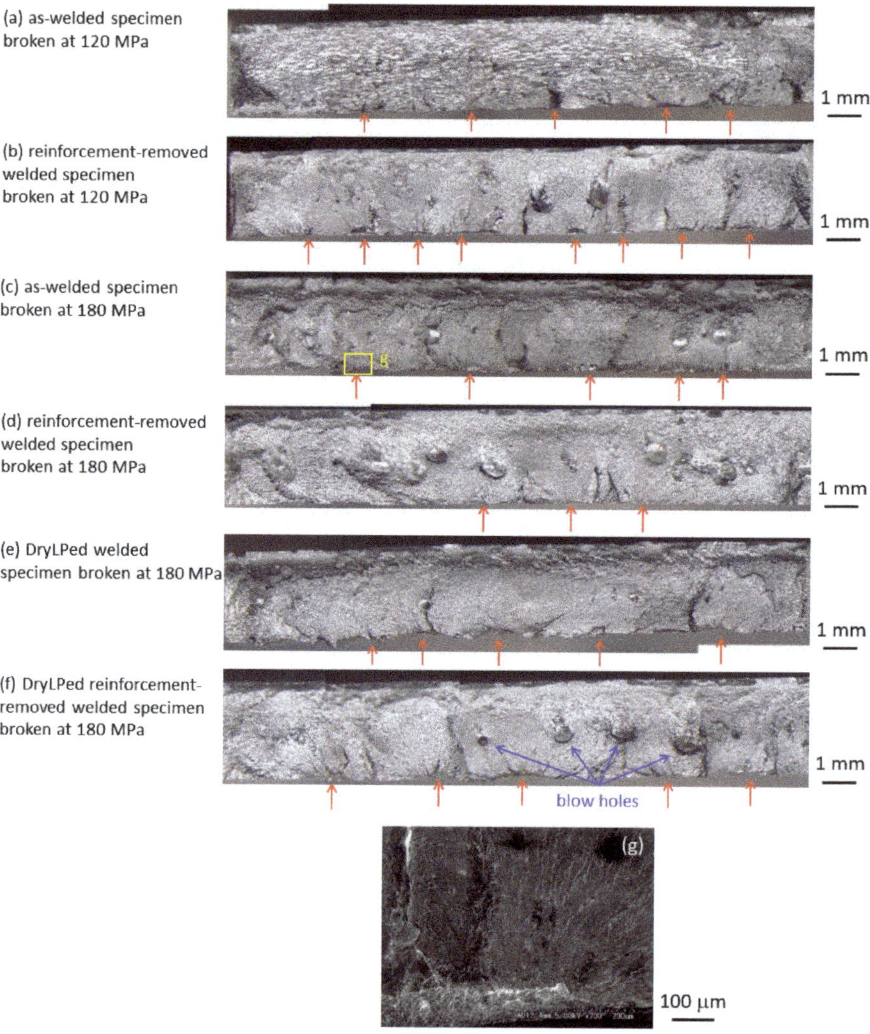

Figure 7. Fracture surfaces of the welded specimens. (**a**) As-welded specimen broken at 120 MPa, (**b**) reinforcement-removed welded specimen broken at 120 MPa, (**c**) as-welded specimen broken at 180 MPa, (**d**) reinforcement-removed welded specimen broken at 180 MPa, (**e**) DryLPed welded specimen broken at 180 MPa, (**f**) DryLPed reinforcement-removed welded specimen broken at 180 MPa. (**g**) Magnified image of a typical crack initiation site, as indicated by the yellow box in (**c**).

3.5. Microstructure in WM

Bright-field TEM images of the region ~10 µm below the surface in the WM of as-welded and DryLPed specimens (with reinforcement) are shown in Figure 8. The incident electron beam direction was nearly parallel to the [110] direction of Al, where the {111} reflection of Al was excited. The dislocations were observed as the darker areas. The dislocation density was estimated using Keh's equation, $\rho = (n_1/L_1 + n_2/L_2)/t$, where ρ is the dislocation density, n_1 and n_2 is the number of intersection points between the dislocation lines and the vertical and horizontal grid lines drawn on the TEM image, respectively, L_1 and L_2 is the total length of the vertical and horizontal grid lines, respectively, and t is the thickness of the TEM sample [39]. The dislocation densities of these samples before and after DryLP treatment were estimated as 1.0×10^{14} m^{-2} and 5.1×10^{14} m^{-2}, respectively. This indicates that DryLP plastically deformed the WM, resulting in hardening and inducing compressive residual stress.

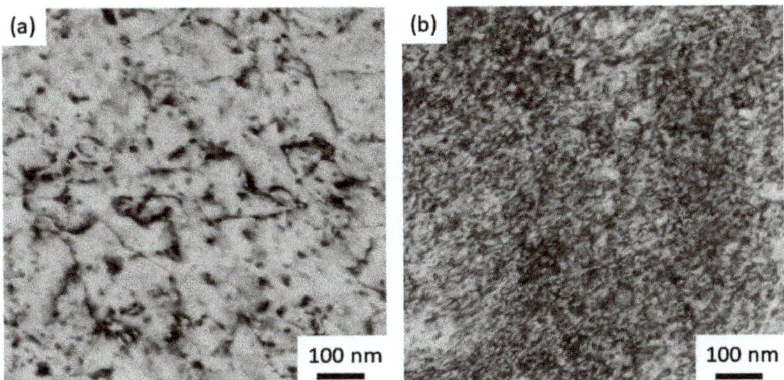

Figure 8. TEM images of weld material (WM) microstructures in laser-welded specimens (with reinforcement) (**a**) before and (**b**) after DryLP treatment.

3.6. Effect of Welding Defects on Fatigue Performance

During fatigue tests, cracks initiated from undercuts at the weld toe, owing to the reduced hardness and tensile residual stress which remained after welding. The fatigue performances of the as-welded specimens with and without reinforcement were comparable at a stress amplitude of 180 MPa, indicating that the stress concentration at undercuts has a greater influence on fatigue performance than stress concentration at the weld toe. The fatigue life of the specimen with the reinforcement removed was shorter than that of the as-welded specimen at a stress amplitude of 120 MPa. Small blowholes were observed in the fracture surface of the specimen without reinforcement, which were expected to influence the fatigue performance at lower stress amplitudes. Welding defects, such as undercuts and blowholes, in addition to softening or tensile residual stress, decreased the fatigue life of the welded specimens.

The fatigue performances of the specimens with and without reinforcement after DryLP treatment were improved compared to the equivalent specimens before DryLP treatment, attributed to hardening of the WM up to the value of the original BM and the introduction of compressive residual stress. The fatigue lives of both specimens after DryLP treatment were significantly increased compared to that of the unpeened welded specimens at lower stress amplitudes. Blowhole defects can lead to stress concentration. However, their contribution is very small, as these features are generally spherical. In addition, the stress concentration at undercuts is smaller at lower stress amplitudes. Therefore, the effect of positive factors induced by DryLP, such as hardening and compressive residual stress, was larger than that of the negative factors, such as stress concentration at undercuts and blowholes at lower stress amplitudes. In addition, for gas metal arc welding lap fillet joint in GA

590 MPa steel sheets, the blowholes in the WM did not significantly affect the fatigue life at relatively lower stress amplitudes, although the fatigue life decreased in the presence of blowholes and surface pores [40]. Overall, DryLP effectively improved the fatigue performance of laser-welded specimens containing welding defects at lower stress amplitudes.

3.7. Plastic Deformation Induced by Femtosecond Laser-Driven Shock Wave

When a peak pressure of a shock wave exceeds a threshold that depends on a material, the pressure increases as a function of the time or the travel distance exhibits a single structure, where the plastic component overtakes the elastic component. The threshold stress for aluminum when the single structure of the shock front is clearly formed is 25 GPa [41]. It was reported that the single structure was observed in the surface layer of 500 nm in pure aluminum, which was irradiated using the intensity of 8.7×10^{12} W/cm^2 with the pulse duration of 150 fs [42]. Therefore, the shock wave with a single structure over 25 GPa should be driven and propagated in the 2024 aluminum alloy, which was irradiated at the intensity of 1.2×10^{14} W/cm^2 with the pulse duration of 130 fs in this research.

It was empirically observed that the strain rate η of the shock wave with the single structure was proportional to the fourth-power of the shock stress σ [43,44]. For the aluminum alloy, $\eta = 9100\sigma^4$ has been reported [41]. Therefore, the strain rate η of 3.5×10^9 s^{-1} was obtained for the shock stress of 25 GPa. The dimensionless Bland number B = $3hs\eta/8c$ was defined [43], where h is the sample thickness, c is the bulk sound velocity under normal pressure, and s is the slope of the u_p-u_s relation, $u_s = c + su_p$, where u_p is the particle velocity and u_s is the shock velocity. When B is greater than 1, steady-wave conditions are expected [41]. The thickness h was estimated to be 3.0 µm for B = 1, $\eta = 3.5 \times 10^9$ s^{-1}, $s = 1.338$, $c = 5.328$ km/s [45]. Therefore, the shock wave with the single structure propagates in the surface layer of 3.0 µm. The single structure splits into two structures, elastic and plastic waves, at the depth of 3.0 µm, and the shock wave with the two-wave structure propagates into the deeper region.

Figure 9 shows the hardness in the BM region in the DryLPed 2024 aluminum alloy as a function of the depth measured using nanoindentation (ELIONIX, ENT-1100a, Japan) with an applied load of 1 mN and loading time of 2 s. The increase in hardness is more significant at a depth of 3 µm from the surface rather than depths of 3–20 µm, although the hardness increased in the surface layer with 20 µm thickness. The thickness of the significantly hardened layer of 3 µm corresponds to the thickness of 3.0 µm where the shock wave with the single structure propagates. This implies that the single structure induces plastic deformation more effectively, thereby increasing the hardness. A high-density-dislocation structure, shown in Figure 8a, was formed in a layer where the shock wave with the single structure propagates, because dislocation generation, rather than dislocation multiplication, was dominant [9,10,46].

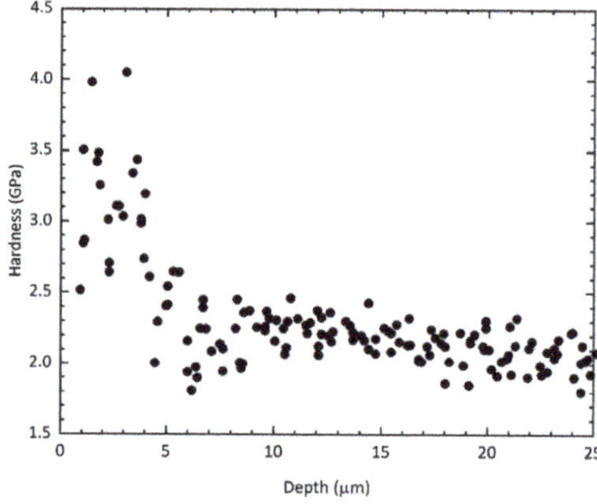

Figure 9. Depth profile of hardness in the BM region in the DryLPed 2024 aluminum measured using nanoindentation.

4. Conclusions

The effects of DryLP on the hardness, residual stress, and fatigue performance of laser-welded 2024-T3 aluminum alloy were investigated. After DryLP treatment, the hardness of the softened WM recovered to that of the original BM, while tensile residual stress in the WM and HAZ changed to compressive stress. DryLP treatment improved the fatigue performances of welded specimens with and without reinforcement almost equally. Positive factors (hardening and introduction of compressive residual stress) induced by DryLP had a larger effect on the mechanical properties than negative factors (stress concentrations near defects at lower stress amplitudes). Therefore, DryLP is expected to be more effective in improving the fatigue performance of laser-welded specimens with weld defects at lower stress amplitudes. Combining high-speed laser welding with DryLP is expected to be a suitable strategy for replacing other welding processes, resulting in high productivity. This combination could be applied in various industrial fields, such as the automotive, rail, aircraft, and space industries.

Author Contributions: This work was supervised by T.S. (Tomokazu Sano). T.S. (Tomokazu Sano) and T.E. conducted all experiments. Y.K. and S.K. supported laser welding experiment. K.A. supported TEM observation. K.M. supported fatigue test. A.S. and T.S. (Takahisa Shobu) supported residual stress measurement. A.H. advised laser welding experiment. Y.S. advised laser peening experiment. T.S. (Takahisa Shobu) wrote the manuscript. All authors discussed the results and commented on the manuscript.

Funding: This work was supported in part by MEXT Quantum Leap Flagship Program (MEXT Q-LEAP) Grant Number JPMXS0118068348, and JSPS KAKENHI Grant Numbers JP16H04247, JP16K14417, and 19K22061. This work was funded in part by ImPACT Program of Council for Science, Technology and Innovation (Cabinet Office, Government of Japan).

Acknowledgments: The laser welding experiment was performed under the Joint Usage/Research Center on Joining and Welding, Osaka University. The synchrotron radiation experiments were performed under the Shared Use Program of JAEA (Proposal Nos. JAEA 2015B-E12, JAEA 2016A-E18, JAEA 2017B-E11, and JAEA 2019A-E07) and QST (Proposal Nos. QST 2016B-H14 and QST 2017A-H18) facilities with the approval of Nanotechnology Platform project supported by the Ministry of Education, Culture, Sports, Science and Technology (Proposal Nos. A-15-AE-0034, A-16-QS-0010, A-16-QS-0026, A-17-QS-0016, A-17-AE-0031, and A-19-AE-0007). The synchrotron radiation experiments were performed using a JAEA experimental station at JAEA beamline BL22XU, SPring-8, with the approval of the Japan Synchrotron Radiation Research Institute (JASRI) (Proposal Nos. 2015B3782, 2016A3783, 2016B3786, 2017A3788, 2017B3737, and 2019A 3737).

Conflicts of Interest: The authors declare no conflict of interest.

References

1. Anderholm, N.C. Laser-generated stress waves. *Appl. Phys. Lett.* **1970**, *16*, 113–115. [CrossRef]
2. Fairand, P.; Wilcox, B.A.; Gallagher, W.J.; Williams, D.N. Laser shock-induced microstructural and mechanical property changes in 7075 aluminum. *J. Appl. Phys.* **1972**, *43*, 3893–3895. [CrossRef]
3. Clauer, A.H. Laser Shock Peening, the Path to Production. *Metals* **2019**, *9*, 626. [CrossRef]
4. Fabbro, R.; Peyre, P.; Berthe, L.; Scherpereel, X. Physics and applications of laser-shock processing. *J. Laser Appl.* **1998**, *10*, 265–279. [CrossRef]
5. Montross, C.S.; Wei, T.; Ye, L.; Clark, G.; Mai, Y.W. Laser shock processing and its effects on microstructure and properties of metal alloys: A review. *Int. J. Fatigue* **2002**, *24*, 1021–1036. [CrossRef]
6. Tenaglia, R.D.; Lahrman, D.F. Shock tactics. *Nat. Photonics* **2009**, *3*, 267–269. [CrossRef]
7. Sano, Y.; Mukai, N.; Okazaki, K.; Obata, M. Residual stress improvement in metal surface by underwater laser irradiation. *Nucl. Instrum. Methods B* **1997**, *121*, 432–436. [CrossRef]
8. Sano, Y.; Obata, M.; Kubo, T.; Mukai, N.; Yoda, M.; Masaki, K.; Ochi, Y. Retardation of crack initiation and growth in austenitic stainless steels by laser peening without protective coating. *Mater. Sci. Eng. A* **2006**, *417*, 334–340. [CrossRef]
9. Sano, T.; Eimura, T.; Kashiwabara, R.; Matsuda, T.; Isshiki, Y.; Hirose, A.; Tsutsumi, S.; Arakawa, K.; Hashimoto, T.; Masaki, K.; et al. Femtosecond laser peening of 2024 aluminum alloy without a sacrificial overlay under atmospheric conditions. *J. Laser Appl.* **2017**, *29*, 012005. [CrossRef]

10. Kawashima, T.; Sano, T.; Hirose, A.; Tsutsumi, S.; Masaki, K.; Arakawa, K.; Hori, H. Femtosecond laser peening of friction stir welded 7075-T73 aluminum alloys. *J. Mater. Process. Technol.* **2018**, *262*, 111–122. [CrossRef]

11. Trdan, U.; Sano, T.; Klobčara, D.; Sano, Y.; Grum, J.; Šturm, R. Improvement of corrosion resistance of AA2024-T3 using femtosecond laser peening without protective and confining medium. *Corros. Sci.* **2018**, *143*, 46–55. [CrossRef]

12. Strickland, D.; Mourou, G. Compression of amplified chirped optical pulses. *Opt. Commun.* **1985**, *56*, 219–221. [CrossRef]

13. Evans, R.; Badger, A.D.; Falliès, F.; Mahdieh, M.; Hall, T.A.; Audebert, P.; Geindre, J.-P.; Gauthier, J.-C.; Mysyrowicz, A.; Grillon, G.; et al. Time- and space-resolved optical probing of femtosecond-laser-driven shock waves in aluminum. *Phys. Rev. Lett.* **1996**, *77*, 3359–3362. [CrossRef] [PubMed]

14. Sano, T.; Mori, H.; Ohmura, E.; Miyamoto, I. Femtosecond laser quenching of the ε phase of iron. *Appl. Phys. Lett.* **2003**, *83*, 3498–3500. [CrossRef]

15. Tsujino, M.; Sano, T.; Sakata, O.; Ozaki, N.; Kimura, S.; Takeda, S.; Okoshi, M.; Inoue, N.; Kodama, R.; Kobayashi, K.F.; et al. Synthesis of submicron metastable phase of silicon using femtosecond laser-driven shock wave. *J. Appl. Phys.* **2011**, *110*, 126103. [CrossRef]

16. Tsujino, M.; Sano, T.; Ogura, T.; Okoshi, M.; Inoue, N.; Ozaki, N.; Kodama, R.; Kobayashi, K.F.; Hirose, A. Formation of high-density dislocations and hardening in femtosecond-laser-shocked silicon. *Appl. Phys. Express* **2012**, *5*, 022703. [CrossRef]

17. Matsuda, T.; Sano, T.; Arakawa, K.; Hirose, A. Multiple-shocks induced nanocrystallization in iron. *Appl. Phys. Lett.* **2014**, *105*, 021902. [CrossRef]

18. Matsuda, T.; Sano, T.; Arakawa, K.; Hirose, A. Dislocation structure produced by an ultrashort shock pulse. *J. Appl. Phys.* **2014**, *116*, 183506. [CrossRef]

19. Matsuda, T.; Sano, T.; Arakawa, K.; Sakata, O.; Tajiri, H.; Hirose, A. Femtosecond laser-driven shock-induced dislocation structures in iron. *Appl. Phys. Express* **2014**, *7*, 122704. [CrossRef]

20. DeWald, A.T.; Rankin, J.E.; Hill, M.R.; Lee, M.J.; Chen, H.L. Assessment of tensile residual stress mitigation in alloy 22 welds due to laser peening. *J. Eng. Mater. Technol.* **2004**, *126*, 465–473. [CrossRef]

21. Hatamleh, O.; Lyons, J.; Forman, R. Laser and shot peening effects on fatigue crack growth in friction stir welded 7075-T7351 aluminum alloy joints. *Int. J. Fatigue* **2007**, *29*, 421–434. [CrossRef]

22. Hatamleh, O. A comprehensive investigation on the effects of laser and shot peening on fatigue crack growth in friction stir welded AA 2195 joints. *Int. J. Fatigue* **2009**, *31*, 974–988. [CrossRef]

23. Sano, Y.; Masaki, K.; Gushi, T.; Sano, T. Improvement in fatigue performance of friction stir welded A6061-T6 aluminum alloy by laser peening without coating. *Mater. Des.* **2012**, *36*, 809–814. [CrossRef]

24. Bussu, G.; Irving, P.E. The role of residual stress and heat affected zone properties on fatigue crack propagation in friction stir welded 2024-T351 aluminium joints. *Int. J. Fatigue* **2003**, *25*, 77–88. [CrossRef]

25. Liljedahl, C.D.M.; Brouard, J.; Zanellato, O.; Lin, J.; Tan, M.L.; Ganguly, S.; Irving, P.E.; Fitzpatrick, M.E.; Zhang, X.; Edwards, L. Weld residual stress effects on fatigue crack growth behaviour of aluminium alloy 2024-T351. *Int. J. Fatigue* **2009**, *31*, 1081–1088. [CrossRef]

26. Nandan, R.; DebRoy, T.; Bhadeshia, H.K.D.H. Recent advances in friction-stir welding—Process, weldment structure and properties. *Prog. Mater. Sci.* **2008**, *53*, 980–1023. [CrossRef]

27. Kulekci, C.; Ik, A.S.; Kaluc, E. Effects of tool rotation and pin diameter on fatigue properties of friction stir welded lap joints. *Int. J. Adv. Manuf. Technol.* **2008**, *36*, 877–882. [CrossRef]

28. Garware, M.; Kridli, G.T.; Mallick, P.K. Tensile and fatigue behavior of friction-stir welded tailor-welded blank of aluminum alloy 5754. *J. Mater. Eng. Perform.* **2010**, *19*, 1161–1171. [CrossRef]

29. Dursun, T.; Soutis, C. Recent developments in advanced aircraft aluminium alloys. *Mater. Des.* **2014**, *56*, 862–871. [CrossRef]

30. Schubert, E.; Klassen, M.; Zerner, I.; Walz, C.; Sepold, G. Light-weight structures produced by laser beam joining for future applications in automobile and aerospace industry. *J. Mater. Process. Technol.* **2001**, *115*, 2–8. [CrossRef]

31. Katayama, S.; Nagayama, H.; Mizutani, M.; Kawahito, Y. Fibre laser welding of aluminium alloy. *Weld. Int.* **2009**, *23*, 744–752. [CrossRef]

32. Shobu, T.; Tozawa, K.; Shiwaku, H.; Konishi, H.; Inami, T.; Harami, T.; Mizuki, J. Wide band energy beamline using Si (111) crystal monochromators at BL22XU in SPring-8. *AIP Conf. Proc.* **2007**, *879*, 902–906. [CrossRef]

33. Shobu, T.; Konishi, H.; Mizuki, J.; Suzuki, K.; Suzuki, H.; Akiniwa, Y.; Tanaka, K. Evaluation of subsurface distribution of residual stress in austenitic stainless steel using strain scanning method. *Mater. Sci. Forum* **2006**, *524*, 691–696. [CrossRef]

34. Little, R.E. Estimating the median fatigue limit for very small up-and-down quantal response tests and for S-N data with runouts. In *Probabilistic Aspects of Fatigue*; Heller, R., Ed.; ASTM International: West Conshohocken, PA, USA, 1972; pp. 29–42.

35. Ahn, J.; Chenb, L.; Heb, E.; Daviesa, C.M.; Deara, J.P. Effect of filler metal feed rate and composition on microstructure and mechanical properties of fibre laser welded AA 2024-T3. *J. Manuf. Process.* **2017**, *25*, 26–36. [CrossRef]

36. Ahn, J.; Heb, E.; Chenb, L.; Deara, J.; Daviesa, C. The effect of Ar and He shielding gas on fibre laser weld shape and microstructure in AA 2024-T3. *J. Manuf. Process.* **2017**, *29*, 62–73. [CrossRef]

37. Hu, B.; Richardson, I.M. Autogenous laser keyhole welding of aluminum alloy 2024. *J. Laser Appl.* **2005**, *17*, 70–80. [CrossRef]

38. Cross, C.E.; Olson, D.L.; Liu, S. *Aluminium Welding, Handbook of Aluminum*; Dekker: New York, NY, USA, 2003; Volume 1.

39. Keh, A.S. *Direct Observation of Imperfection in Crystals*; Interscience Publishers: New York, NY, USA, 1962; pp. 213–233.

40. Kim, D.Y.; Hwang, I.; Jeong, G.; Kang, M.; Kim, D.; Seo, J.; Kim, Y.M. Effect of Porosity on the Fatigue Behavior of Gas Metal ArcWelding Lap Fillet Joint in GA 590 MPa Steel Sheets. *Metals* **2018**, *8*, 241. [CrossRef]

41. Crowhurst, J.C.; Armstrong, M.R.; Knight, K.B.; Zaug, J.M.; Behymer, E.M. Invariance of the Dissipative Action at Ultrahigh Strain Rates Above the Strong Shock Threshold. *Phys. Rev. Lett.* **2011**, *107*, 144302. [CrossRef]

42. Ashitkov, S.I.; Agranat, M.B.; Kanel, G.I.; Komarov, P.S.; Fortov, V.E. Behavior of Aluminum near an Ultimate Theoretical Strength in Experiments with Femtosecond Laser Pulses. *JETP Lett.* **2010**, *92*, 516–520. [CrossRef]

43. Swegle, J.W.; Grady, D.E. Shock viscosity and the prediction of shock wave rise times. *J. Appl. Phys.* **1985**, *58*, 692–701. [CrossRef]

44. Grady, D.E. Structured shock waves and the fourth-power law. *J. Appl. Phys.* **2010**, *107*, 013506. [CrossRef]

45. McQueen, R.G.; March, S.P.; Taylor, J.W.; Fritz, J.N.; Carter, W.J. *High.-Velocity Impact Phenomena*; Kinslow, R., Ed.; Academic: New York, NY, USA, 1970; p. 293.

46. Meyers, M.A. A mechanism for dislocation generation in shock-wave deformation. *Scr. Metall.* **1978**, *12*, 21–26. [CrossRef]

MDPI
St. Alban-Anlage 66
4052 Basel
Switzerland
Tel. +41 61 683 77 34
Fax +41 61 302 89 18
www.mdpi.com

Metals Editorial Office
E-mail: metals@mdpi.com
www.mdpi.com/journal/metals

9 783039 367986